THE
HOUSE RESTORER'S
GUIDE

THE
HOUSE RESTORER'S
GUIDE

HUGH LANDER

David & Charles

FOR JUNE
my comrade-at-arms

Drawings by Eric Berry

A DAVID & CHARLES BOOK

First published 1986 ISBN 0-7153-8386-8
Second impression 1988
Third impression 1989
New edition 1992 ISBN 0-7153-0003-2
Reprinted 1996
Updated edition 1999

A catalogue record for this book is
available from the British Library.

ISBN 0-7153-0003-2

Printed in Great Britain by
Redwood Books, Trowbridge
for David & Charles
Brunel House Newton Abbot Devon

CONTENTS

1

INTRODUCTION

The restoration of old buildings is no work for the faint hearted; but the results may be enjoyed by dreamers. Efficient and strong-willed people can achieve much in the way of repairs and so-called improvements without ever opening a book on architecture. The half-understood odds and ends of historical styles which they have absorbed over the years confirm their self-image of sensitivity. 'Of course I'd never want to do anything which spoiled the character,' they tell themselves as they cart in the new uPVC replacement windows with their sham glazing bars.

Tender optimists, on the other hand, spend many happy hours picking the mud and cow's-hair plaster off stone walls to reveal the wealths of texture. Standing back amidst the rubble they see in their mind's eye the 'quaint old room' with its stripped pine clock and brightly coloured folk-weave rugs.

Do-it-yourself men of sanguine disposition and not a few builders (who should know better) burrow out the floors of elderly cottages to gain extra headroom, and retire to conveniently sited caravans and bungalows for blameless evenings before their television screens. Meanwhile, the night winds howl and summer rains of unexpected ferocity beat upon the forsaken shell which they have left

exposed. Returning next day they are astonished and dispirited to find that the meagre foundations have collapsed bringing those apparently impregnable walls with them.

In schools of architecture, determined students with mugs of instant coffee sweat over the intricacies of modular building systems and wish themselves Corbusiers overnight . . . but not Palladios. The books on the orders of architecture, recommended by reactionary dons, lie abandoned beneath heaps of dirty jeans and socks.

In the empty wind-swept squares of market towns, holidaymakers fortified with steak and chips peer into the windows of the estate agents, with their revolving stands of colour prints. 'If we sold the boat and didn't go to Spain,' she says wistfully, 'couldn't we buy some grotty little cottage up on the edge of the moor?' Certainly they could – and probably will; and it is from that moment onwards that their joys and afflictions will begin. But what of the cottage itself and how will they treat it? What should they all do, these ever-hopeful house restorers, and how should they go about it? Since this is to be a very practical book it is all the more necessary to understand some of the thinking which underlies people's attitudes to old buildings and how they should be altered, repaired and restored. Before you touch an old building, you must have some idea of what it is you are hoping to enjoy by living in it. If it is merely the sight of green fields and cows grazing up to the garden wall, you could equally settle for a bungalow.

The truth is that you are beguiled by the sense of age, the atmosphere, the beauty of old walls, mellow textures,

(opposite)
You don't get much earlier wall hearths than this one in the gatehouse of a medieval castle. The stone hood has very little projection and sits on a lintel supported by engaged columns with stiff-leaf capitals. The fireplace, which must have billowed smoke into the room on a bad day, has a wide internal gather, terminating in a meagre flue which pops through the wall into the battlements. This is the thirteenth-century progenitor of those classical fireplaces which were not to become commonplace in domestic buildings for another three or four hundred years.

7

(opposite)
Here is a typical small stone cottage of the sort found in
Cornwall or any other granite area of the country.
Mellow, pretty and dilapidated, it is obviously worth
saving. The character of the building is apparent and
should never be forgotten at any subsequent stage of
repair and restoration. There is no need to despair at
the sight of decayed windows and doors, missing slates
and cracks in the walls. An old building must have very
serious faults indeed before it becomes either too
expensive or too difficult to repair.

The cottage is shown after thoroughly informed and
sympathetic repairs and extension. The chimneys have
been rebuilt using the original materials. The slate roof
has been re-laid on new common rafters and battens.
The twelve-pane horizontal sliding sash windows
upstairs have been replaced with identical new ones
and so have the delightful sixteen-pane sashes on the
ground floor. The stonework has been unobtrusively
repointed and a bathroom extension has been built
above the lean-to dairy: it has been built up in concrete
blocks and slate-hung to disguise it. The doors have
been properly designed and the garden and low walls
have retained their unassuming but robust flavour. It
would be difficult for anybody who had not seen the
cottage before to know that so much work had been
carried out.

(above)
This is the same cottage – so much modernised that it
is hardly recognisable as an old building. The roof pitch
has been flattened out and the chimney partially
unbuilt and hard-rendered. The slates have been
replaced with interlocking composition tiles. The
extension, as in the good example, has been built up on
the walls of the old dairy, but the roof line has not been
stepped down. Sharp and relentless rendering has been
used down to a bell-cast. The pointing of the old
stonework is crude and unsuitable. Shiny plastic box
guttering has been used and the downpipe has been
placed in a conspicuous position. Horrendous windows
have been inserted in the old openings – some of obviously modern design and others, like the ground-floor
bay, hideous pastiches of Georgian ones with all the
glazing details misconstrued. Bull's eye panes have
been used and the front door, with its incorporated
fanlight, is in the old doll's-house or Wendy-house style
so often seen today. They are normally made of hardwood and few people even soften the blow to the extent
of painting them. A vulgar modern coachlamp has
been fixed on the wall, and a graceless garage with a
steel up-and-over door has been tacked on the end of
the building. The tree has been savagely pollarded, the
garden walls unnecessarily pointed, and a nasty little
wrought-iron gate replaces the old one. The area
around the house has been laid with concrete paths and
suburban lawns, edged with garden-centre blocks. This
kind of modernisation work bears no relationship to
genuine repair and restoration. It may be seen in count-
less versions all over Great Britain; and local authorities
continue to give planning permission.

dipping roofs and elegant windows. Benefits of this kind are not easy to analyse; but if you are to stand a fighting chance of retaining them after you have repaired and restored the building, you have to make the attempt. So let us consider what it is that makes a building attractive. These are the elements involved:

1 The architectural quality of the design – scale, proportion, detail and decoration.
2 The visual qualities contributed by the materials used to construct the building – slate and brick, stone and timber.
3 The setting of the building – how it integrates with the landscape, whether rural or urban.

These three points could have been assessed when the building was brand new. A Georgian gentleman, standing in front of his fine brick house, tucked into a fold of land at the edge of the meadows, could admire his handiwork with complacency on all three scores.

But there are other factors which greatly influence our minds today when we look at an old building. First, the materials have weathered and changed in both colour and texture. Brick which was once smooth has become pitted and spalled; timbers have sagged; lichen has grown on roofs and stonework; subtle and undreamed-of colours have blended and amended the original substances so that in many ways we could claim to be looking at a different building.

The house or cottage has developed its own unique atmosphere and character. This is partly a matter of the ageing process mentioned above, which can never be other than entirely individual to each building. It also has to do with what may be described as the 'psyche' of the house and its surroundings.

To describe the 'psyche' of a building is really the business of the poet or philosopher; but it is of the greatest importance. We do not have to know exactly what it is – and it will be different for each house and for each person. What we must do is be keenly aware of its supreme importance to our feelings about a building and the practical measures we eventually employ to repair or restore it.

The 'psyche' of a building (and the more I repeat it, the more pretentious the word becomes) is made up of the feelings and experiences which we bring with us and direct towards it, and the atmosphere of the place itself. Our tastes, and phobias, loves, dreads and suspicions are all subconsciously affecting our appreciation. The building, for its part, is busily sparking off memories in us – some will remain subconscious while others will surface. It is vital that nobody who reads this book should emerge with the idea that the practical techniques of construction are all they need bother about. This is not just a book for earnest swots with an ability to apply theories to the task in hand.

To deal satisfactorily with old buildings you must first have a feeling for them and an understanding of them. The feeling for old houses and cottages is something which many knowledgeable people never achieve. It is rather like having a good ear for music: some are virtually born with it, others develop it, and many can find little sense in the business whatever well-meaning steps they take in that direction.

Understanding, on the other hand, can be sought by training the eye. If you look at enough old houses with some degree of educated attention and curiosity you will learn both to understand and to enjoy them. For anybody who actually has to restore and repair old buildings this can prove a mixed blessing. Such tastes can be expensive and if you have to gratify them with your own money you will need to be prudent, resourceful and ready to make financial sacrifices elsewhere in your budget.

There are two more requirements. Firstly, you need imagination; and secondly, you must read books.

Without imagination, it is doubtful whether you would want to live in an old house at all; any old building can only be appreciated in the most academic way without it. There is a strong element of romanticism in the whole matter of enjoying old buildings and it is this which makes it so difficult to devise legislation to protect them. Where the romantic imagination dictates one course of action, practical considerations or business interests may dictate another, and the two are often diametrically opposed.

When the Town and Country Planning Act of 1990 or some Department of the Environment circular talks about the 'character' of a building without actually defining the word, the legislation is frequently found to mean all things to all men, or sometimes nothing at all. For 'character' may reasonably be taken to imply some nebulous aura conveyed by the architectural features of a building, in a consort with an agreeable patina of age, as perceived by the fevered imagination of the beholder. Alternatively 'character' may be narrowly interpreted as 'style'. That is to say, the introduction of gold and white Louis Quinze panelling to a sombre medieval open hall would provide a shocking mixture of styles which are 'out of character' one with the other.

Those who periodically refresh themselves with the astringent *Maxims* of the Duc de La Rochefoucauld may recall that he says: 'No people are more often wrong than those who cannot bear to be.' It is probably fair to use that as the basis for another epigram: 'No people more often need to read books than those who cannot bear to do so.' Anybody who doubts this statement should ask the members of his or her local planning committee how often they read books on architecture, or even look at pictures in them. It would be interesting to find out how many planning committee members throughout the country actually own a basic book on architectural history of any kind. While it would be thought usual for a member of a finance committee to know something of money and the operation of business, it is obviously considered unnecessary for a planning representative to read about an important subject upon which he or she continually makes decisions.

The moral of all this is that only by reading widely on old buildings can you hope to understand and enjoy them properly. Naturally, the study of old houses and their historic development must lead on to other kinds of reading: painting, sculpture, antiques, social history, costume, not to mention novels and poetry, all play a part.

Finally there is a kind of reading matter which few old-house enthusiasts will need much encouragement to tackle: numerous splendid books on the techniques of building construction. Some are listed in Further Reading. You can find out about plastering techniques, roof carcasing, joinery, how to lay drains and all the tried and tested ways of jointing one component to another or building this on that.

Our book is not intended to be a basic primer in building methods; but rather to give detailed advice on how to encounter the problems of old buildings, often using very ordinary and straightforward means. On the whole, it is assumed that readers either know the basic techniques of construction or will take steps to find out what they are. I do try to explain practical systems by which certain ends may be achieved, and the general forms of construction from which to choose. Space is not wasted on showing, for example, how to construct a stud partition, when the modern method is little different from the old one. But, if a partition has to be constructed using an early technique, which differs from modern practice, then I describe it in more detail. Generally, both illustrations and descriptions are meant to make clear the principles involved, not to be interpreted as working drawings.

Now for some important terms, which we should try to define.

Conversion

This is technically what a district planning officer would think of as a 'change of use'. To turn a barn into a dwelling is a 'conversion'. To make the downstairs room of a terrace town house into a shop is also a 'conversion'. They are 'material' changes of use and thus constitute 'development' for which you would seek planning permission.

However, I shall use the term conversion in a non-technical sense, to apply to a broader range of changes, not necessarily involving an official 'change of use'. An example would be the 'conversion' of a farm dairy to a downstairs parlour, or a powder closet to a bathroom. Indeed it has been usual to talk about 'converting' an old cottage – by which people readily understand that you are repairing and restoring it and, perhaps, putting in some modern services. I have often used the term in that sense; but since it is slightly vague it will be more carefully employed in these pages.

Repair

All old buildings require repairs from time to time. Here I use the expression for the process of mending any major or minor architectural feature, structural element or other component.

Replacement

Old sash windows may sometimes need repairing, while at other times, they have to be 'replaced', albeit with windows of identical design. Replacement, therefore, is simply a matter of exchanging one door for another or one roof truss for another, whether you match the originals or not.

Restoration

This term means that you are returning a building or part of its structure or decoration to what it used to be in former times. In its broadest sense it may involve major works of repair and replacement, or even the rebuilding of entire elements which have been demolished at some earlier period. In its narrowest sense it may merely require some operation like cleaning and retouching a gilded plaster ceiling band. Because the word tends to be imprecise, many old-buildings experts dislike using it. However, it has its place as a catch-all for a number of connected activities.

Alteration

If you move a partition wall into a new position, in which no partition formerly existed, you are making an alteration. If you insert a new dormer window in an old roof-slope or provide a new staircase against the back wall of the house when it previously ran inside the gable, that again is an alteration. It is in the matter of alter-ations to old buildings that the greatest degree of taste, knowledge, discretion and imagination is required. It is also the area in which you will encounter the most violent disagreements and hottest debates.

Extension

As the word implies, you are adding to the building, whether a wing which is bigger than the original structure itself, or a tiny lean-to for a downstairs bathroom. Planning laws allow for a certain amount of 'permitted development' in the way of extensions, but Listed Building Consent may still be required. See the section on planning permission in the next chapter.

Maintenance

Anything you do to maintain the fabric of a building and ensure that it is working efficiently can loosely be described as maintenance. However, it is normally taken to mean such minor running repairs as easing sash windows or the fit of doors, replacing tap washers, oiling hinges and refitting wooden floor-blocks which have lifted. It also involves repainting and redecorating, as well as various cleaning operations, including unstopping choked gulleys and gutters.

Modernisation

Modernisation is a term which I should like to expunge from our vocabulary as far as old buildings are concerned. You may provide a house with a bath and a lavatory, put insulation in the roof space, devise an inconspicuous system of central heating. None of these comforts should amount to modernising the appearance of the building, which is nearly always what people mean when they use the expression. If somebody told you that they had 'modernised' their kitchen you might expect to see banks of flush fitted units, a stainless-steel sink, a new picture window and fluorescent lighting overhead. Nine times out of ten you would be right – so neither modernise nor imply that you have done so. It makes people who love old houses nervous.

2

BEFORE WORK STARTS

Many people reading this book will be about to buy an old house or cottage, or will recently have done so. Others will be proposing changes or alterations to property which they have owned for some while: they too will find that much of this chapter concerns them.

Perhaps it is fortunate for vendors and their agents that all too many house-hunters have only the vaguest idea of what sort of building they are seeking. How many times have you heard somebody say: 'We fell in love with it the moment we saw it!' As with all affairs of the heart, you may fall out of love with distressing rapidity, when that first euphoria evaporates and you are faced with the terrible reality of the building's shortcomings. Before we turn to the vital preliminaries which precede the purchase of an old house, and the subsequent months, or even years, of physical work which may be involved, it is necessary to think briefly about basic attitudes of mind – about having a sense of discrimination and clarity of purpose.

Though it appears to be a statement of the obvious, experience suggests that one cannot over-emphasise the fact that house-hunters should decide at an early stage whether they want to live in a cottage or a castle, a mill or a mansion. They could save themselves, and numerous frustrated vendors and estate agents, much time and nervous energy. These decisions should be settled *before* they trundle off to make exhaustive inspections of properties which they must know are unsuited to their needs. It is a pointless exercise to spend an afternoon looking at a one-bedroom Victorian lodge with honeysuckle around the door if you have a family of three school-age children waiting rebelliously in the car outside. It is

best to resist the temptation to while away a golden summer day peering into the attics and cellars of a dilapidated farmhouse at the end of a bad cart-track, with no prospect of mains water and electricity, if an invalid mother has to be housed and looked after. The househunter must know what he wants, how much he might conceivably pay for it, and whether he has the time and the resources to give the building the attention it deserves.

It is vital to work out exactly what sort of old building will satisfy your tastes and requirements. Write down the specification, item by item. Be firm with yourself about the area within which it is worth searching. Decide the amount of land or garden you consider a minimum necessity. How many rooms must the building ultimately have, and for what purposes will they be wanted? Would you consider putting up with the absence of certain mains services, such as water, drainage, electricity or telephone? If you would, it may widen the scope.

This book is full of counsels of perfection. The chief of these is to resolve upon a plan and *stick to it*. That does not imply being rigid about minor matters, or that there can be no flexibility over methods and aspects of design. It *does* mean that major factors must be thrashed out early and counted from then on as fundamental.

If your happiness depends upon having mature trees, then avoid buying a benighted and windswept house in the high fields. If you have a thing about privacy, don't even look at dear little terrace cottages in cosily intimate rural villages. The neighbours may be lovely people; but you could soon grow to hate them. If you wish your outside surroundings to be clean, tidy, well lit and organised, never buy a country cottage. We all know of retired people who move to cottages and subsequently spend their days complaining that the cows leave mess on the roads, the leaves block up their gutters, the farmer has a 'dangerous bull' in a field where they like to walk the poodle and that there are no street lights in the lane.

In this book, old buildings are divided into two main groups: the Rough and the Smooth. I define 'rough' buildings as those which have few pretensions to elegance or symmetry. The materials from which they are built lack fine finish; their design is informal and often haphazard. Obvious examples include cottages, mills, crofts, many small farmhouses, certain industrial buildings, barns and the more primitive schools and chapels.

Smooth buildings are mainly those which were originally intended as dwellings – usually for the more prosperous part of society: anything from a little Georgian rectory, or a mid-Victorian stuccoed villa, to a gracious stone Palladian manor, or the well-appointed late seventeenth-century pub in the High Street. A smooth building will have panelled interior doors, cornices, chair rails, properly designed fireplace surrounds and a layout of what an estate agent might list as 'reception' rooms.

Less easy to place in a broad category are the numerous small terrace houses in towns, which are neatly and smoothly built, forming a repetitive visual unity. Their interior details are simplified and scaled-down versions of their grander neighbours in the square around the corner. Where the latter have six panels in the doors, these perhaps have four. Their cornice mouldings are rudimentary and confined to the hall and the front room. The staircase looks like one you might see in the bigger house, but it is meaner in its proportions, and the detail is cruder or less intricate. However, there are no open joisted ceilings (a 'rough' feature) and the place feels like a house rather than a cottage, despite its size.

The other obvious exception to the 'rough' and 'smooth' division is the early building which is 'rough' in its materials and layout, but abounds in fine craftsmanship and even elaborate detail. Many of the older manor houses and the grander houses of town merchants predating the mid-seventeenth century, fall within this group. It is really a question of flavour.

Buildings must be assessed on their individual merits, and we should be faithful to the spirit of the building, or the part of it with which we are dealing. If the hall, for example, displays a strange mixture of rugged textures and sophisticated detail, then it should be approached with that in mind. We should neither try to smooth it out nor rough it up. The idea of having 'rough' and 'smooth' categories is not some arbitrary rule of law, but just a method of guiding one's approach in a suitable and sympathetic direction.

In this book terms like 'historic' or 'ancient' applied to buildings are avoided where possible, since they defy adequate definition. Old buildings are normally referred to as such, and where necessary the actual period is stated.

Dating Buildings

Detailed advice on dating buildings is outside the scope of this book, which is mainly concerned with the aesthetic and technical principles involved in works of repair and restoration. All the same, the problems and techniques discussed provide a number of pointers to age. It is, of course, essential to have some notion of a building's age, for everything to be done will hinge upon an understanding of period methods of construction, detail and decoration. Many readers will already have studied the basic elements of English architectural history, and those who have not, will need to do so. The history of architecture and building practice is endlessly fascinating, and the more you learn, the more aware you become of how much you still need to know. One thing is certain: many owners of cottages and similar vernacular buildings have the most fanciful beliefs about how old their buildings are. 'It's very old,' they tell you with conviction.

'How old is that?' you ask politely.

14

Tumble-down cottages, like this one in east Cornwall give very little away about their origins. The tin roof sometimes suggests that a building may have once been thatched, but in this case the pitch looks too shallow. The most interesting feature is the hollow-chamfered and stopped lintel over the main door. Is it one of the original jamb posts or something salvaged from another building? Certainly, the stonework of the existing jambs is very ragged. There are no early houses close at hand so the suspicion is that this moulded 'lintel' makes the apparently modest cottage into a seventeenth-century farm of some social status.

'Oh, at least three hundred years. It's sixteenth century.'

'Seventeenth century, do you mean?'

'Yes, yes – built in sixteen something.'

When you finally see the building it turns out to be mainly mid-eighteenth century, but contains earlier features which suggest seventeenth century; and it is built on the site of an Elizabethan house.

There are four main ways of dating a building. They are:

1 From visual points of style.
2 From materials, structural details and techniques.
3 From archaeological evidence.
4 From written records, references,

maps, plans and pictures.

The first and second of these methods often prove most useful. Visually, note the proportions, the shapes of the window openings, the kind of windows, the pitch of the roof and its form of construction. the eaves details, the type of chimneys and all the multitudinous interior fixtures and decorations. The grander the house, the more can be told about its date from points of style. This is because such houses display a greater degree of fashionable embellishment and decoration typical of their period. Small, old and exceedingly plain buildings may convey almost nothing which can be confidently assessed to the nearest hundred years. A granite upland cottage in Wales or Cornwall could often date from the late sixteenth century or the late seventeenth century, and despite our best endeavours we may end up none the wiser. What you can look for in these instances is some giveaway feature which will help to narrow the time scale within which you are working. It is a tricky business, requiring knowledge and a keen eye for detail.

This also applies to dating by the third method, that of archaeological evidence. Here, it is a question of analysing the plan of the building and its possible development – trying to work out whether some part cannot logically be earlier than this, or later than that. The position in which artefacts are discovered, if they have obviously not been disturbed, will offer clues. Proper archaeological evidence should not be a matter of conjecture. It is painfully acquired and must be rigorously tested.

Timbers in an old building may sometimes be dated by a technique called *dendrochronology*. This involves the analysis of the annular rings of the components in the house with those of a specimen of the same age. Advice to beginners: first find your specimen of known age. *Carbon dating* is another expensive and highly specialised method, which is normally employed for dating only the earliest and most important structures. It is done by measuring the radioactive decay of carbon 14 in a sample timber.

A method of dating which overlaps with the archaeological one, is that of analysing materials, structural details and techniques. Clearly the type, shape, size and condition of such components as timbers, tiles, stones and bricks, all tell a story. Likewise, the ways by which they are fixed in place or jointed together often indicates their age.

An obvious example might be found in the roof space of a timber-framed southern farmhouse. You stand on a ladder, open a hatch and peer into the roof space with a powerful torch. You see a huge cambered tie-beam with a moulded crown post rising from its centre. This supports a stout collar purlin running the length of the roof. From near the top of the crown post project four curved braces, two of which pick up the load of the purlin, while another two help to sustain the collar itself. Here is a type of roof construction which tells you at once that the building is likely to be fifteenth century and is possibly earlier still. That, of course, is an easy one, but there are many other constructional details to look for, though few of them are conclusive when taken on their own as evidence of a building's date.

Written records as an aid to dating can be extremely helpful and are rightly beloved of scholars. However, they too have their shortcomings. The difficulty lies in deciding whether the building or feature mentioned is actually the one which you have seen. All too frequently the references are inconclusive. Old sketches may be wildly inaccurate, either because they were drawn from memory, or for reasons of romantic caprice and artistic licence. Early maps and plans are frequently too small in scale to show a building as more than a blob, and the cartographers sometimes confused the positions of buildings or misrepresented their basic shapes.

In the final analysis, dating of a building will be achieved by a judicious mixture of all the above methods. Ideally, you try to get three or four pieces of dating evidence to cross-check with each other.

But let me strike a warning note. Three things must be avoided. Firstly, you must remain objective and not allow yourself to believe evidence which is not really proper evidence at all. It is easy to be beguiled by a preconception and then try to make the evidence fit. Secondly, the pitfall associated with dating by style is that many stylistic details remained in constant use for centuries, or were revived at intervals as fashion dictated. Thirdly, allowances must be made for considerable variations of style between one region and another. A feature which had long been abandoned in a southern town, for instance, might still have been used in the rural North a hundred years later.

Here are a few suggestions for helping to date a building. Listen to local hearsay, checking it against the building itself and any available records. If a village elder stoutly avers that a cottage was once the blacksmith's forge, it is more than likely that at some point in its history it did indeed serve that function, even if no obvious signs remain.

Ring up the county, city or borough records office, tell them what you are looking for and make an appointment to spend an afternoon studying whatever maps and other documents are available. These could include wills and probate inventories, tax returns, copies of glebe terriers, tithe maps (dating around 1840), Ordnance Survey maps, estate maps and records which have been lodged with the local authority by the owners, and parish registers.

Glebe terriers are lists of the property owned by each particular church living and can contain useful details. Sometimes there are drawings of the rectory or vicarage and other buildings associated with the benefice. A tithe map shows all the buildings and land upon which fixed charges were made after tithes proper were abandoned. Perhaps even more useful is the accompanying tithe apportionment, which gives the names of owners and occupiers at that date.

Where various records will be found depends upon the area in which the building is located. The church-related documents may either be in the diocesan registry or the records office. Maps and deeds connected with properties of big estates may also be in the records office or in the estate's own administrative office. Parish registers should also be remembered, but they are not very helpful if you do not already know the approximate dates to be researched; it can be a long and tedious business to check through all the births, deaths and marriages.

Before going to these lengths, you obviously check the deeds of the building; but in many cases they only date back to the late nineteenth century and are of little assistance. The offices of the local planning authority can tell you if the building is listed by the Department of the Environment as of 'special architectural or historic interest'. If it is – and a great many old buildings *are* listed – there will be a brief description and an idea of the possible date. I say 'possible date' advisedly, because many such buildings have only been assessed from the outside, and even then mistakes can be made by the original listers.

Yet another source of documentary evidence are the manorial court rolls. The system of recording the tenancies of 'copyholders' in a written document pre-dates the ordinary leasehold agreement. One special advantage, if the house is a really old one, is that such evidence may go right back to medieval times. However, the writing may be meaningless to anybody who has not learned the archaic Latin hand of the notaries.

As far as tax returns are concerned, it is worth recalling that there were numerous levies connected with building at different periods. To name a few: Window Tax (1696–1851); Brick Tax (1784–1850); Candle Tax (1709–1831); Peter's Pence – a pre-Reformation contribution to the Pope in the form of a Hearth Tax (up to 1534) and a proper national Hearth Tax (1662–89). The ways of gathering these so-called taxes varied considerably and not all were truly taxes, but rather excise duties.

The most obvious means of dating a building is to read the information off a date stone or other inscription. If this appears to be in its original position, it can be invaluable. All the same, it should be treated with extreme caution, if not suspicion. It may come from another part of the building altogether. It may merely have been placed to celebrate a marriage or some important event and thus does no more than suggest that at least that bit of the building is no later than the stone. It may even be an outright fake.

Finally, do not forget to visit the local reference library. Some authorities have specialised ones which deal with local history and have staff who are knowledgeable and experienced in searching out information which helps with dating. They may have old maps, parish histories, topographical books and useful biographies, or even collections of old prints and photographs. Some libraries have early newspapers on microfilm which go back to the eighteenth century. Newspapers can include records of sales, fires, advertisements, births, deaths and marriages and various other items which might mention the building or give some other clue.

Grants
There is an inherent danger in discussing grants in a book like this one, since the systems and criteria are subject to fairly regular changes by central and local government.

The allocation of grant money is decided on the basis of a means test. Secondly, the concept of providing fixed amounts for specific works has been replaced by the idea of a local authority paying out on the actual cost of the works.

In the end, you will have to see the grants officer at your local council offices and discuss the possibilities in detail. However, these are the main kinds of grant available:

1   House renovation grants.
2   Common parts grants.
3   Houses in multiple occupation grants.
4   Disabled facilities grants.

House Renovation Grants
Many people repairing and restoring old buildings will consider a house renovation grant, since it is the most generally applicable. Second homes or holiday cottages are *not* eligible. Should a conversion have taken place less than ten years before your application, you are again debarred, and grants are not available for internal decoration. Most importantly, extensions to provide more bedroom or living space do not attract grant aid.

House renovation grants are discretionary. The local authority may decide to make a grant if it wishes. The purpose may be to bring the property up to an acceptable standard of fitness for human habitation or you could be trying to make the standards rather better than the basic requirements.

Insulation, replacements of things like rotten windows, damp-proofing, wiring and so on, together with repairs to roofs and timbers, walls and foundations, floors and staircases may all be grant aided in some circumstances. If the building is listed, the grant-aided works, or any works for that matter, may need listed building consent.

Conversions are another discretionary matter and will more often apply to buildings which are thought to be underused and capable of making a contribution to the stock of available dwellings in an area.

Common Parts Grants
These grants apply to the shared areas such as access passages, staircases and halls of buildings divided up into flats or maisonettes. Grant money is again provided on a discretionary basis.

Houses in Multiple Occupation Grants
These grants are normally provided only to make a house fit to live in, and only the landlord can apply. Houses divided up into bed-sitters would be in this category.

Disabled Facilities Grants
Largely directed at improving facilities for those registered as disabled, or those who could reasonably expect to be registered, these grants are intended to be used to improve access, heating, lighting and other controls, and other provisions directed towards a disabled person's needs.

The Means Test
The means test applies to all types of local authority grant and is assessed as follows.

To your average weekly income is added any revenue from savings in excess of £5,000. Then by various formulae of their own, they work out how much is required for your basic needs, taking into account disabilities, commitments such as children, whether you are a single parent or advanced in years.

If these sums show that you have less money coming in than you are thought to need, you may not have to pay anything towards the grant-aided works. If your income is more than your assessed needs, a proportion of the notional surplus is designated as your contribution to the works. This sum is called the *affordable loan*. The final equation therefore looks like this:

Amount of grant = cost of works minus affordable loan.

Professional fees, by the way, for people like architects, surveyors and historic buildings consultants may be included in the costs.

This relatively brief summary of the local authority grants guidelines may be followed up by reference to the Department of the Environment's booklet *House Renovation Grants*; but the district council's grants officers are the ones who will supply chapter and verse.

Other Grants
English Heritage can also make grants towards the repair and restoration of Grade I and II* buildings but owners lucky enough to get anything from the limited kitty may have to contribute large sums from their own resources.

District Councils may allocate funds for the repair of listed buildings, according to the state of their finances. Any money available will be related to overall government funding and the way your councillors feel about the importance of old buildings and their protection. Normally, there will be no money for buildings which are eligible for grants from other sources. Second homes are also usually excluded and a property should have been owned for more than two years. Rules can include a condition which allows the authority to reclaim a proportion of the

grant if the property is sold within five years. Most grants are restricted to a percentage of a maximum sum. Grants for roof thatching can attract grant aid. Commercial properties are rarely eligible for district council funding.

There are also various privately funded charities which raise money for the purchase, repair or upkeep of worthy old buildings. Some, like the Landmark Trust, buy threatened buildings, repair them, and let them out to short-stay tenants. There are other local preservation trusts doing much the same thing on a county or town basis. Lottery funds are sometimes available for particularly high profile projects.

Planning and Listed Buildings
The summaries of parts of the law on planning given in this book are merely intended as an informative outline of a complicated subject which has many legal ramifications, exceptions and interpretations. Readers are therefore advised to go to a lawyer for expert guidance before taking an important action which hinges on these definitions. In some cases it will be enough to consult the local planning department. Readers are also recommended to study:

The Town and Country Planning Act 1990
DoE Planning Policy Guidance notes PPG15
Town and Country Planning (General Permitted Development) Order 1995
Planning (Listed Buildings & Conservation Areas) Act 1990
Planning and Compensation Act 1991

Listed Buildings
The Department of the Environment has lists of buildings which are considered to be of 'special architectural or historic interest'. There are two grades. grade I includes those which are of major importance; but most are Grade II buildings. Some particularly favoured Grade II buildings are designated Grade Two Star (Grade II*).

The lists for each area, copies of which are held by local authority planning departments, together with identifying maps and written descriptions, constitute a legal document giving certain degrees of protection to the buildings. The main protection is against unauthorised demolition. The second, and a highly contentious one, is concerned with 'character'.

Nobody may carry out any work on the exterior or *interior* of any part of a listed building which could affect the 'character' without obtaining *listed building consent*. This is sought from the local planning authority, which may decide on its own about alterations to Grade II buildings, but must inform the Secretary of State for the Environment about requests to demolish, regardless of the grade (no fees for listed building consent).

Proposed alterations to Grade I and Grade II* buildings cannot be decided unilaterally by a local authority. The Statutory Consultees (see page 21) must be informed so that they can comment. In certain rather rare instances, the Secretary of State may 'call-in' an application and hold a public inquiry. This procedure usually takes place when matters of wide public concern are at stake. Sometimes a 'call-in' is made because the local authority has a vested interest which makes it impossible for them to determine an application impartially.

The usual grounds for an inquiry are either that the planning authority and its professional officers agree that a building should be altered or demolished and there is opposition from amenity societies or individuals; or that the applicant wishes to have an impartial hearing of his own appeal against the authority's refusal. It may be that he wishes to change the building's use and the planning authority claims that his ideas are contrary to its stated policy.

In most instances, the local authority's planning committee relies heavily on the policies, reasoning, tastes, knowledge and advice of its professional planning officers when deciding whether to approve or refuse an application. In places where the planning department has keen staff with a good understanding of architectural history, it often makes recommendations which an ignorant or philistine committee chooses to ignore. Conversely, far-sighted and informed committee members may occasionally overrule their own badly informed officers.

Taking into account the frequently disingenuous assurances from owners who swear that they 'would never dream of spoiling the character' of their buildings, it is easy to understand why the listed-buildings law can be described as contentious. Owners, planners or mischievous groups and individuals insist that reasonably sound buildings are dangerous or past salvation. Speculators, who often do not even have a legal title to a listed building, may apply for consent to demolish, knowing it could pay them handsomely to procure the valuable site. Equally, they may be struck with the idea of renovating or rebuilding so brutally and so extensively behind the façade that they provide themselves with a new office or shop in this way.

'Listing' a building does not, alas, protect it for all time. It merely ensures that procedures are observed which may serve to protect it, if firmly and imaginatively operated. Many owners do not realise that they are not automatically sent a copy of the *listing description* when their house is listed. This is held by the local authority and they can ask for a copy. The description is not an inventory of all the important features of a building, but merely indicates some that are especially noteworthy. Because something is not mentioned in the description it does not mean that it can be changed without consent. Everything is deemed to be listed.

It is frequently argued that the State should refrain from interfering in the citizen's control over his or her own property; but this has seldom prevented governments, of all persuasions, from interceding in other matters touching on our supposedly inalienable rights. The trouble with old buildings is that they do not engender quite the same degree of instinctive and sometimes sentimental response as causes like child welfare or the protection of animals. Even within the field of conservation, buildings are the poor relations of such fraught topics as pollution. You do not have to be a professional chemist to appreciate that swimming in sewage-infested waters is a disagreeable experience. It would be idle to pretend that sensibility, even sentimentality, is not sometimes an influence in our passion to protect old buildings. But that passion is also geared to our degree of knowledge, aesthetic judgement, sense of history and personal taste. None of this is likely to bring tears to the eyes of the average 'caring' householder.

The purpose of this book is to help readers understand not only the technical options available, but also some of the legal, procedural and philosophical issues which are involved. Anyone who thinks that restoring an old building is always, or even usually, just a matter of setting the builder on, or a glorified exercise in do-it-yourself, will be quickly disabused of those ideas.

During the late 1980s the DoE carried out a massive programme of rural relisting which has vastly increased the number of old buildings which are protected. New lists of buildings in towns have also been prepared. Whatever our self-interest may seem to dictate, we should never forget that the listing system is the principal bastion against the destruction and degradation of old buildings.

Conservation Areas

Under the Civic Amenities Act of 1967 a system was devised by which a local authority, or the DoE, designates selected parts of towns or villages as 'conservation areas'. These gather together everything from the Georgian butter-cross to the ragtag and bobtail of modern post-war development. The purpose is to give additional protection, and encourage a more sympathetic use of the buildings and their surroundings. The key duty of the local authority is to 'preserve and enhance'. No building in a conservation area may be demolished without local authority consent, and action may be taken to withdraw various forms of otherwise 'permitted' development by making Article 4 Directions (see below). This valuable power, capable of preventing the introduction of uPVC windows and doors in unlisted buildings, for example, is too seldom applied by local authorities.

Statutory Consultees

In nearly all cases, applications to demolish or partly demolish a building in a

The ghost of a seventeenth-century house survives in this gable wall. The pitch of the roof appears to have been flattened out and built up in brick at some later date; and sockets for early joists may be seen at first-floor level. A fine chamfered fireplace lintel and jambs suggest the date of the demolished building. Was it listed? Was it in a Conservation Area? Did it really have to go?

conservation area must be referred for consultation to the following bodies by the local authority:

English Heritage
The Council for British Archaeology
The Ancient Monuments Society
The Society for the Protection of Ancient Buildings
The Georgian Group
The Victorian Society

For the purposes of recording, the Royal Commission on the Historical Monuments of England must be given at least one month's notice of any decision to demolish a listed building or part of it.

Applications to alter or extend any Grade I or Grade II* building must also be referred to the above consultees for their comments. This is done by the local authority.

An important element of control which can assist the enhancement of conservation areas and protect those outside such areas as well concerns business premises, like shops, offices, warehouses and flats. Owners of these need planning permission for any extension or alteration which would materially affect the external appearance of the building. Business premises do not have the permitted development rights enjoyed by dwellings.

Article 4 Directions
A local authority's powers of control within a Conservation Area are surprisingly limited where buildings are not listed. Owners of such buildings can change all kinds of important features for the worse by claiming rights of permitted development. However, authorities can obtain Article 4(1) directions to remove these rights by application to the Secretary of State. More usefully, they can, after various public consultations, impose Article 4(2) directions which have the same effect. These do not require the Secretary of State's approval.

Repairs Notices
These may be served on owners of listed buildings who allow them to deteriorate. The notice requires the owner to carry out work to rectify the situation. Many owners of buildings which they find inconvenient let them fall apart in the hope of procuring a useful site for a new structure, or to rid themselves, perhaps, of a liability. Local authorities are loath to serve repairs notices, which may involve them in aggravation. Sometimes the action even leads them into the expense of compulsory purchase, as this is the only legal remedy if an owner refuses to comply.

Compulsory Purchase
If a listed-building owner either cannot or will not carry out repairs, the local authority can either demand to buy the property, with the approval of the DoE, or it can sometimes do the more urgent repairs itself and charge the owner the cost. Both courses cost the taxpayer money. Where the authority repairs the building and charges the cost to the owner, the building must be unoccupied. Local authorities are not keen to purchase buildings for which they have no special use. Compulsory purchase has, of course, other applications, which do not concern us here.

Closing Orders
A closing order can be served on a listed-building owner to stop anybody living in the house until the work which the council feels necessary has been done. However, listed-building consent may still be required first.

Demolition Orders
Owners of buildings thought to be dangerous can, under the Public Health Act, be obliged to demolish them forthwith, or the council may decide to do the job itself. Again, listed-building consent may be needed.

Unlawful Demolition of Listed Buildings
Some owners may decide for their own advantage that a building is unsafe, demolish it, and then seek retrospective consent. These actions may be defended under Section 9, paragraph 3 of Planning (Listed Buildings and Conservation Areas) Act 1990, which says that you may seek to prove that the works were 'urgently necessary in the interests of safety or health, or for the preservation of the building, and that notice in writing of the need for the works was given to the local planning authority as soon as reasonably practicable.'

This is one of the most damaging loopholes which the planning laws afford. Since it is almost impossible to prove the previous condition once the building is demolished, a case is likely to hinge on any other possible motives which the owner might have for demolition, evidence of witnesses as to condition and whatever impression of integrity the various parties can muster in court. Under Section 9 of the Listed Building Act 1990 a defendant may be imprisoned on summary conviction for up to three months or heavily fined, or both. It is notable that courts are required to consider any financial benefit which might have accrued from the offences. In some circumstances a defendant may have to rebuild what was there before.

Breach of Condition Notices

In many cases a local authority will give planning permission or listed-building consent on the basis that certain specific conditions are met. For example, they might insist that all details of internal and external joinery should be submitted for their approval. If the authority decides that the conditions are not being complied with, they can serve a Breach of Condition Notice which takes effect in not less than twenty-eight days of it being served. Failure to comply with a Breach of Condition Notice can result in a fine or summary conviction.

Enforcement Notices

If a local authority believes that an owner has carried out development which is a breach of planning control, it may serve an enforcement notice, which must state what it considers the breach has been, and what should be done to rectify the matter. The notice does not normally take effect for twenty-eight days, and the date by which the breach of control should be remedied must be stated. However, when a breach of control, such as demolition or works affecting the character of a listed building, or an unlisted one in a conservation area, are undertaken without consent, it is a criminal offence. The authority may then serve an immediate enforcement notice without any taking-effect period of grace. The owner then has to reverse whatever he or she has done wrong, or comply with the terms of the listed-building consent, if there is one. The authority can, however, prosecute at once if it wishes to do so. In the case of a listed building a Listed Building Enforcement Notice is issued. The usual rights of appeal are available.

Stop Notice

In the case of development or breach of control *not* connected with a conservation area or a listed building, there is, as described above, a twenty-eight-day period which elapses before an enforcement notice takes effect; and this means that the contravention may continue unabated. In this case, the council should serve a 'stop notice' to insist that the operations are halted. The notice normally requires compliance in not less than three days from the time it is served, but where an authority thinks it necessary, it may require earlier compliance, but must give its reasons. Failure to comply can result in up to £20,000 fines of the same order as those relating to Enforcement Action (see above).

In the case of a listed building, a local authority will normally write a fierce letter to those responsible and then, if they persist, must seek a court injunction.

Rights of Entry

If a local authority believes that a breach of planning control is being carried out on land or within a building, it may demand right of entry to check up on what is happening. The authority may authorise any person they think fit to enter the premises but this must be in writing. In the case of entry to a dwelling, the occupier has to be given twenty-four hours' notice.

Building Preservation Notices

Local authorities may serve a 'building preservation notice' on the owner of a building which is *not listed* which they believe is endangered. This is a method by which demolition, or work which is detrimental to the building's character, can be halted or pre-empted for six months breathing space, while steps are taken to have it included in the DoE lists. It has to be subsequently confirmed by the Secretary of State. Councils can if they wish give their officers powers to serve building preservation notices without waiting for the approval of a committee, so that action can be taken instantly.

Planning Permission

If you wish to carry out work which constitutes 'development', you should apply to your local authority for planning permission. The definition of development is vital. It is defined as: 'the carrying out of building, engineering, mining or other operations in, on, over or under the land, or the making of any material change in the use of any buildings or other land.'

Not classified as 'development' are

'works which affect only the interior of the building or which do not materially affect the external appearances of the building'. But in the case of listed buildings it must be remembered that such works will still require listed-building consent.

For other exceptions, see relevant legislation listed above.

Planning permission may be sought on property which you do not actually own; but unless you are a tenant, you must notify the owner. It may sometimes be important to seek 'outline planning permission' for possible future works to establish your chances of being allowed to go ahead with your ideas. Even if outline planning permission is granted, you will still have to put in detailed plans later. The local authority may approve your plans *in toto* or it may make conditions, or it can refuse your proposals entirely.

Detailed planning permission expires after five years if work has not started. Outline permission requires that you lodge detailed plans within three years of it being granted. The work must actually start not later than five years from the grant of outline permission, or two years from the final approval of detailed plans – whichever is the later date.

In all these matters you should consult the planning officer. If you are doing your own drawings and applications, get him or her to spell out what forms, plans, sections and elevations are required, and how many copies of each. Fees are demanded.

Appeals and Inquiries

There are several situations in which an applicant for *planning permission,* or *listed building consent* can appeal against a local authority refusal. It is also possible to appeal against an authority for not determining an application within the required period of time. What you cannot do, unfortunately, is appeal against a decision to give some other party a consent which you dislike. An obvious example is the conservation society which deplores the fact that their council has given listed building consent to some owner who proposes to ruin the interior of a Georgian house.

You can choose from three different types of appeal procedure:-

1. *Written representations.* These involve presenting a case in written form only, with plans, maps, diagrams and photographs supporting your case. The government's appointed inspector will assess the evidence on both sides and visit the appeal site. You and the local authority can attend his visit but may not discuss the case with the inspector.

2. *Local Inquiry.* In this instance, the inspector acts rather like a judge in a court of law. Both sides can be legally represented, witnesses called and cross-examined. Proofs of evidence must be compiled in advance and made available to both sides. Members of the public can attend and make statements with the inspector's consent. You can conduct the case yourself if you wish, but unlike your lawyer, you will be required to undergo cross-examination. The inspector provides a written decision or recommendation to the Secretary of State some weeks after the inquiry.

3. *Informal Hearing* before an inspector. This procedure gives the parties most involved a chance to discuss the bone of contention with an inspector acting as chairman. You probably sit around a table with the planning officer, your architect and one or two others. Again, the inspector makes a site visit and subsequently writes a decision.

Permitted Development

Under the terms of a statutory instrument called the General Permitted Development Order 1995 you may carry out various types of development without planning permission. These exceptions cannot be listed here, but they relate, amongst other things, to small extensions, garages, conservatories and porches.

It is of particular interest that an extension may be constructed without planning permission provided that it does not exceed 70 cubic metres as calculated from external measurements, or is not bigger than 15 per cent of the original building, whichever is the greater ($50m^3$ or 10 per cent for conservation areas, areas of outstanding natural beauty and for terraced houses). It must not finally be more than $115m^3$.

The extension should be no higher than

the highest part of the original house. No part of the extension should project beyond the 'forwardmost part' of any wall of the original house which fronts on to a highway, or beyond 20m, whichever is nearest to the road. If the extension is within 2m of a boundary wall of the property, it should not be over 4m high.

The area of ground covered by buildings within the curtilage (the land attached to the house) should not exceed 50 per cent when the extension has been added. The area covered by the original house itself, however, may be excluded from the calculations.

You may also erect a porch, so long as the total ground area does not exceed 3 square metres (about 5ft 8 in x 5ft 8in). The porch's maximum height should not be over 3 m from the ground, and no part of it should be nearer than 2 m to any boundary with a highway. Many good buildings have been spoilt by the addition of horrid porches, but they can be both pleasing and practical if the style of the building is respected. *Remember that you still need Listed Building Consent for anything which could affect the character of your house or cottage, even though it is 'permitted development'* and may be exempt from Building Regulations.

The Building Regulations

In previous editions of this book, I have attempted to describe the presentation and the various sections of the Building Regulations together with the options open to those applying for them. I now believe that the whole subject is so complex that to do so would merely further mystify the reader. The rules are constantly changing and only those professionals who deal with such matters day by day can hope to keep up with the endless stream of requirements which pour forth from the European Union and our own government offices.

What can be said is that Building Regulations approval must be sought from the Building Control Department of your local authority for works involving structural alterations, fire spread, means of escape, insulation, chimneys and hearths, drainage, plumbing, staircases and ventilation, to name some of the main headings.

For anybody dealing with the repair and faithful restoration of old buildings, the regulations can be an unmitigated curse, since they are designed to provide standards for new housing, but are often applied to old ones with various degrees of ferocity. If you are replacing like-with-like you can avoid regulation, but should what you are doing be counted as new work, all sorts of difficulties may arise.

Extensions to old cottages are treated as if they are new bungalows on a modern housing estate. The conversion of a barn to a dwelling is a change of use and thus fully regulated. If you widen a door or window opening, the necessary lintels must be approved and backed up by structural calculations. A new open-joist or beamed ceiling for a cottage kitchen will require approval in respect of structural stability and fire-spread, if it is thought to be a *material alteration* which adversely affects the building.

If you can claim that you are only repairing what was there already, you may be free of this bain; but once your Building Control Department is involved, you may find yourselves subject to more and more restrictions. The sensible course of action is to discuss proposals with the building inspector before finalising your plans. Some inspectors are vastly more forgiving and sensitive to old buildings than others. If fully annotated application drawings have to be submitted, you would be wise to employ a professional architect, surveyor or architectural designer. However, such a professional must be one who understands and cares about old houses and be ready to argue the case for essential detailing which departs from the norm.

It is not enough that your professional accepts the control officer's standard solutions. Alternative methods should be examined which relate to the aesthetic and historical authenticity which you are trying to achieve. It is worth noting that methods shown in the *approved documents* are not mandatory, but alternatives will need to be proved satisfactory. In many cases this can be a difficult and tedious operation.

You must not be too cowed by the prospect of complying with regulations. The authorities cannot make you change an existing cottage staircase or floor however far it falls short of their requirements. Indeed it may remain dangerously inadequate, provided you leave it alone, or merely carry out selective repairs. If you are unable to reach a compromise with the building control department, you can request a dispensation or relaxation of some part of the regulations. This means writing a clear, well reasoned application to the local authority for consideration by their appropriate committee. With luck, they may come down on your side, especially if a *listed building* is involved and the conservation officer backs you up.

It would be wrong to suggest that the Regulations do not serve many useful purposes. It is probably a very good thing that a control officer checks the way in which your builder means to underpin a wall and ensures that a load bearing beam inserted to knock one room into another will do its job satisfactorily.

If you are trying to do everything yourself, you should buy the complete regulations from HMSO through your local bookshop or a book such as *The Building Regulations Explained and Illustrated* by Vincent Powell-Smith and M. J. Billington.

A final note of warning is this. The planning department's *conservation officer* may be delighted with what you are proposing for *listed building consent*. His colleague, the *planning officer* may see no obstacles regarding *planning permission*. Despite all that, the *building control officer* may have very different ideas indeed.

An alternative to the submission of *full plans*, attracting fees based upon the estimated cost of the works, is the *building notice* procedure. In this case you tell the control department what you are about to do, not less than 48 hours before work starts. You are charged on a set scale of fees. For minor operations it is probably the best system.

Amenity Societies
There are numerous groups and societies at both national and local level to which owners of old houses may turn for advice and assistance. These range from those which are nationally known and have great prestige, like the Society for the Protection of Ancient Buildings and the Georgian Group, to the small town conservation society.

The SPAB, as it is generally known, the Georgian Group, the Victorian Society and the Ancient Monument Society all have a special status as 'statutory consultees' in the framework of planning procedure. At certain stages local authorities are obliged to inform these groups of proposals which affect listed buildings, so that they can make comments. Their members include many professionals with expertise and experience in matters concerned with old buildings. It is well worth belonging to at least one of these organisations. They may be able to support you in a planning dispute about an old building, and there is also the altruistic aspect – if you are keen enough to repair an old house for yourself, you may want to do something for the cause of saving old buildings in general.

Local societies can be immensely helpful in fighting battles, in giving advice about conservation and restoration work, and helping to date houses. A library can often supply addresses. English Heritage (The Historic Buildings & Monuments Commission) plays a strange dual role in conservation. It is a watchdog, able to fight in the interests of preserving 'the heritage', but also acts as a specialist arm of the DoE, with various statutory powers and duties too complicated to described here.

3

OFFICIALS AND PROFESSIONALS

Anybody buying or repairing an old building will immediately become involved with a host of professionals and officials. Their business is to sell him the benefit of their expertise, carry out work for him or vet his activities. An intelligent or knowledgeable enthusiast may dispense with some of the professionals, but he will still have to accustom himself to dealing with officialdom, which means deploying as much tact and sang-froid as he can muster.

The temptation to tar all officials with the same brush should be resisted. Both diplomatically and psychologically it is a mistake to allow yourself to become obsessive. Usually I have found them to be kindly and helpful, if not always entirely sympathetic to the needs of old buildings. As in life generally, everything finally comes down to individual personalities; you may discover more architectural sensitivity in the man who comes to install the telephone than in the architect you commission to draw up your plans.

Estate Agents
Most people are well acquainted with the ways of estate agents and much has been written of their love of hyperbole. Praise coming from some agents for the way in which a building has been altered or restored is almost tantamount to handing the owner the 'black spot' – and though this view will not endear the writer to some members of the profession, so long as they continue to describe old houses which have been brutally modernised as 'restored to a high standard, maintaining the period character', the accusation will be justified.

There are, of course, some agents in most areas who do understand and appreciate the qualities of old buildings; but even they are placed in a dilemma. New rules on accurate description of matters like period detail can sometimes terrify an estate agent into not mentioning some of a building's best features at all.

There may be the occasional eccentric who makes a virtue of overwhelming honesty – 'Nice little lock-keeper's cottage, foully converted to a chic studio. Not beyond redemption in the hands of some discerning nut-case' – but most agents find that this approach fails to please the client. If the aesthetic and architectural shortcomings of all the old buildings at present on the market were frankly pointed out, sales would slump beyond redemption. Agents understandably feel themselves obliged to sell the goods, not to conduct an historical critique. In recent years they have vastly improved the way in which they present the details of houses. Many of us recall the days when no prospectus included a photograph – now a standard practice. House-hunters must read between the lines, play fair, ask the right questions and act with despatch.

Do not, if you can avoid it, look at a building with the agent on the first occasion. It is distracting, however discreet he or she may be. You need to concentrate. Make considerable allowances for the effect of furniture and decorations in occupied buildings; they can be either off-putting or encourage you to be dangerously optimistic. If you are not employing a surveyor, remember what astonishing defects may lurk beneath a turkey carpet or behind a walnut escritoire. Steel yourself to ask the owner the impertinent questions which his haughty mien suggests he would not care to answer. Conversely beware the lively and imaginative occupant who is both amazingly frank and yet at the same time suspiciously vague.

'Where is the septic tank?' you enquire, peering gloomily through the leaded lights towards a sinister-looking pond, covered in green scum.

'Oh, everything just runs away down there. We never have any trouble with the drains,' comes the answer, with a touch of impatience.

The best advice for dealing with estate agents is to be concise, businesslike, consistent and considerate – and demand the same qualities from them. Do not accept fudged answers to questions, and, if you are a buyer, withdraw an offer at once if you suspect you are being used as a stalking-horse to provide some other client with a sense of urgency. If you make an offer, insist that it is accepted or rejected within a definite time-limit. Do not leave it on the table as a permanent bid.

Surveyors

The particular skills and qualifications of surveyors are directed towards assessing the condition of buildings. They may belong to the specialist professional institutes, like the Royal Institute of Chartered Surveyors, or hold qualifications issued by other relevant bodies, and have training and experience which fit them to do the job. A member of the Chartered Institute of Building, for example, is well qualified to undertake domestic surveys, if this is his chosen field of activity. However, anybody may call himself a 'surveyor' and carry out the duties with whatever competence his training, temperament and ability allow. It is best to play safe by paying fees to somebody who has relevant training, insurance cover, and is responsible to a professional body with established standards.

A surveyor will vet your building's structural and decorative condition and report in writing. Agree with him ahead about his charges and how detailed the report is to be. You might want him to prepare a full measured survey with plans, elevations, sections, photographs and a report, but this would be an unlikely demand unless you had already exchanged contracts with

the vendor; and a full survey would be a costly white elephant if the deal fell through.

Prospective purchasers usually require a written report of condition after they have made an offer 'subject to contract' or 'subject to survey'. The contents of the report obviously influence the buyer's attitude to the final sum he or she is ready to pay. All the same, old buildings are unique and are seen to be worth what you are willing to pay for them, unlike a production-model bungalow. Thus an owner will normally accept your final offer only if it suits him to do so regardless of the contents of the surveyor's report, which will be either no news to him or of academic interest. The report's chief value is more often as a manoeuvre in a complicated saraband where timing is all. The buyer is saying: 'I won't pay that much' and the owner or his agent wonders 'Does he mean it? And if he does mean it, what reason can he put forward to justify such a miserable offer?'

The buyer's answer may be: 'I won't pay it because this report says there are signs of foundation movement in the gable wall.' The owner thinks: 'I want his money although more would have been better. Anyway, this prize fusspot may actually *believe* in this thing about the gable wall, though it's been like that ever since I can remember.'

The point is that the buyer could just as well say, 'I won't pay that much for a house with so many stairs for my aged mother to climb when she goes to bed.' If the owner genuinely wants to sell, and there is no better prospective purchaser in sight, he will accept your quaint reasoning with a straight face.

What a buyer really wants to know from a surveyor's first written report is whether the building has costly defects which would prove too expensive, too much nuisance or too time-consuming to rectify. (It is worth remembering that given *sufficient* time, expertise and money, almost any fault in an old building can be overcome.)

Once contracts have been exchanged, a full structural survey, with accurate plans and sections, may prove invaluable. This survey can be used to work out your strategy for all that needs to be done. It gives you something concrete to refer to at all times and, as far as planning applications are concerned, the drawings will serve to show the 'existing building'. If the survey is a good one and sufficiently detailed, it could be the basis for the plans required under the Building Regulations, and it will give an architect what he needs to work from.

It is helpful to have numerous copies of the drawings, on which you can sketch in the proposed alterations and repairs, drainage runs, the layout of wiring and electric sockets, the positioning of temporary supports and the layout of any extension.

It is particularly important that amateurs who are going to work on a building with their own hands should have expert advice on structural matters before they start knocking through walls, digging out floors or removing roofs. Far too many pleasant old buildings collapse as a result of the ministrations of enthusiasts who do not realise the structural significance of what they are doing. Here again the surveyor or similar professional can help to avert disaster.

Architects

Architects can, or should, be able to provide most of the expert services which surveyors undertake, but will normally do so on different terms. More importantly, they should have additional abilities which, at best, will make them the ideal professionals to employ when you are repairing or restoring old buildings. A good architect who specialises in conservation work should have a comprehensive knowledge of basic architectural history, great sensitivity, discrimination, imagination and the talent to design new work which matches or blends with the old.

Unfortunately, many architects, while honest and competent, have very limited knowledge of how to treat old houses. They all know about procedure, how to submit the right plans and applications for approval, and read the professional journals and trade information. They are often excellent draughtsmen themselves or employ such people They can produce a beautifully presented dossier on a building. With the self-monitoring emphasis of the new Building Regulations (1985), they could prove most helpful by interpreting the requirements and providing technical expertise. Where many architects appear to fail is in their aesthetic standards – taste and feeling for old buildings. Anybody who doubts this should take a good look at a dozen architect-designed extensions or restorations of old cottages in his or her locality.

An architect charges a percentage fee based on the cost of the final building contract, or ordinary professional fees for his drawing and other services.

It could be argued that, where an architect contributes nothing more than might be provided by the draughtsmen and technicians in a conventional firm of building contractors, he is not the person for the job. It is just those qualities which go beyond mere professional know-how that you must demand of an architect. Exhaustive enquiries should be made before picking one, and look at examples of their work on repair and restoration of old buildings. In some architectural practices there may be considerable differences of style and opinion between one partner and another. A client may be surprised to find that having chosen sensitive Mr A, it is philistine Mr B who actually does most of his work.

The right architect – one who is on your wavelength – can prove a boon. He will do surveys, plans and specifications, deal with applications and officials, obtain competitive quotations, research and design authentic details, supervise the operation of the contract and save money by making wise choices of methods and materials. Should relaxations of the Building Regulations be needed, the architect will apply for them. He may persuade obdurate planning officers of the merits of your case and take your application to appeal when all else fails. If he is highly regarded in the field of conservation, his suggestions will carry weight with the better planning officers and their committees.

Those who have the misfortune to pick the wrong architect are beaten before they have even started. No matter how efficient he may be, there is no antidote which will assuage the poison of ill-chosen windows, heartless modern finishes and a wilful determination to contrast rather than blend the old and the new. As far as unsuitable architects are concerned, prevention is better than cure, and failing that, amputation rather than physic.

Architectural Designers

There are numerous architectural design practices, run by people of varied qualifications and abilities. They range from the man who, though he never got around to taking his final examinations, had years of architectural experience as a senior assistant in good firms, to the most talentless, untrained, badly informed and acquisitive cowboy.

It is fashionable in many circles to speak harshly of architectural designers, but some of them may be excellent; there is no fundamental reason why they should not be as competent as an architect. Indeed, on occasions such designers might have additional skills or qualifications which make them a first-rate choice. All the same it is well to be careful since they do not have to observe the same professional constraints as members of the Royal Institute of British Architects or the Royal Institution of Chartered Surveyors.

With regard to old buildings, however, it is not difficult to imagine the advantages of putting together a team the members of which, although coming from a variety of disciplines, all entirely devote their time to that kind of work. The combination of a trained or experienced builder, an accountant, an architectural historian and two or three resourceful technical draughtsmen, for example, could be a winner, and such people would be no more or less likely to fall out with each other than any similar group of people working together in a common cause.

Structural Engineers

These are the specialists to call in when there are complex structural designs to carry out, or faults to diagnose. An example of the former might be a system of temporary steel supports for the façade

of an eighteenth-century town house; the latter could involve assessing the significance of cracks in the walls of an old barn and analysing the bearing capacity of the subsoil on which it was built.

Structural engineers are usually consulted by architects, planners and other professionals when they want to play safe, or prove a point, or are technically out of their depth. A structural engineer's services should seldom be needed when dealing with the smaller kind of old domestic building, unless particularly difficult problems are involved or design calculations are required by the council's building control department.

An engineer is often employed in connection with ordinary Grade II listed buildings when an owner, local authority or prospective purchaser wants an authoritative voice to back the idea of demolition. Conservationists, on the other hand, can rarely afford engineering opinions of equal weight to claim that no such measures are necessary. Two firms of engineers, or even two members of the same engineering practice, can make different deductions from the available evidence. Likewise, they can put more or less emphasis on the structural importance of one set of circumstances rather than another. Finally, according to the tenor of the client's instructions, they can put forward vastly different solutions to the same problem.

To point this out is not to denigrate consultant engineers. They must be highly qualified and often come up with the most ingenious schemes for countering the structural faults in old buildings. It entirely depends upon the circumstances in which their advice has been sought: to this may be added their own natural desire to play safe in the knowledge that most old houses abound in structural shortcomings which may eventually prove the sanguine engineer's undoing. When the conservationist, the lawyer, the planner, the architect and the district councillor has each had his say, the buck stops with the consultant engineer.

The Builder

For most people this is the really important professional in their lives when they start working on an old building. The right choice of builder can be the making of an entire scheme. This seems obvious, but it is astounding how many thoroughly bad builders are around and how many owners end up employing them.

Builders are remarkably varied in abilities and training, and the scope of their organisation can be equally diverse. They may have a concrete mixer, a van and a shed in the back yard; or a large modern headquarters employing contracts managers, surveyors, quantity surveyors, estimators, site managers, foremen and hosts of craftsmen and labourers.

Some builders are extremely well-qualified people with expertise of the highest order. The Chartered Institute of Building is the professional organisation concerned: the letters FCIOB or MCIOB after a man or woman's name signify that you are dealing with somebody who is expected to be competent.

Whether a builder, however qualified or experienced, will feel for and understand the architectural subtleties of old buildings is another matter. Since few members of other professions do so, it is no special slight on builders to say that the majority fall short of this ideal. It is normally useless to depend upon your builder when it comes to making choices of materials, components and finishes – the visual rather than the technical elements of design. Many builders have a first-rate knowledge of traditional methods of construction, but little judgement about how and when to use them.

Good craftsmen in the building trades may know all kinds of ways of pointing a stone wall, for example – most of them abhorrent – and have a knack of choosing the wrong one. Reading the more technical sections of this book, you will understand that it is not a question of selecting a technically sound way of doing a job, but of finding one that is both sound and looks in character.

The size of a building firm may be more or less immaterial when it comes to handling smaller old buildings. The high overheads of a large concern can mean higher prices, but the equipment available and the capacity to draw on a sizeable work-force, coupled with good management, might allow it to make a very competitive tender. It might even cut profits to avoid sacking good tradesmen during a slack period. Small contractors can cut estimates to the bone to get work, may own all the equipment required for a relatively small job – or be able to hire it – and can have more involved and committed personnel and better day-to-day supervision.

What really counts is the experience, craftsmanship, integrity and imagination of the people who finally appear on site to do the work. These qualities can be found in teams from firms of almost any size. It is no good expecting the man owning two ladders, a van and a mixer to be able to draw up your plans and produce polished and detailed estimates or specifications, but even the very small-time operator will usually know somebody who can supply these services if they are required. The principals of many smaller building concerns started out as tradesmen in one of the crafts, like carpentry, bricklaying, plastering, masonry or plumbing. Some of them still work regularly 'on the tools' themselves or help their men out from time to time. It is also essential to find a builder with whom you can work, especially if you are dealing with him direct. You may want to do a number of the jobs yourself and use the brawn and skills of his men for selected tasks: such arrangements require tact, good planning and a sympathetic contractor.

He should be experienced and conscientious, and employ people who are the same. The sort of craftsmen he can command is of the greatest importance. Old buildings need men who are not only skilful but flexible in their approach, who do not have to be constantly supervised or driven, and who can show initiative and a high degree of care. Many small teams of building workers on conservation projects include representatives of the main building crafts but rely considerably on semi-skilled labourers who can turn their hand to almost anything. These men do not have their City and Guilds certificates but are the backbone of restoration and repair work. They are accustomed to old houses, are not fussed about union rules, and can lay bricks, plaster walls, do first-fix carpentry, slate a roof and throw in a bit of simple plumbing for luck. They may be asked to spend the day digging and wheeling out barrows of spoil, and drive the truck too.

They are often the men who will be on site during most of the contract, while the skilled craftsmen come and go from one job to another. Look for the builder who has good men of this calibre who have been with him for years. No decent builder keeps men who are work-shy or dishonest for long. It is getting increasingly difficult to find highly skilled specialist craftsmen – joiners who know the old methods, plasterers who can cast or run mouldings, and masons who are not just block-layers but can really handle stone. Sometimes it will be necessary to bring in a specialist subcontractor or have particular components made elsewhere.

Specialist firms or those employing very skilled craftsmen tend to be found in areas where the number and quality of old buildings has kept up the level of demand. Thus you are more likely to find a really competent plastering company which can replace cornice mouldings in Bath than in say Cornwall. Similarly, you will discover more thatching firms in Devon than Durham. The cathedrals, some major public buildings and great private houses retain their own teams of skilled craftsmen; but even they call in outsiders for specialised operations. Having said this, remember that many ordinary craftsmen, and labourers too, can rise wonderfully to a challenge. It is for the architect, builder or building owner to encourage the development of skills by asking for them.

The best plan, when you start to search for the ideal building firm I have described, is to ask around and look around as well. It is no use turning to the Yellow Pages directory and hoping for the best. Note local works in progress which you think are being done well and find out the name of the firm. Ask the planning department if they know of any contractor who has done a lot of work on listed build-ings which has subsequently been commended by people who know what they are talking about. Telephone the secretaries of amenity societies: they often know somebody suitable or can tell you where to find out.

However, because a firm has done a good job for, say, the National Trust, it cannot just be left to get on with it. That success may have been wholly dependent on the supervision by the architect, the tightness of the specification or the vigilance of the client.

Specifications and Contracts

Once a suitable builder has been found, the architect or surveyor will deal with the negotiation of the contract, unless the owner of the building decides to do it himself. In the case of repair and restoration work on old houses, this is frequently the case. So it is worth explaining one or two points for those who are not professionals.

It cannot be too strongly emphasised that anybody contracting with a builder to carry out work should make certain that his or her intentions are crystal clear and laid down in writing. There are endless rows and legal disputes because builders claim they have agreed to do one thing and owners say they were expecting another.

Without going into all the ramifications, these are the essentials to bear in mind.

Specifications

The specification is a written explanation of the work which must be done, giving sufficient information for the builder to cost it. The writing of really good specifications is tantamount to an art form, and few without professional expertise can hope to achieve good results for complicated proposals.

Generally speaking, a specification is the usual contract document employed when instructing builders who are quoting for lesser works of repair, restoration, alteration and extension (drawings apart). However, many small building firms seldom see a detailed specification for these operations, especially if they are dealing direct with the owner.

The more likely course of action is for the builder to visit the site, inspect, assess, measure up the work and quote a fixed price for the whole contract. While this method constitutes a proper legal contract – once the price has been accepted – it is unsatisfactory, since it will be lamentably lacking in detail. Another possibility is that the builder will look at the job and price it from a set of working drawings which he has prepared or the owner has given to him. Alternatively, the builder may be asked to price certain parts of the work separately, the total amount being agreed both in part and as a whole.

Here again, in the absence of a specification, the detail will depend upon the builder's probably sketchy descriptions. Undoubtedly numerous contracts are successfully carried out on this basis; it is when things begin to go wrong that owners start wishing that they had been specific.

It is also possible, but very unwise, to agree that the work should be undertaken without a fixed price. In this instance the costs are measured on the basis of a schedule of prices which has been agreed ahead.

With all lump-sum contracts the builder is obliged to complete all the agreed work, which means that the extent of that work must be clearly determined. That again emphasises the need for a proper specification.

If the work is to be tendered for by a number of builders, as local authorities sometimes demand for grant purposes, a specification is essential. Without it, each builder will probably have a quite different idea of what he is quoting for.

The specification may be accompanied by measured drawings – plans, sections and elevations. Where works involving the Building Regulations are to be undertaken, drawings must be submitted for approval. However, these may not be sufficiently detailed from the builder's point of view, particularly if carefully worked-out items of joinery are to be made and fitted, or relatively complex processes of repair undertaken.

There is another form of contract that

owners or their agents may agree – one based on the work being measured and paid for at intervals as it proceeds. It is paid for at quoted rates.

An excellent method of carrying out repair work with a reliable and honest builder is the time-and-materials contract. Officially known as a 'prime-cost' or 'cost plus percentage' contract, this merely means that you pay for the hours of work done at quoted rates and for the materials used. For a genuine prime-cost contract the builder adds sums for overheads and profit. When he is working directly for the owner, these are often built into his quoted hourly rates. Since work on old buildings is so involved, and full of such unexpected complications and additional items, the time-and-materials system often prevents the builder over-quoting in order to play safe. Provided that his men can be relied upon to keep at it, or you are on the spot to see that they do, it can prove a good system.

Naturally, owners of old buildings or their agents need to make sure that the materials used in a contract are of the type and quality which have been specified and that the builder does not dispose of timbers, panelling, doors or any other salvageable items without first getting permission.

You may occasionally be astonished by an astronomical quotation from a builder. This is often his way of saying 'Thanks for asking, but I'm really too busy to bother with this, unless you agree a figure I can't refuse.'

Contractors
Although this term can be applied to a great variety of people and establishments (including a building company), I am talking about the businesses which hire out excavating machines and similar plant, by the day or the hour. If the job merits it, they will also enter into a fixed-price contract. Most such firms also have lorries with drivers available to take away earth or building rubble, or to bring in sand, gravel and topsoil.

Anyone with a full scheme of reclamation on an old country cottage is likely to need to hire a digger. Suitable firms are listed in the Yellow Pages telephone directory, under 'Contractors' plant and machinery hire'. It is usually best to hire the machine by the hour. I have rarely seen a man at the controls of one of these who did not get on with the job, when working on a small private site. There is something compulsive about operating a digger which is a great antidote to lassitude. What is important is to check the driver's natural impulse to attempt everything short of combing his hair with the digging bucket. Unless the customer gives absolutely clear instructions, and preferably sets out pegs or other markers, it is astonishing how much damage a digger can do in a few minutes. It can carve through the water main, scrabble up the drainpipes and haul down the electricity cable with a kind of terrible inevitability. So insist that the driver listens to your instructions, turning off the engine to do so. If you are prepared to shout at him over the din he will seldom think of switching off and paying proper attention.

Digger drivers can be very skilful and it becomes second nature to them to use the machine as an extension of their own arms. This sometimes makes them over-confident. Near to old buildings, anything tricky should be done by hand, the brawn of the hydraulic arm being used only in a carefully planned manner. It is so easy to knock loose an ancient quoin stone or undermine a wall by scraping out that bit too much from a trench. It is well worth requesting that the digger driver brings a narrow bucket with him when he first comes on site. To dig wide trenches for pipe-runs and other works that do not need them is disruptive and messy. Be careful to measure the access to your property and any awkward places where the digger has to go before the contract begins.

Excavation contractors are very keen about payment from small private customers since this helps their cash flow. They frequently go bust, but not as a result of slow payment from persons converting cottages. It is the big and apparently lucrative contracts which are their undoing (or the lack of them), tied up with the huge capital cost of their machines. Expect to pay on the nail the moment the job is done.

If hardcore or similar material is being hauled in, remember that a 16 ton modern tipper lorry is not only wide but often too long to manoeuvre round tight corners. A big excavator may be too wide. Measure up and ask for the smaller lorry or machine if necessary. (They send the big ones just the same, if they are the only ones available, but at least you tried.)

A 16 ton lorry may be 8ft 6in (2.6m) wide and 20ft (6.1m) long, and a 10 tonner 8ft 6in (2.6m) wide and 15ft (4.5m) long. A 'small' JCB3 excavator is about 7ft 2in (2.2m) wide and 9ft 6in (2.9m) high. A crawler-type mini digger is about 4ft 8in (1.4m) or even 3ft (0.9m) wide.

Other equipment that house repairers may need includes pumps, drills, compressors, powered barrows, ready-mixed concrete, scaffolding and skips.

VAT and Listed Buildings
Government rulings on the payment of VAT for works connected with listed buildings are far from helpful. The whole matter is dealt with by Customs and Excise and there are grey areas which must be decided in certain cases. How it seems to work is that alterations to listed buildings may be zero rated by the contractor, but repair and maintenance of what is already there attracts VAT.

The inference is that you may ruin an old building by extensive and unsuitable alterations and get the VAT knocked off for your trouble; while the faithful replacement of old features or repair of original ones will attract the duty. Your fate will hinge on whether you can convince all concerned that what you are doing is an alteration. It could be argued that replacement of an unsuitable window with one of the correct period detail is an alteration; but the VAT men may say: 'A window is a window'. Convert a barn (however badly), and new partitions and joinery will be zero rated, while repairs to the old roof structure will not.

All this seems positively to encourage people to modernise and abuse old buildings, when the emphasis should clearly be on informed repair and restoration of what is there. If a Government minister should ever chance to read this book, I earnestly implore him or her to devise something more forgiving.

Skip Hire

When repairing or altering a town or village house, it may be sensible to hire a rubbish skip to keep on site, or in the road outside. Then you can get rid of the vast amount of rubble which accumulates as the job goes on. Do not, however, waste skip space by throwing in worthless chunks of wood or old doors that could as easily be stacked or burned. The skip is often exchanged weekly and can be rented at an agreed rate with an additional charge for the changeover. While the rent is fairly low, the changeover may cost five or six times as much. So on sites with good access and plenty of space to make a rubble heap that can be cleared periodically or at the end of the job, hiring a skip all the time is not worthwhile.

Lawyers

Conveyancing follows a fairly set pattern, whether the building is old or new, though more legal complications can arise with an old one. With any house sale or purchase, you never know whether it is your lawyer or the other party's who is making it all move so remarkably slowly. But sometimes the layman's natural impatience should indeed be curbed, especially if there is no reason to believe that the other party is about to renege. Solicitors tend to be cool and amused by their clients' hot-blooded attitude towards the peccadilloes of the 'other side'. They are trained to think this way, since temper tends to cloud the judgement and promote mistakes. Yet there are times when a client would do well to instruct his lawyer to tell the other side that he will be pushed around no longer.

When to get angry in the midst of a deal is a matter for nice timing, and should never be anything but a calculated display to show that you mean business. The act should only be played out after the odds have been weighed and the possible results assessed. Not all vendors or pur-chasers are guided by sweet reason, and it is pointless to lose a house, or its sale, in order to get something off your chest.

Questions of condition can loom large when you buy or sell an old house and there can be weird demarcations of ownership, such as 'flying freeholds'. These, which usually occur in town properties, establish possession of parts of one building which overlap bits of another. A good example would be the room over a coach entrance. This might belong to one building, while the entrance itself does not.

Use of the system of 'enquiries before contract', which are sent to the vendor's lawyer, should reveal matters affecting boundaries, outstanding disputes which a purchaser might inherit, the position regarding gas, water, electricity and drainage, rights of way, ownership of roads and which fixtures and fittings are included in the sale. But there are vendors whose understanding of honourable behaviour in these matters is tenuous. They have no compunction about de-camping with fixtures, fittings and mate-rials which were meant to be included in the sale. Their reasoning is that the cost and bother of reclaiming the goods by process of law will decide the purchaser to let it pass.

Some old buildings may be the subject of 'restrictive covenants'. A former owner may have decided to safeguard by this means the future of some part of the building or its land. He may, for example, have wished to prevent visual damage being done to the building by the removal of a statue in a niche on the stairs, or he perhaps anticipated an unsuitable use which he was determined to prevent. Someone who has moved out of a fine old house into its converted coach-house may 'covenant' that no development should take place on the land which now consti-tutes his remaining view of the ancestral acres.

Although a lawyer should turn up these important facts during the searches, it is safer to ask him or her about them, just in case a maverick item has not been covered by the standard questions sent to the vendor.

It is now possible for people who are not solicitors to carry out conveyancing, which may be cheaper and even quicker, but I should not care to bet on the procedures being any more effectively pursued.

Boundaries
Above all, buyers should ensure that the ground plan and boundaries of the property are clear beyond dispute. It is essential, for example, to find out whether the sycamore tree is part of the property; or whether the *whole* kitchen-garden wall, including the half-hidden corner swathed in buddleia, is within the 'curtilage', as the area in which a building stands is called.

If there is the slightest vagueness in the demarcation of boundaries, the time will come when what seemed a thing of trivial importance becomes a major issue. Purchasers should insist that properly fixed markers are erected wherever necessary and that they are accurately transferred to the plan which goes with the contract documents. Faintly or crudely delineated boundaries, where coloured lines fail to make clear exactly what is going on, can prove fatal to an owner's rights on some distant day.

There are few commoner causes of litigation than disputes over boundaries and peripheral corners of property. Even when buying from the nicest people, it is useless simply to say 'Do we get that bit of the stream that runs along the side of the barn?' and leave it at that when they reply 'Oh yes, that should all be part of it'. Months later, on looking at the deeds, you might notice with a sinking heart that the vital line of red ink has left the stream outside your domain.

The Lawyer also makes 'local searches' with the district or borough council, which will supply certificates showing whether there are any claims against the property, such as the cost of making up an unadopted access road. There could be a 'repairs notice', or some other demand, (see previous chapter) which the local authority has served on the vendor, the terms of which have never been met. Important planning decisions may have been taken which could greatly affect the

future of the property. These might be so serious that the buyer no longer feels he can exchange contracts. There may be plans afoot which the purchaser feels need further personal investigation at the local planning department's offices. Some proposal which sounds innocent enough could prove horrendous on closer examination.

The guiding principle with lawyers is to ask them the questions you know should be asked, and insist on unequivocal answers, even if they give the impression that you bore them by fussing over trifles.

Officials

Planning and Conservation Officers
The planning officer is the person who conducts the local authority's business regarding the matters of planning law which have been outlined in the previous chapter, and implements local and nation planning policy. There are usually not enough officers for the work in hand, and many worthy schemes and investigations connected with conservation are set aside while urgent statutory duties take priority.

Gradually, more enlightened local authorities are employing Conservation Officers to deal with listed buildings and conservation area concerns. The best of these are truly knowledgeable and have a thorough understanding of architectural history. Never fear to consult them at an early stage – the more they know of old buildings, the easier it will be to get sensitive proposals accepted. Unfortunately, some councils take on conservation officers without specialist knowledge, failing to give them even the most rudimentary test in the identification of period styles and building methods. Such persons sometimes lack the taste and confidence to give good advice and make sound decisions.

National policy, as interpreted by many experts, demands the preservation of much that is wholly devoid of architectural merit. They work on the basis that almost everything which shows the historic development of a building tells a story which is worth relating. Soon, a new generation of officials will be demanding the retention of beige tiled fireplaces in

Georgian houses on the grounds that they speak for the ethos of the 1950's.

The opinions of old-buildings specialists of one kind or another are sometimes contributed at Conservation Area Advisory Committee meetings, seminars with amenity societies and so on; but none of it can ever make up for the absence of thoroughly knowledgeable planning staff who have studied and read widely in the field of architectural history. Nor is architectural history enough. Everything to do with planning, and especially the conservation of old buildings, requires a very wide cultural background if good decisions are to result.

Most planning officers I have met have been friendly people whose company you could enjoy in any other context than that in which you are both imprisoned. It is the planning officers who are going to decide whether your proposals for planning permission are acceptable or even lawful. However much you disagree with them it is well to show sweet reason, while yet remaining persistent. The majority of planning officers are not rude or boorish; as public servants, they are obliged to remain civil if they can.

The Department of the Environment periodically advises on policy relating to old buildings and planning matters by publishing 'circulars'. One or two of these have been mentioned in Chapter 2. Circulars provide useful guidelines but they tend to be self-contradictory or to lack sufficient definition just when it is most needed. The emphasis is supposed to reflect the views of the current Secretary of State, whose true interests may be whippets and wurlitzers. In fact, circulars are mainly influenced by the ideas of the senior civil servants who draft them. Above all, circulars and policies relate to current government priorities. Beware of using a circular as a stick to beat a planning officer, for the one which he then brings from behind his back may refute everything that you have been claiming.

If you feel you have a planning point which requires some official support, buy from the council, or study in the local library, a copy of the County Structure Plan and its explanatory memorandum.

District councils also produce local plans containing policies for almost everything. These contain the agreed line which will be taken on all kinds of development and you may find useful information in them. However, peruse them with care, seeking references to your point under different policy headings, to ensure that the authority has not left itself some vital loophole, or stopped up the one through which you were meaning to dive.

Building Inspectors and Approved Inspectors
Despite the arrival on the scene of the 'approved inspector', it is still the local council's building control officer you will normally deal with when handling an old house or cottage. The part he plays in your life will remain vital to the success of the enterprise.

It seems to me that the approved inspector is an esoteric creature who will be unlikely to cross the path of the average house restorer. At present, the National Housebuilding Council and one or two other big private enterprises have been given the go-ahead to carry out control duties; but other professionals are waiting to join their ranks. Their role is almost always in connection with new housing schemes. However, the necessity for demanding forms of insurance is proving to be a powerful disincentive. The criteria for approved inspectors are set out in the introductory *Manual* of the Building Regulations.

The main thing to remember when dealing with building inspectors is this. They are individuals with the same degree of human fallibility, variety of character and vulnerability to the pressures of a demanding job as the rest of us. They have to administer the regulations with a fine mixture of common sense and precision.

The advisory methods shown in the *approved documents* of the regulations are not the only answers available to the problems, but clearly it is much safer and easier for all concerned if they are followed to the letter. All the same, the inspectors know that it is the brief and enigmatic legal requirements stated at the beginning of the sections which really count. Section K, about stairs, merely says:

'Stairways and ramps shall be such as to afford safe passage for the users of the building.'

The inspector is obliged to assess any well-thought-out alternative you have to offer. If he takes a common-sense approach he may agree your method, if he is a man who always goes by the book, he will fight you to the end to follow the approved methods in every detail. Some inspectors take a tolerant line on old buildings: 'Is the new staircase going to be at least as safe as the old one?' Don't boil over when an inspector is obstinate, it is his head on the block if something awful results from his decision to relax or dispense with part of a regulation, or to accept alternatives you have put forward.

Fire Officers and Others

An officer from the local fire brigade will inspect your building if it comes within the provisions of the Fire Precautions Act so that you need a 'fire certificate'. The legislation is complex and there is no room here to quote chapter and verse. The duties of the Fire Brigade overlap those of the Public Health and building control departments. Old buildings which are to be used as hotels, guest houses or offices with more than twenty people occupying them need fire certificates. The requirements of the Fire Precautions Act and similar legislation are some of the most architecturally damaging to old buildings. This matter will be referred to elsewhere. The council's building control department liaises with the Fire Brigade which issues a completion certificate.

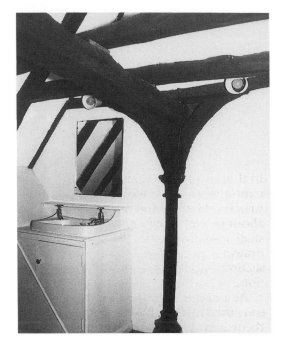

Clear evidence of a really early domestic building, this late fifteenth-century crown post stares accusingly out from a later partition. The hall of the timber-framed building was obviously floored in at tie-beam level in the seventeenth century when new hearths and chimneys were introduced.

On closer inspection you can see the smoke-blackened timbers of what must have been an open hall with a central hearth. The crown purlin, supporting the collars of the paired rafters can be seen peg jointed to one of the braces which curve out from the top of the crown post.

4

ASSESSING THE BUILDING

In this chapter we shall discuss the examination of an old building for evidence of structural failure and other faults brought about by bad design or wear and tear. We shall also be looking at the method of drawing preliminary sketch plans and sections, and how to record the information.

As a matter of principle, you cannot be too thorough in your inspection and recording techniques. No detail is too insignificant to be worth jotting down as a note, even if it is not marked on a plan or photographed. Although you may need expert assistance in assessing the building, it is important to know the kind of thing the professional will be looking for.

Equipment for Survey
Use a clipboard with sheets of graph paper for sketching plans and plain paper for making notes. Use a pencil with a rubber on the end. (One never gets a line right first time and a ball-point sketch can turn into an illegible mess.) Take a surveyor's 20m tape, marked in imperial and metric divisions. You should also have an imperial/metric 3m steel retractable tape.

A fairly fine screwdriver is important for digging into timbers to check for decay, as is a large tough screwdriver for wrenching back boards and so on. A sharp penknife will be required, and a length of bricklayer's line with a plumb bob. It is also helpful to carry a lump hammer and mason's chisel for assaulting suspicious-looking areas of plaster or blocked-up wall. A very powerful torch is a must, for looking in roof spaces, up flues and in cellars.

I carry a small mirror attached to a 2ft length of very springy batten with a string attached to the end. With this you can see

around corners, often using a small torch which can be inserted into such awkward places as gaps in lath and plaster or under floorboards. A pair of iron lifting hooks, rather like boot jacks, are a help for heaving off manhole covers.

If you mean business, it is essential to have a pair of household steps for inside the building, and a long ladder for outside. A garden trowel is frequently needed for digging out around wall footings and other investigations of that kind. An adjustable template is important for taking profiles of mouldings.

If you can beg or borrow a damp-meter from a professional it will be useful for reading the moisture contents of various components like joists, skirtings and wall plaster. However, these gadgets should be interpreted with discrimination, since the points at which you choose to dig in the electrodes are crucial. It should also be remembered that moisture contents vary according to weather conditions, the time of the year, and a host of other factors.

Binoculars are useful to inspect bits of roof or wall which are inaccessible by ladder. A camera should be carried as a matter of course: ideally have a wide-angle lens for photographing interiors and use very fast, black-and-white film (ASA 400). You also need an electronic flashgun.

There are now relatively cheap gadgets for locating pipe-runs and studs under surface finishes. Depending on what stage your investigations have reached, you may need a spirit-level to check the run of pipes, gutters, roof-flats and so on. To find out differential levels between the ground outside and floor inside, or the fall of land between one point and another, a dumpy level is good if you can get hold of one and know how to use it. I find that the relatively cheap and easily used Cowley level is adequate for this purpose. It is a non-magnifying, split-image instrument mounted on a light aluminium tripod. Unlike a dumpy it does not have to be properly levelled before use. With it goes an aluminium graduated staff in two sections with a horizontal fluorescent timber cross-piece at which you aim the level.

Your assistant holds the staff while you look through the eyepiece of the level. The cross-piece on the staff appears to be split in two, one side seeming higher than the other. The staff-holder moves the cross-piece up and down the staff until the two images coincide. When they do you have a level line of site and fix the cross-piece with a thumbscrew before reading the graduation on the staff. By taking a series of readings without moving the position of the level and tripod, you can find out whether the floor inside the building, for example, is higher or lower than the ground outside, and by how many millimetres.

You will discover that there can be astonishingly deceptive variations in level between one end of a building and another. By knowing what these variations are you can determine whether there will be problems with drainage, what digging-out may have to be done at the back of a building to prevent ingress of damp, or how much deeper it may be necessary to dig foundation trenches for an extension at point A rather than B.

Most builders would never take even a small part of this equipment with them when looking at a building, but that does not mean that it is futile to do so. It is better to have too much equipment with you than to be short of an essential tool, torch or ladder, and to have to make yet another thirty-mile drive to investigate a point which could have been settled before.

The Survey
First make a preliminary tour of the building and its curtilage to give you a rough idea of the layout and the general state of repair. Then make a systematic exterior tour, noting everything of impor-

tance about the house's condition and construction, together with any special architectural features.

Next, photograph every elevation from the most straight-on viewpoint you can manage. You photograph the setting to show the way in which the building lies in relationship to the land, plus any outbuildings. Lastly take pictures of road, lane or drive access points and special boundary markers.

Inside the building, make a similar tour, noting and photographing as you go. Take photographs of all the defects, and all points at which you anticipate operations of some kind may have to be carried out in future. Then work around the building on the outside, taking detailed photographs in the same way. In other words, photograph the corner between the two wings where it has crossed your mind you might have to add a staircase extension; and photograph the cracks in the gable wall.

If you have reached a stage at which it is not a waste of time to measure up the building in preparation for drawing a full set of survey plans, you should go about it in the way outlined below.

Fig 1 This survey sketch plan relates to the semi-derelict cottage illustrated on page 8. Although roughly drawn, it is more or less to scale and clearly shows the information from which more accurate drawings might be prepared. Diagonal measurements are recorded, the positions of joists indicated and heights are ringed. Running measurements are taken along walls. Thus, the left-hand corner of the plan is marked O. The left-hand side of the first door opening is 2.3m from the corner; and the right-hand side is 3.31m, giving a door opening 1.01m wide. The overall length of the building is 13.46m.

Plan of the cottage shown on page 8, as it would be accurately drawn from the rough sketch-plan measurements and details shown in Fig 1. Draughtsmen do not normally block the walls in black when preparing plans for planning permission or Building Regulations approval. For survey drawings of old buildings, however, it gives contrast and makes the plan easier to understand and interpret. The broken line marked A–A shows the points at which the draughtsman has taken his section through the walls. Since a plan is really a horizontal section through a building, it must be taken at a level where the window openings will show. Broken lines for the stairs indicate that the steps in question are above the level of the section.

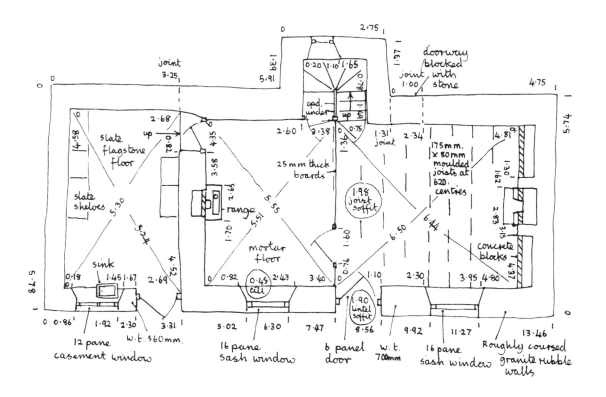

12 pane
casement window

slate
flagstone
floor

slate
shelves

sink

joint

25mm thick
boards

cpd.
under

range

mortar
floor

16 pane
sash window

w.t. 560mm.

6 panel
door

w.t.
700mm

doorway
blocked
joint with
stone

175mm.
× 80mm.
moulded
joists at
620
centres

concrete
blocks

16 pane
sash window

Roughly coursed
granite rubble
walls

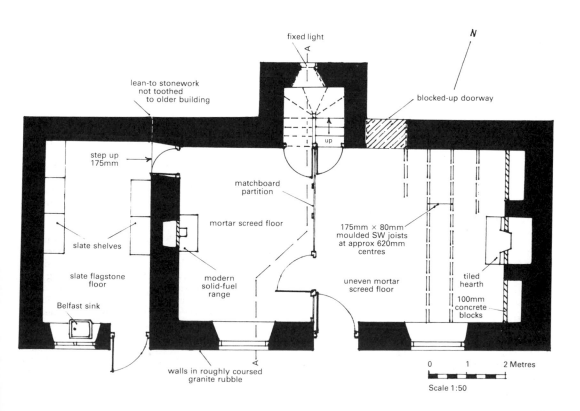

fixed light

lean-to stonework
not toothed
to older building

blocked-up doorway

N

step up
175mm

matchboard
partition

mortar screed floor

175mm × 80mm
moulded SW joists
at approx 620mm
centres

slate shelves

slate flagstone
floor

Belfast sink

modern
solid-fuel
range

uneven mortar
screed floor

tiled
hearth

100mm
concrete
blocks

walls in roughly coursed
granite rubble

0   1   2 Metres

Scale 1:50

41

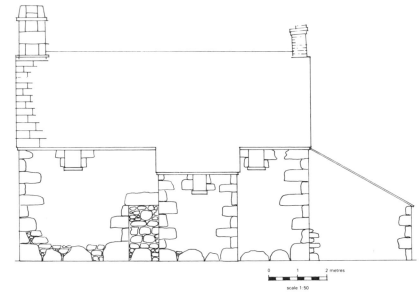

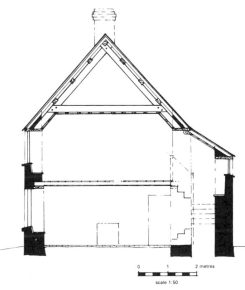

Fig 1b The north (rear) *elevation* of the specimen cottage shown on page 8 is just an outline drawing of the building taken flat-on, without any kind of perspective and preferably to the same scale as the plans and sections. Some of the stonework is sketched in and also a portion of the roofing slates – to show that the former is roughly coursed rubble with big granite quoins, and that the slates are in diminishing courses. Any windows should be carefully drawn to show their type and glazing pattern. In the case of this building the frames are all that remain on the north side.

Measured Drawings
Take full exterior measurements, starting with the front façade. The measurements should be running ones so that you start at the corner made by the front and gable walls, and read off the figures from the tape as they appear at each opening or feature. This process is repeated for each wall around the outside. (If you can dragoon somebody into holding the other end of the tape for you it will enormously speed up this tedious chore.) Fig 1 shows a survey sketch plan of the building on page 8.

Then measure the drop from the eaves to the ground (in as many positions as necessary) and the heights of the window openings, including those upstairs. It should not be assumed that windows are the same size because they look as if they are. Disastrous errors in ordering joinery or proportioning window-panes can be made by not measuring windows individually. Remember to check the actual thick-

Fig 1a Section at A–A of the same cottage. By referring to the plan, it can be seen that the section takes in the staircase at the back; and the broken line A–A on the plan is cranked so that the section of the front wall is through the windows instead of a blank piece of wall. Depending upon how detailed you want the section, you may, as in this case, merely outline the position of window frames, or you can draw a true section through the various components such as sills, rails and glazing bars. Although strictly not within the terms of the section, it gives much greater value to the drawing if a typical roof truss, a floor joist, the chimney and the position of important door or fireplace openings are also shown in elevation.

ness of the sills; and to note the thickness and width of the window sashboxes or frames, and those of the doors.

Take the height of the chimneys, using the ladder if it is long enough, and their girth as well. The pitch of the roof is better calculated from inside the roof space by triangulating measurements off a roof truss. If the collar component is level, a right-angle triangle can easily be obtained. If the collar runs at a marked angle, pencil a true horizontal line on it using the spirit-level. The pitch may then be drawn on paper to scale, or calculated by trigonometry. I have an excellent little adjustable spirit-level with a protractor scale from which rafter angles can be read.

Old buildings are far more difficult to measure than modern ones, since so many walls are out of plumb or not at right angles to each other on plan. The walls themselves, especially if built in random rubble stone or cob, tend to vary considerably in thickness. This means taking far more tape readings of measurements through door and window openings.

The fact that a building has been constructed in various phases over a number of centuries will produce revealing anomalies of plan which, while difficult to survey, may tell a lot about its history.

To obtain the correct shape of the rooms on plan, measure the diagonals from corner to corner, so that again, by triangulation, you can see what angles are formed by the walls. Heights of rooms should be measured to both joist and ceiling soffits. The figures should be circled on your rough sketch plan and clearly marked as to what they represent.

Do not forget to measure inside the recesses at windows, those flanking fireplace projections, and also areas containing staircases. Include the interior measurements of hearths and record the materials with which floors are finished. Mark in all steps and changes of level, drawing an arrow showing whether the steps lead 'up' or 'down' from that floor. Mark in, with a dotted line, any blocked doors, windows or similar openings and note any architectural features of interest.

It is important also to record mouldings

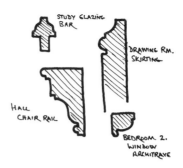

Fig 2 When recording a building, profiles or sections of details such as glazing bars, chair rails, window architraves and skirting mouldings should be sketched freehand. Later, it may be necessary to transfer accurate profiles of some of them from adjustable templates or squeeze moulds.

or architraves, glazing bars, fireplace surrounds, panelling and so on, using quickly sketched profiles (Fig 2).

Go systematically around the house marking in existing electrical sockets, light switches and pendants, the cooker point, distribution board, earth leakage circuit-breakers, meters and main fuses. Note whether the circuits are ring mains or radial circuits and how many fuseways there are in the fuse box. Some buildings may have a great unwieldy array of iron-boxed switched fuses, each with a separate radial circuit to control. All these may need renewing, since such a system is likely to be at least forty years old and the wiring faulty.

Establish the positions of the rising main and stop valve, together with the taps, waste-outlet pipes, soil-and-vent pipes, roof storage tanks, header tanks for central heating and radiators, and the pipework serving them.

Go outside and discover the positions of the drain-runs and manholes. Follow these through to the public sewer, if there is one, or to the septic tank or cesspit. If there is no main drainage, try to establish what the private treatment system is, if any. A septic tank will have two manhole covers, either one over the inlet dip-pipe and another over the wall dividing the two chambers, or separate ones over each chamber. These days many septic tanks are preconstructed units made from glass-reinforced plastic. They are usually in the shape of a flask. The tank overflows partially treated effluent into a soakaway and this, in turn, seeps away though a pipe or land drain.

If there appears to be only one chamber, it is almost certainly a cesspit, which has to be emptied at regular intervals. You will often find that it has been built with an overflow linked to land drains, so that waste either seeps into the surrounding subsoil, or discharges into a stream through some form of pipe. If you find a patch of marshy-looking ground somewhere on the far side of the so-called septic tank you will be able to hazard a guess at what is happening.

Some irresponsible characters construct a watertight cesspit, designed to be emptied by the local council tanker or that of a private contractor and then, once the pit has been passed by the building inspector, knock a hole in the bottom to allow the future contents to drip away steadily into the ground. Apart from being a bad potential health hazard, polluting fresh springs, wells or other water supplies, this is a blatant and self-defeating infringement of building control.

Check the position at which the electricity supply cable enters the building and whether it runs from underground, from ordinary 240 volt overhead lines, or from a more substantial pole carrying an 11,000 volt to 240 volt step-down transformer. The transformer is normally a grey metal box into which the two (level and parallel) supply cables are fed via porcelain insulators. The 240 volt supply is then taken from the pole to the building by a fairly substantial insulated cable.

Find the main gas supply pipe, if there is one, and check where it enters your land and the building itself.

Finally measure any entrance gates, coach arches or other strategic openings and mark them on your plan.

Faults of All Kinds

There are only a limited number of headings under which faults in an old building can be listed, but there are endless manifestations of them. It may be assumed that an old building will have a certain number of weaknesses, and some of them are quite likely to be fairly serious. By serious, I

mean that, while curable, they are either complicated or costly to put right.

The first way in which a building is likely to become vulnerable is through faulty design. The second way is through neglect, abuse or misuse. The third is through the effects of wind and weather and the fourth is as a result of infection or infestation.

Faulty Design

If the original builders failed to provide adequate foundations, undersized important structural components, omitted to weather abutments between one surface and another, or used unsuitable materials, these basic design faults will have detrimental effects on the building as time goes on. Gradually, some form of structural collapse will occur, or the more rigid building components will fracture, allowing the ingress of moisture. Flexible components, such as oak structural timbers, will sag and distort. In turn various types of decay and infestation will occur.

Neglect, Abuse or Misuse

If routine repairs are not carried out, surfaces painted, holes or cracks blocked, gutters cleaned, and earth kept back from outside walls, trouble will ensue. All that is neglect. Abuse and misuse may be said to include the overloading of floors, the botching of repair and alteration works, and a tendency on the part of owners to knock a building about – to handle it roughly. Huge fires are lit in hearths which cannot withstand them, tie-beams and collars of roof trusses are cut out, weakening the structure. Hard renderings are applied to surfaces which should not have them. Openings are knocked through walls without thought of the effect they will have on the building's ability to transfer loads to the ground.

There is literally no end to the ways in which people will damage old buildings, through either thoughtlessness or callous disregard.

Wind and Weather

Even a well-looked-after building, which is not misused, must suffer continuous deterioration as a result of wind and weather.

You may paint and repair it with almost craven zeal, and still it will run from bad to worse. Anybody living in an old house has to accept this fact – and it does not only apply to old ones.

How well a building can stand up to the elements, maintenance apart, depends upon the soundness of its design. The most important point to remember about old buildings is that many of them depend for their survival upon a reasonably high degree of flexibility in their structure. All materials try to move, settle, expand and contract. If they are sufficiently flexible both in themselves and in the way in which they are put together, they can prove surprisingly durable. Trouble often starts when attempts to repair or renew parts of a building are made with rigid forms of jointing, materials and finishes.

If this book does nothing else, it will have succeeded if it can convince owners and builders of the need to *allow a building to be flexible*. This will be a recurrent theme when we discuss methods of repair.

Infection and Infestation
Infection can be defined as the introduction into the building of micro-organisms which multiply in the form of fungi and similar growths. The obvious examples are dry and wet rot, and black moulds formed on wall surfaces as a result of damp and lack of adequate ventilation. There are also chemically formed moulds which result from the interaction of incompatible materials and the presence of substances such as salts in mortar, or sulphur in the atmosphere.

Infestation can be summed up in one word – 'beetles'. You may have rats or mice, bugs or cockroaches – and unpleasant they may be – but, apart from leaving mess about, nibbling timber, blocking up places which should be open and opening places which should be closed, they do only limited damage. They can be eradicated, albeit with some occasional difficulty, and afterwards whatever damage has been done remains more or less as they left it.

Beetle is another matter altogether. It is a menace and few old houses do not have some signs of its presence. Houses containing many hardwood components may have suffered from death-watch beetle, and almost every old building has undergone some attack by woodworm. These are the main beetles the house restorer needs to worry about, although mention should be made of the house longhorn beetle and certain weevils.

Having said that, it should be remembered that there are numerous other types of beetle, all with Latin names and nasty habits. If you include such esoteric creatures as masonry bees, which bore holes in soft stonework, you have a formidable array of bugs to beware.

Identifying Faults
We consider in the remainder of this chapter the danger points to inspect for faults of neglect. Remedies, where they are not self-apparent, will be dealt with under their appropriate chapter headings. It is a temptation to deal with building faults under the main headings of faulty design, wind and weather, neglect, infection and infestation, but it is not really possible to do so; you find that one fault promotes another, and the pattern of cause and effect varies from case to case.

For example, did the valley gutter between the roof-slopes deteriorate because it was badly designed in the first place, with insufficient up-turn of lead under the slates and fall towards the outlet? Or did it fail because it was abused and neglected? Was it never cleaned out or was the lead holed by workmen stamping over it during other repair work?

The simplest approach is to discuss the visual evidence of faults, pointing out what is indicated by each of the superficial signs that things are not as they should be.

Roof Faults
It is not unusual to see the ridge of an old roof sagging in various places along its length. This shows that there has been some failure of the ridge purlin, pole or board, or of a batten, or a movement of the rafters or the trusses. When you climb into the roof space, you can discover whether the fault has already been dealt with or is still active. The rotten timbers

What could look more Victorian than this Welsh farm-house in the mountains not far from Machynlleth? The fretwork barge boards, rustic porch and brick chimneys all remind you of workers' cottages on some thousand -acre estate. The interior tells another story.

Inside the simple stone farmhouse is a stylish mid-eighteenth century staircase with turned balusters, moulded handrail and solid, square newel posts. You might suppose that it had been salvaged from some nearby gentry house, but the flights fit snugly in the hall and there are no obvious signs that treads and risers have been adapted to new requirements.

Even more surprising is the evidence of timber framing in upper-floor rooms. What looked at first like a run-of-the-mill nineteenth-century building turns out to have much earlier origins. The framed farmhouse might have been built in the mid-seventeenth century. The place was then 'done-over' in the eighteenth century and finally Victorianised.

may have been replaced long ago, rafters may have been doubled up, or other remedial measures taken. The sagging roof-line may only mean that the roof was repaired from inside without the tiles or slates being stripped. If the roof is stripped for repair, any deflection is often adjusted, since it is much easier to lay roof coverings on a flat surface.

The slope of the roof itself may also run in a series of waves or hollows, as a result of rotten or sagging rafters or purlins. Here again the roof covering may be fairly sound and even quite recent, since many contractors re-tile or slate using the old carcass timbers, merely laying on new battens and felt. Some attempts are made to level up the battens with chocks of wood; but all too often the workmen let the roof follow its own contours. From the aesthetic point of view that is not always a bad thing. From the constructional angle it is, since it prevents the roof tiles or slates from sitting down tightly, one on top of another.

Look at the abutments between the roof slates and the chimneys. This is a constant area of concern, since many old roofs have no lead flashing and soakers to weather this important connection Frequently there are slate, brick or tile listings which protrude from the stack to cover the junction. These may have broken or decayed. The usual method of repair – a very unsatisfactory one – has been to run a hard mortar fillet down the sides and along the front of the brick or stone chimney. As a result of thermal movement, this nearly always cracks away, allowing water to get into the roof and rot the timbers.

Another obvious fault is missing, broken or slipping tiles. Here again, the cause can be rotten tiling laths or battens, or broken, decayed and twisted fixing pegs. Sometimes there is a condition described as 'nail sickness' which means different things to different people. Some use the term to denote widespread rusting and breaking of slating nails. Others say nail sickness is when sarking boards or battens have been so frequently nailed and re-nailed that most nails are loose, and there is no firm timber into which you may hammer new ones.

Obviously signs of vegetable growth in valley gutters make you suspect a danger of damp building up without any means of ventilation – a certain method of rotting any nearby timber which is vulnerable. Eaves gutters too may be obstructed so that they overflow – causing damp problems.

The chimneys themselves may lean, sometimes in a most pronounced manner, usually away from the prevailing wind. This fault is normally found in brick chimneys and is caused by constant wetting and drying over a period of years. The mortar joints on the damp, windward side expand through the action of sulphate salts. The fault is particularly associated with hard Portland cement mortars used to bed the brickwork of chimneys serving coal-burning slow-combustion stoves. If the inside of the flue is unlined, the chimney is especially prone to suffer expansion of the mortar joints, caused by the dilute acids condensed from the escaping gases.

The flaunching around chimney pots is another weak point to check. This form of mortar capping, which is struck off to a fall around the foot of the pot, often cracks, allowing water into the brickwork. In the absence of a damp-proof course in the stack (a normal omission in old buildings), the water finds its way to the inside of the wall upon which the chimney is built.

Grouted roofs are also likely to cause trouble. A quick brush-over with a cement slurry is a traditional means of blocking leaks in the slatework. However, it stops the roof breathing and seals in unventilated damp; it is also likely to crack and draw in further moisture by capillary action.

Carefully inspect the line of the eaves for signs of broken gutters, blocked downpipe hopper-heads, rotten fascia boards or cornices, rotten rafter ends, decayed wall plates and the absence of doubled-up slate or tile cloaking in the eaves course. It is in the nature of slating and tiling that an extra row must be laid at the eaves to provide the same double thickness at this point that will be the automatic minimum over the rest of the roof slope. Badly designed roofing sometimes lacks this vital course of slates, allowing moisture to seep into the cracks between those in the first row. The moisture soaks the top of the wall and rots the wall plate, rafter ends, lintels and even joist ends or panelling.

Ridge tiles are frequently broken, slipped, or in need of pointing or re-bedding. The verges of the roof (ie the edges which run down the gable walls) may also need pointing, both under the slates or tiles and between them.

Georgian buildings often have parapet walls which mask from view the bottom part of the roof-slope and its lead-lined gutter. An inherent fault in parapet walls is the absence of a damp-proof course, and the inadequacy of the coping for preventing rainwater finding its way into the wall. It is surprising what floods of water can channel down through the joints in the stone or brickwork to the inside of the rooms below.

Look for decaying barge boards and purlin ends. These are especially liable to decay in gable roofs where the purlins are carried across the rubble stone walls to be cut off flush with the wall. The trimmed ends of the purlins are weathered with a cloak of slates or tiles, which crack or slip away, leaving them exposed. Mortar is smeared over the faults as a form of repair and this exacerbates the situation. The purlin, which is usually buried in stone and earth bedding material, is unventilated and soon begins to rot away.

Dormer windows should also be checked with care. Cheek tiles, boards or slates may be faulty, or the abutments between the roof covering and the cheeks may be poorly designed. Secret gutters are particularly vulnerable, since they become blocked with wind-borne earth and moss, allowing water to build up and spill over on to the rafters and batten or lath ends.

Flat roofs are the source of endless tribulations. Well-designed and well-executed lead flats may survive the centuries, but they still require careful maintenance. Modern felted or asphalted flat roofs, however, are always in need of attention. One of the worst faults derives from a lack of adequate fall in the roof timbering so that rainwater does not drain

away. Look for puddles of standing water on such roofs, especially in dry weather when they should have long since drained down the appropriate outlet. Also consider the possibility of damp having penetrated the felting, or found its way into cracks and similar weaknesses in the asphalt. This may have caused decay in the roof timbering and a subsequent dishing of the roof in that area.

Wall Faults

It requires considerable experience and know-how to diagnose the meaning of cracks in walls, and we shall consider these in a moment. But first let us discuss the other types of faults which will be apparent to any keen observer with some knowledge of the way in which a building is put together.

One of the first points to look for is whether the building has a damp-proof course of any kind at the base of the walls. Very few old buildings do, since this was a feature which did not develop until the early twentieth century. However, there may be extensions to an old building which do have damp-proof courses, or the building may have had some form of barrier to rising damp introduced in fairly recent times.

If there is a damp-proof course, the classic cause of trouble, as Fig 3 shows, is 'bridging'. This means that the impermeable damp-proof course, which is intended to prevent moisture from the ground

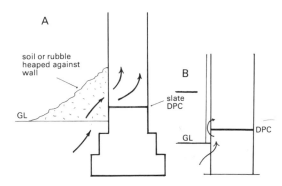

Fig 3 The arrows in A show how moisture from the ground, below the damp-proof course, can find its way into the wall by bridging through earth or rubble which has been heaped at its foot; B Shows how damp may bridge the damp-proof course via rendering.

rising up the walls, has been by-passed. The commonest form of bridge at ground level is the banking-up of earth or rubble against the wall. If there are signs of damp inside from that cause you may breathe a sigh of relief, because the remedy is about the simplest that this book contains – the use of a shovel and wheelbarrow.

Look at the condition of the window frames and any other joinery. Timber sills, particularly, are very prone to decay and should be prodded with your fine screwdriver. Dig into the frame of the sash-box at the point where it abuts the window reveal, another danger spot. Double-hung sash windows are suspect at the closing rail of the upper sash, where sheeting rain-water continually gathers and drips off. Also look at the condition of the putty, which may have cracked and flaked away in chunks, allowing moisture to linger in the glazing rebates.

Door frames often rot at the ends resting upon the threshold, especially if they are made of softwood. The timber plugs, which provide a nailing ground for fixing back frames in openings, often decay, allowing the frames to pull away. Broken sash-cords may have allowed top sashes to stay open, admitting driven rain.

All joinery may suffer from lack of paint. If the paint has peeled down to exposed timber, this is sometimes a result of the omission of a priming coat. Hardwood frames stand a great deal of neglect without decaying (special points in this connection are mentioned in Chapter 9).

The area of ground at the foot of the walls should be inspected to ensure that there is an adequate fall, away from the building, to drain surface water. If the surrounding area is of permeable material like earth, clinker, hardcore or gravel, its height in relation to the interior needs to be considered.

Where you notice any kind of garden wall or buttress adjoining a wall, anticipate damp inside the building at that spot. Correctly, such walls should have a vertical damp-proof course to stop them bridging or channelling damp into the building; but naturally this concept was either unknown or unconsidered by builders in

former times. They might have argued, with some reason, that the garden wall was no damper than the wall of the house itself, but it is only partly true, since the walls of the building are protected from above by the roof, while a faulty garden wall is considerably more exposed. Buttresses are considered elsewhere, since their structural significance is far more important than their potential for promoting damp.

An obvious sign of trouble in plastered walls is the presence of cracked, flaked areas which show the stone or brick beneath. In some cases, through a mixture of poor adhesion and ingress of moisture there will be a definite sponginess to the touch and it may be possible to depress areas of the plaster with your fingers. Faults in exterior renderings, when not caused by structural movement, are often the result of inadequate key, bad materials or unsuitable mixes. Hard renders which have separated from their background key will often ring hollow when you tap them.

Timber-framed buildings are most vulnerable in places where the panels of infilling abut the structural posts and beams. Here again is proof that a *flexible* building is to be preferred. Traditional lath or wattle panels rendered with clay daub or plaster have an inherent ability to absorb the relatively small amounts of thermal movement which occur. Brick nogging between structural timbers is less amenable and should be inspected with a beady eye. Partly for aesthetic reasons, and partly because the weight of the bricks permits a tighter joint against the sides, these panels will often be seen in a diagonal form of bonding.

Any brick infill panels bedded or pointed with Portland cement mortar are likely to promote difficulties; they require very soft lime mortars. As a general rule be suspicious of any part of a building where mastics have been used either to repair cracks in walls or to form a seal between components. Although some speak well of mastics it is my own belief that they are only effective as part of a properly designed joint. On their own they are at best unreliable.

In areas where granite is the traditional

building material for walls, there should be little deterioration of stonework, although the occasional soft and crumbling piece may need replacing. This exercise of replacing individual stones is easier recommended than done.

Walls composed of freestones such as sandstone or limestone are prone to all kinds of deterioration. They spall (splinter), weather unevenly, and suffer from wind erosion, staining and numerous other problems. They depend for their durability on the geological structure of the rock from which they were quarried. Some, like Portland stone, are relatively hard, but there are sandstones so soft that you can quickly form a depression by merely twisting the edge of a coin against the surface.

Staining can be seen on walls, dribbling down from rusted ferrous-metal fixtures like iron balconettes, brackets or hinge-pins. Where iron cramps have been used to hold coping stones together, for example, you again find staining from rust. Worse is the expansion effect of rusted iron, which tends actually to split the stone around it.

In towns, walls may badly need cleaning, after standing through generations or even centuries of soot deposits from the palls of chimney smoke which once enveloped all urban areas. The stones may also show signs of chemical attack from sulphur in the atmosphere, which causes the formation of crusts, scaling and flaking – especially on limestones.

Brickwork too can be badly worn by erosion or spalled by moisture freezing below the surface. Bricks can also crumble when bedded or pointed in hard mortars. It is a general principle, which will be referred to again, that mortar should never be harder than the stone or brick for which it is used.

The pointing itself may be loose or cracked, or have fallen out altogether. Soft lime mortars traditionally used for brickwork can lose both structural strength and adhesion over a period of many years. Some brick-built houses have suffered so badly as a result of the original builders' meanness with the lime component that you can literally pull bricks out of the walls with your fingers.

Timber lintels in walls decay, and as they cease to provide support the brickwork or stonework may visibly sag over openings. Major structural disaster has sometimes been averted by the prudent incorporation of a relieving arch which transfers most of the loading to the sides of the opening. In that case it will be observed that the area of stone above the lintel and

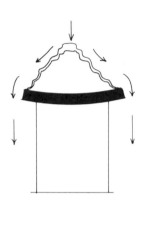

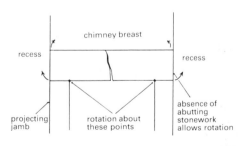

Fig 4 When an inner timber lintel sags badly (*top drawing*) or fractures, it drops slightly and the triangle of masonry above falls away. The remaining stonework above acts as an arch carrying the load to either side of the opening. A cracked stone lintel will drop fractionally and may then lock (*middle drawing*), forming an arch within its own depth. Bottom drawing shows detail of lintel not locking when unrestrained.

below the relieving arch has dropped, no longer serving any structural purpose.

Tile- or slate-clad walls suffer in rather the same way as roofs: rotten battens or laths, rusted fixing nails, and the effect of head-nailing can all help faults to develop. Since cladding is frequently used in conjunction with timber framing, any structural faults or decay will often be manifested by collapse or distortion of the wall itself. This obviously brings about a general loosening, lifting or cracking of the cladding material. However, the flexibility of timber, and the fact that slate or tile cladding is capable of quite a lot of movement before any noticeable damage is done, mean that you should investigate such walls with care. All may not be as it seems.

Old buildings are continually settling and moving for various reasons. Much of this movement is harmless. It either takes place when new work is carried out and then stops, or it may be seasonal or occasional. A crack will open up in a dry summer and close again during the winter rains. Some quite severe faults may be seen to have developed within a few years of the building's original construction, but stabilised themselves perhaps two or three hundred years ago and have not moved since.

Any new building settles down on to the subsoil which its floors and walls are compressing for the first time. If all is well, that movement ceases when the subsoil has finally taken up its permanent load. Trouble arises when one bit of a building settles further than, or at a different rate from, another. Flexible buildings absorb some of this movement and rigid ones tend to crack up.

The most serious wholesale movement of a building results from general subsidence of the land upon which it rests. The obvious examples are houses which disappear into holes in the ground after the collapse of mineworkings; and buildings which slip, tilt or break up when massive erosion or landslides take place on the sides of mountains or at the edges of cliffs. Houses constructed on made-up ground which is inadequately consolidated can suffer similar problems.

In these cases, the entire building may move, especially if it is built all of a piece, as is a sixteenth-century timber-framed house for example. If it is of brick or stone it will probably break up as it does so.

However, far more likely than full-scale subsidence are the various effects of differential settlement. Clay soils are particularly prone to expand and contract, and the foundations move in consort. Outside areas of wet subsoil dry and shrink causing movement of the foundations, while those inside the building retain a more constant moisture content and remain relatively stable. Leaking drain-pipes, water mains, underground springs and so on can wash away the subsoil beneath the foundations, causing them to move or collapse.

Weaknesses in the ground beneath the foundations may have allowed hollows and pockets to form, so that masonry either sinks or remains suspended until some additional load or disturbance causes collapse. Sometimes originally adequate foundations are subjected to loads for which they were not designed, and they suffer accordingly.

When you are looking at the walls of an old building be wary of areas in which large new openings have been made at a later date. It is often the redistribution of the loading which proves dangerous, by concentrating additional weight on piers and other elements which are inadequate. Buildings which have been adapted to contain wide shop-windows have frequently suffered in this way (Fig 5). An extension can crack away from walls of an earlier building, either because the bricks or stones have not been properly bonded in with the original work, or because of differential settlement. Ideally, movement should be absorbed by some kind of vertical slip joint at the junction between old and new.

It is important to find out the kind of subsoil upon which a building stands, especially if the walls seem a bit unstable. Although this is really a specialist matter (soil mechanics), you can often discover all you need to know by digging a few trial pits and sampling the soil at different levels. First dig neatly and carefully at the base of the suspect wall until you reach a

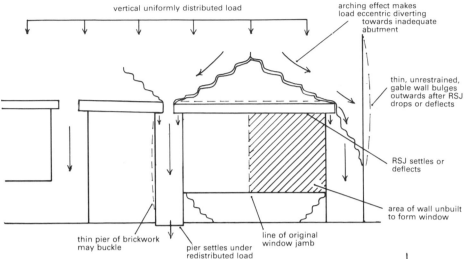

vertical uniformly distributed load

arching effect makes load eccentric diverting towards inadequate abutment

thin, unrestrained, gable wall bulges outwards after RSJ drops or deflects

RSJ settles or deflects

area of wall unbuilt to form window

line of original window jamb

pier settles under redistributed load

thin pier of brickwork may buckle

Fig 5 Widening a window opening to form a shop-front, for example, can have adverse effects on the adjoining walls. If the new steel beam or lintel drops or sags, even minimally, the universally distributed load of the wall above becomes eccentric and thrusts any weak abutments outwards. A thin pier may sink or buckle under the new loading conditions; and an end wall may bulge. Diagonal cracks also appear, especially around doors and windows.

Fig 6 Inspection pits may be dug to examine the subsoil beneath the foundations of a suspect wall. Dig away from the foot of the wall to discover the depth of the foundations; but the hole should be made close to the apparent point of maximum settlement, rather than *below* the more stable area of the wall, from which the crack rotates. Standard practice is to keep subsoil inspection pits at least lm from the foot of the wall; but some experts believe they may safely be dug against the foot of the wall provided they are only 750mm square and are backfilled immediately after inspection.

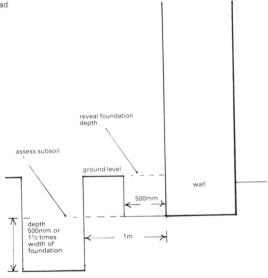

reveal foundation depth

assess subsoil

ground level

wall

500mm

depth 500mm or 1½ times width of foundation

1m

point where there are no further signs of foundation masonry or brickwork. This hole should be about 500mm square. Opposite this, and about lm from the base of the wall, dig another 500mm² hole, to a depth of about 600mm below foundation level. For a new building a depth of at least one and a half times the width of the foundations is normally recommended (Fig 6). The pit needs to be proportionally greater in area to allow free movement when digging to greater depths.

Some people will dig the initial hole right down against the foot of the wall below foundation level. I am exceedingly cautious about this sort of thing, preferring to look at the nature of the subsoil at a little distance from the wall to begin with, especially if the walls are poorly bonded and the pits are not going to be immediately backfilled.

Having dug the trial hole you can take samples of the soil at different levels and decide to what broad category or categories they belong. In the most simplified classification, there are three main kinds of soil you will encounter: non-cohesive, cohesive, and rocks. Non-cohesive soils are mainly sands and gravels. Cohesives consist of various sorts of clay. Rocks may be anything from granite to a clay shale or chalk. The soil you are investigating may turn out to be a mixture of these types – gravelly clay, for example. Space does not

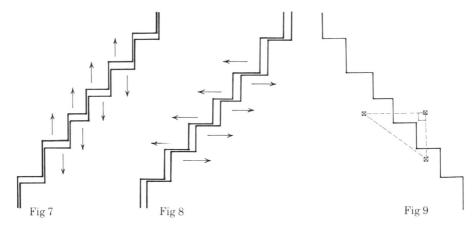

Fig 7   Fig 8   Fig 9

allow us to consider the different properties of these soil types and how to test them; but it is enough to say that the two main points to be considered are how hard the sample was to get out and how it handles when you squeeze some in your fingers. Clays can be moulded in the hand, rocks have to be drilled and broken up to excavate. Sands can be dug straight out with a spade, unless very compact.

Types of Wall Fault
The several types of fault which may be found in the walls of an old house can be more or less classified by their causes. Some occur because the foundations have settled for whatever reason. Others are the result of thermal movement – expansion or contraction due to heat. There are others again that arise from similar forms of expansion and contraction set up by wetting and drying, frost or chemical processes. Yet a further group come under the heading of poor construction, inferior materials and bad design. Finally, there are the faults which derive from over-loading or the application of weight (thrust) in a way which unbalances the structure.

Foundation Settlement
Cracks indicating foundation settlement are normally diagonal; and the maximum settlement has generally taken place directly below the highest point of the crack. By measuring or inspecting the width of the crack at top and bottom, and whether or not the more vertical parts are wider than the more horizontal ones, you

can get some idea of what has been happening (Figs 7 and 8).

Although you can learn the general principles for analysing cracks in walls from a book like this, it cannot be too strongly emphasised that knowledge and experience are still needed to come to sound conclusions. The guidance and information given here is not an infallible set of rules which an amateur may depend upon without recourse to professional advice.

Having made that clear, let us consider how to monitor the movement which may be taking place in a wall. We want to find out whether the crack is 'live', 'dead' or 'dormant': is it still opening up and, if so in what direction? Has it developed as far as it is going to, because the foundation or other movement has stabilised? Or is it subject to periods of seasonal or intermittent activity? It might not be caused by foundation movement at all, but result from one of the other causes mentioned above.

Tell-tales (strips of glass) should be fixed across the cracks in suitable positions so that changes can be observed. Bed them in pats of mortar, well keyed to the masonry or brickwork. A pat of mortar (say 1:5) can be similarly applied over the crack and, like all these tell-tales, clearly dated. The mortar must not be too strong since it could then crack through shrinkage and give a false impression. Or a sophisticated system of non-ferrous pegs can be drilled and mortared into place in the form of an accurate right-angle triangle (Fig 9). The lengths of the sides are very finely measured with vernier slide callipers and noted. Future movement can then be assessed, together with the direction, by further measurements and trigonometry.

Another way is to use transparent perspex tell-tales, one marked with cursor threads, the other with calibrations. They are fixed to the wall, one bridging the crack, and the other at right angles on one side of it. Then the cursor threads and calibration marks are brought into exact register, one above the other (Fig 10). The important thing is that the two pieces can move independently; they are not stuck together. Any movement of the markings in relationship to each other can thus be noted.

When assessing cracks in walls, look also at the colour and condition of the cracked material. If the crack is very dirty and discoloured, it has probably been there for some time. Old cracks will also be filled with far more particles of grit and dust than a fairly new crack.

In general, a crack which is more open at the top than the bottom suggests the end of the wall is sinking in relationship to the centre (Fig 11). If it is more open at the bottom than the top, then the centre is probably sinking (Fig 12). A fine crack

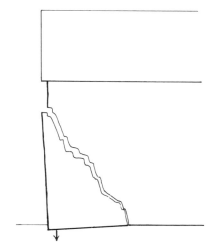

Fig 11 When one end of a wall settles in relation to the rest, a diagonal crack often appears. It starts at the corner of the building and tapers down to the point at which the foundation movement effectively ceases. The principle involved is that of rotation. The position of maximum foundation movement is normally below the highest point of the crack.

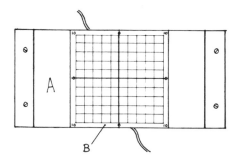

Fig 10 Although relatively expensive, specially manufactured tell-tales for monitoring wall cracks are easy to fix and give great accuracy. The lower plate (A), which has an opaque area calibrated with a grid marked in millimetres, is fixed across the crack. The upper plate (B) is entirely of transparent plastic and has an intersecting hair-line cursor, which is fixed to overlap plate A so that the cursor exactly registers a nil reading, both vertically and horizontally. Any subsequent movement will then be shown, as the intersection point of the cursor lines moves to a new position in relation to the calibrated grid.

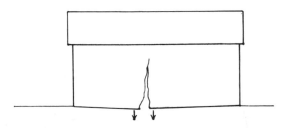

Fig 12 A crack which is more open at the bottom than the top, suggests differential settlement at the centre of the wall, while the ends remain stable.

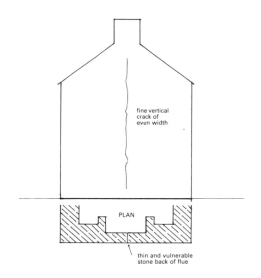

fine vertical
crack of
even width

PLAN

thin and vulnerable
stone back of flue

Fig 13 A fine vertical crack of even width in a wall containing a flue is often caused by thermal movement of the stone or brickwork. The crack may result either from a fire in the flue itself, or from constant heating and cooling over a long period.

Fig 14 A badly foundationed buttress for a leaning wall may itself begin to settle, thus pulling away from the top and becoming totally ineffective. Damp and fine rubble then enter the crack.

Fig 15 A buttress with inadequate foundations which has been well bonded into a leaning wall, may aggravate the original fault by settling and pulling the wall with it.

Fig 16 When lateral restraint provided by roof timbers and floors is removed, a wall may start to pull outwards, breaking its bond with the gable walls. The condition is made worse by the effects of rain on walls unprotected by a roof covering. For that reason the tops of rubble walls with cob bedding material must be kept dry, if a roof cannot be provided for any length of time.

which is of equal width from top to bottom may usually be attributed to thermal movement or vibration rather than foundation settlement (Fig 13).

External buttresses can bring about further movement of a leaning wall, rather than stabilising it as intended. The dangers are shown in Figs 14 and 15.

If a building has become a roofless shell, a crack may develop between the quoins and the gable wall, due to lack of lateral restraint (Fig 16).

A diagonal crack which runs down the wall from a point fairly high up on the corner indicates settlement at the end of the building, perhaps in the foundations below the crack only, but probably in the return wall as well (Fig 17). A common cause of this type of settlement is the drying out of the clay subsoil because the roots of a tree are drinking up the moisture content. The opposite may happen when a reasonably substantial tree is felled: the clay's moisture content increases, and the subsoil accordingly expands. This 'heaves' the foundations upwards, breaking sills and opening up cracks which are wider at the top than at the bottom.

Where cracks start at the base of the wall some feet apart, and converge diagonally upwards, they suggest that the section of wall below them is sinking (Fig 18). This applies also to cracks above

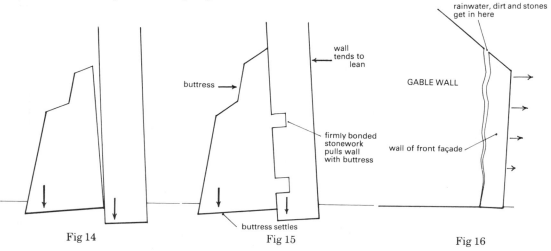

wall
tends to
lean

buttress →

firmly bonded
stonework
pulls wall
with buttress

buttress settles

rainwater, dirt and stones
get in here

GABLE WALL

wall of front façade

Fig 14   Fig 15   Fig 16

Fig. 17 If the subsoil which supports the foundations of a building is robbed of moisture by the roots of a tree, the soil dries out and shrinks. The foundations then drop and cracks appear in the walls. This most frequently occurs in clay subsoils. If the tree is chopped down, the clay may heave the foundations upwards as it swells with added moisture.

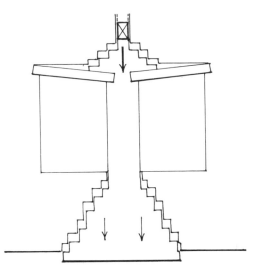

Fig 19 The weight of a beam end bearing down upon a narrow pier between window or door openings can prove too much for the bearing capacity of the foundations; settlement then results. If the foundations can withstand the loading, the pier may buckle or bow instead.

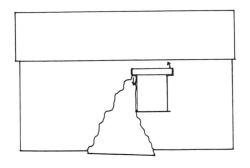

Fig 18 Localised settlement of the subsoil away from the ends of a wall can produce converging diagonal cracks which track up the sides of window openings and around the ends of lintels. Fine cracking and very slight rotational movement may also be seen at the opposite end of a lintel.

window and door lintels, indicating their failure. Here you will see that the crack follows the bedding joints and perpends of the brickwork or stonework, and is widest apart at the former.

This same overall pattern of cracking may also occur to either side of an overloaded pier between wall openings (Fig 19). The load on the thin pier of stone or brick follows the jambs of the openings and then spreads out in the wall below, cracking it diagonally.

It should be noted that cracks will often start near the tops and bottoms of openings, tracking their way by the easiest route to the ground.

A further cause of differential settlement is the uneven distribution of weight between gable walls and those running along the front and back of a building. Tall chimneys, together with high, steep gables, may be transferring much greater or more concentrated loads to the ground than the long walls to front and back. This can bring about vertical cracking at the corners of the building and may involve a number of different stresses. The gable wall may be top-heavy and lean outwards through lack of lateral restraint. It may be sinking, more or less vertically, straight into the ground, or it may combine both conditions (Fig 20).

A constant source of trouble in old buildings is the unequal settlement between bay windows, or similar projections such as porches, and the main walls to which they are attached: cracks run down the junction between the two. The reason for this is that the bay has been

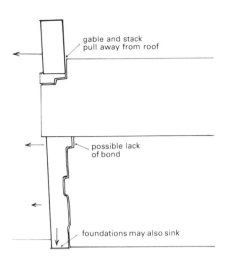

Fig 20 A gable wall containing a chimney stack often develops more-or-less vertical cracks at the junctions with the front and back walls. This can be due to foundation settlement, inadequate bonding at the corners of the building, or the overturning effect of a top-heavy flue and chimney structure.

provided with shallower and more flimsy foundations than the main wall and has thus settled differently from the rest of the house. If the bay has been properly bonded into the brickwork or stonework of the main wall, the bricks or stones at the junction may crack. If it has been added without any mechanical bond, the mortar joint between the two will merely open up so that the bay becomes detached. A decently constructed lead flashing will possibly have succeeded in keeping the abutment between the roof of the bay and the wall of the house weatherproof; or the roof may have stuck to the house and separated from the bay. In any event, water will pour in at the sides.

Another fault arising from badly designed or constructed bays results from the uneven distribution of loads above the wide opening in the main wall. The lintel bridging the entrance to the bay from the rest of the room can sag or collapse through overloading. Brick- or stonework above will be seen to have developed diagonal cracks.

If the lintel or beam is sound, it may be that the pier on one side or the other is itself overloaded and settling. Again, the bay will crack apart from the main wall,

but vertical cracks will also be seen, in many cases, descending from the sill of the opening above, and in line with the side of the bay.

Broadly speaking faults of this sort can be expected at any junction between an original building and later extensions, whether they are porches, staircase projections or wings. Always look for lack of proper bonding between the two, and differential settlement.

Cracks in plaster renderings as a result of poor craftsmanship, design or materials will be dealt with later; here we are looking at structural movement in the building. Similarly, leaning chimneys can be left for the present, since they are less often caused by lack of restraint or bad foundations than chemical changes in the bedding mortar.

Wall Faults Inside the Building

Inside a building the main signs of faults in walls are again cracks, damp, decay and the separation of one structural unit from another. We can consider damp when thinking about methods of curing it, while

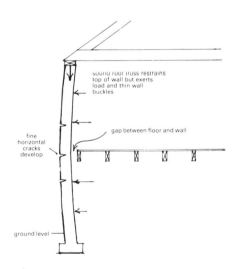

Fig 21 Fine horizontal cracks running along the outside face of a thin or badly bonded wall may be caused by the absence of lateral restraint. In this drawing, the first-floor joists run parallel to the wall and do not help to restrain it. The sound roof truss restrains the top of the wall and the foundations restrain the bottom, so it buckles under the load of the roof. The horizontal exterior cracks must not be confused with those caused by thermal movement, when the face of the wall should remain plumb.

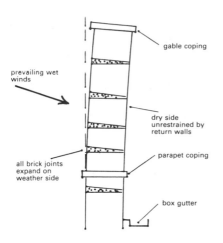

Fig 22 Exposed gable and parapet walls can lean away from the prevailing wet winds. This is caused by expansion of all the mortar joints between bricks as a result of chemical action. Portland cement (in a rebuild, for example) can react with the salts in the old bricks in a way which the original lime mortar did not. Deep pointing in cement mortar can also have this effect. Hard renderings can crack horizontally and chimneys too, are frequently victims of this sulphate reaction with cement.

decay in timber is a subject on its own. Perhaps more important at this stage of inspection are cracks between floors and walls or ceilings and walls.

A sure sign of structural movement in a wall is a gap between the edge of the floorboards and the face of the wall (Fig 21). A gap between the bottom of the skirting and the surface of the boards indicates vertical movement in the joist ends and this is bad news. The trouble has usually been caused by failure of a timber lintel or plate upon which the joists are bearing or by decay in the joists themselves, or sometimes by settlement of the inner masonry upon which the joist ends rest.

Diagonal cracks running down from the point at which a joist bears in the wall are usually the result of some fairly minor settlement of the joist bearing (Fig 23). What happens, especially with newly inserted joists, is that the point load exerted by the joist end compresses the stonework or brickwork below, cracking the plaster. This may be nothing more than a little initial settlement which then stabilises. Indeed, panelled or dry-lined walls may well conceal minor movements of this sort because they are flexible

enough to absorb small changes. If a strong bearing plate or spreader is put under the joist end when it is first positioned, it usually prevents this fault.

If the diagonal crack below a joist end appears on both faces of the wall, the settlement is rather more severe. What counts, as always, is whether it is 'live' or inactive. Scratch the surface of the plaster to key it and apply pats of mortar as telltales.

Some fine cracking where timber stud

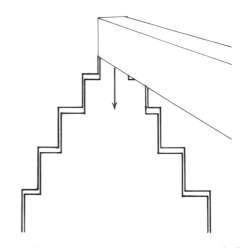

Fig 23 When new joists or beams are inserted, the concentrated load may depress the masonry below the bearing, especially if no suitable pad or plate is used to distribute the weight more evenly. Fine cracks spread diagonally apart in a downward direction. The movement often ceases when initial settlement has finished.

partitions join brick or stone walls is to be expected. This results from differential shrinkage or expansion of the materials; or it can be brought about by the 'spring' in the timber suspended floor upon which a stud partition rests.

When checking the exterior walls with the plumb bob for any outward or inward lean, remember that what may *look* like a quite severe inclination may merely be caused by the fact that one face of the wall has been built to a batter, a gradually receding slope. Measurements of the thickness at ground- and upper-floor levels will reveal whether this is so. As far as leaning walls are concerned, you need to find out how far the structure departs from the perpendicular, and to sketch the wall in section, marking in the notional

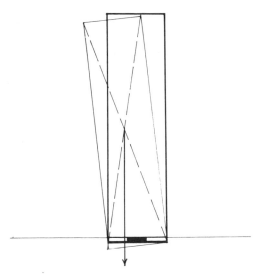

Fig 24 The 'middle-third' rule states that a wall may lean out of plumb and yet be in no immediate danger of collapse, provided that: (a) there is no further foundation settlement, new and excessive loading or disturbances; (b) a notional line suspended from its centre of gravity hangs within the middle third of its thickness at the base. This wall shows the plumb-line arrow pointing *outside* the middle third, and the wall could be unstable.

line dropping from the centre of gravity to the base. If this line 'hangs' outside the middle third of the base, the wall's stability is highly suspect (Fig 24).

Displaced quoin stones, without any signs of those diagonal cracks leading towards the foundations, suggest various possibilities. One is that an eccentric thrust on the quoins is being applied at eaves level, perhaps because a hip rafter has spread through lack of restraint. Another possibility is some very localised settlement under the quoins. Or a stone might be loosened on its bed by some kind of impact: a good example would be damage caused by a projection on a lorry or cart knocking the quoin stones as the driver makes a tight turn into a narrow opening. Yet again, it could be that leaking slates over the top of the wall, at the corner, have brought in rainwater which has washed away the bedding material of the quoins, allowing them gradually to drop and loosen.

Buildings which include some form of arcading or closely spaced openings between relatively narrow piers, can often develop a knock-on effect arising from an eccentric thrust from some distance away. It comes about, as a rule, when the loading of some higher part of the building, such as a tower, fails to direct its forces vertically to the ground. Instead, an eccentric thrust is taken up by the abutment of an adjoining arch, which gradually succumbs to the pressure, as do the other arches in line, one after another.

The whole purpose of buttresses of any kind is to provide an equal and opposite counter-thrust to prevent this kind of thing. Sometimes a sound piece of wall above an arcade or series of closely spaced openings expands and undergoes a fractional degree of seasonal movement along its entire length. The piers or columns below, having insufficient mass to resist, bend away from the source of pressure. Some experts call this type of movement 'drift' (Fig 26). It is more normally encountered in buildings like halls and churches. If you look along the nave of a church towards the east window, and see all the arcade columns leaning in that direction, 'drift' is a possibility. Although the movement may be seasonal, the defect works on the ratchet principle: the wall is unable to return to its original position, because there is no counter-thrust in the return direction as it cools. Also, the cracks fill with small pieces of dust and rubble which prevent them closing up again. Expansion joints are used to accommodate this wall movement.

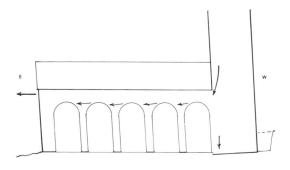

Fig 25 The condition known as 'drift' is usually seen in arcades inadequate to the task of buttressing a tower which has begun to settle and lean. Here, because the foundations are too shallow to the east, due to a sloping site, they have sunk. The tower then leans against the arcade of the nave, creating a knock-on effect along it. Finally, it begins to push the east wall outwards.

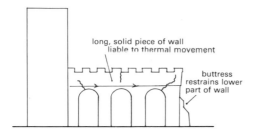

Fig 26 Another definition of 'drift' is that solid walling over an arcade expands in hot weather, so gradually pushing outwards from a very stable tower towards the less well-restrained eastern wall. In this example, external buttresses restrain the lower part of the east wall and cracks appear above that level.

A very obvious sign that there has been movement in walls in the past is the presence on the façade of iron anchor-plates or crosses which denote the use of tie-bars running through the building, probably at upper-floor levels. These have always been an effective method of checking the outward spread of walls. With this in mind, look inside a building for what appear to be mysteriously placed beams or joists running just below the ceiling plaster. They may be boxed-in tie-rods whose anchor plates are concealed outside the building by a plaster rendering.

Sometimes you may need to prise up a floorboard to look for a tie-rod since, ideally, they are run between the joists. At others you may see unexplained patches of discoloration on exterior plaster which turn out to be caused by concealed rusting anchor-plates. Another cause of rust marks on stone ashlar walls is the corrosion of ferrous cramps used to tie the face to the inner core.

Inside the roof space of a building, as well as the timbers, check the condition of the gable walls and chimney stacks. Many faults develop from two causes. The first of these is the deterioration of brickwork or stonework in the area of the flue caused by thermal movement, sulphate attack or disturbance during changes or repairs. Second are the numerous defects which may arise after walls have been built up to accommodate a new roof which runs at a different pitch from the original one. I have seen, for example, the gap between the original gable wall and the new verge

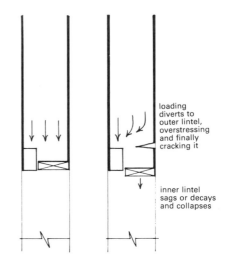

Fig 27 When an inner lintel sags and collapses, the main load of a very well-bonded brick wall may be diverted to the outer lintel. The arrows show the load turning towards the exterior lintel, which in turn could crack.

rafters filled with all kinds of rotting stud-framed panels, sheets of iron, crumbling brickwork, asbestos and so on. It may look decent from outside, but the picture inside can be horrendous.

Be alert throughout for any sign that interior lintels, almost always of timber in old domestic buildings, have decayed, deflected or collapsed. This is one of the most persistent types of structural defect

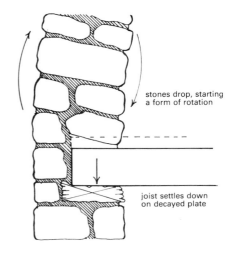

Fig 28 If an internal wall plate or bonding timber decays, the joist ends will settle. Masonry above then drops so that a rotary movement begins. The wall leans inwards and cracks may appear horizontally on the outside face.

60

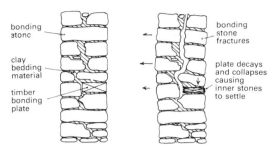

bonding stone

clay bedding material

timber bonding plate

bonding stone fractures

plate decays and collapses causing inner stones to settle

Fig 29 The left-hand drawing shows the timber bonding plate as still sound; it firmly supports the masonry above. On the right, however, the plate has decayed and differential settlement has occurred between the inner and outer skins. This movement has overstressed and cracked the bonding stone which ties the faces together; and they begin to move apart.

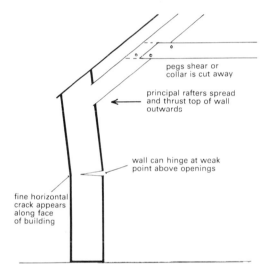

pegs shear or collar is cut away

principal rafters spread and thrust top of wall outwards

wall can hinge at weak point above openings

fine horizontal crack appears along face of building

you will encounter. A cracked stone lintel over a window or door could indicate the failure of the timber inner lintel. This diverts the total wall load to the exterior lintel (Fig 27). Figs 28 and 29 demonstrate other problems caused by rotten lintels or bonding timbers.

Within the roof space it is often found that the only element tying the tops of the walls above window openings is the wall plate that bears the rafter ends. As a result, the partially or wholly unrestrained jambs of the top-floor windows may be pushed out of true by the thrust from the roof, or they may have started to fall apart as the dry and friable clay mortar bedding runs out from the top. Another fault is the hinging effect caused by thrust from unrestrained roof trusses, as the principal rafters spread outwards (Fig 30).

Fig 30 Roof timbers both load and restrain the walls of a building. If, as in this case, the collar ceases to prevent the feet of the principal rafters from spreading, the tops of the walls may be pushed outwards. The upper part of the wall may then hinge where it is weakest – above window or door openings. First-floor joists, bearing in the wall's thickness, can help to restrain any movement in the lower part.

5

DEALING WITH DAMP WALLS

In the previous chapter, we considered the diagnosis of structural problems in a building. These were usually typified by the presence of cracks in the walls. There are three further very important types of defect which must now be mentioned and they are all connected with damp.

The first is rising damp, which finds its way up the walls from the ground. The second is penetrating damp, which may best be described as rainwater which enters the building, taking advantage of any imperfections in either the design or condition of the structure. Third is airborne moisture in the form of vapour which condenses on cold surfaces.

Briefly, rising damp is caused by the capillary action of the moisture in the ground below the building; it is drawn up the walls because no damp-proof course prevents it. No firm rule can be given for the height which rising damp will reach in walls, but you will be unlucky if it comes above 4ft from ground level. It depends upon the capillary structure of the walling material, whether it is sucked high up the wall through the equivalent of thousands of very fine tubes, or whether it stops at a low level. The effect of rising damp is also governed by the degree of ventilation available, especially on the inner face. A well-ventilated surface will more quickly evaporate much of the moisture. Finally, of course, it is influenced by the moisture conditions of the ground, both outside and under the building.

Rising damp often leaves an upper band of efflorescence in the form of powdery white salts. Below that, the wall will be discoloured and dark-looking, often with black mould growths and peeling papers or flaking plaster.

The penetration of rainwater may be through faulty pointing of brick or stonework, structural cracks, hairline cracks in plaster, porous building materials, fissures between walls and door or window frames or a host of other points of entry. One of the weakest places in a building's defences against rain is the junction between roof surfaces and walls. These will be discussed in another chapter.

The overriding principle is this. Rain will penetrate the face of a building to a much lesser degree if the stonework, brick or plaster is relatively soft and porous, allowing an evenly distributed but shallow movement of moisture. Hard, impermeable materials develop cracks which draw moisture in by capillary action. It resembles the difference between a sponge and a straw. The finest of hairline cracks will take in water coursing down the otherwise impenetrable face of a building and draw it deep into the wall as if through a straw. Water soaking *evenly* into the wall, as into a sponge, to a *shallow* depth, will dry out and evaporate when the weather changes, without doing so much harm.

The lesson is that hard building materials must be totally waterproof, while softer ones should be allowed to breathe. This, as far as damp-proofing is concerned, is the sister theory to the principle of flexibility in the structure. Flexible buildings can absorb some movement without collapse. Soft and relatively porous plasters and mortars can absorb some moisture and dry out again without the damp penetrating to the inside.

Having said that, commonsense will tell you which parts of a building must clearly *not* be porous and which may be slightly permeable. A permeable bitumen roofing-felt or solid floor would obviously be useless. Timber components must normally be kept dry unless they are so placed that the wetting and drying process is thoroughly effective as a result of excellent ventilation; but even then no risks should be taken with softwoods, which are highly susceptible to decay. For external renderings, the kind of mix you need is one of say 1:1:6 to 1:3:12 of cement, lime and coarse sand. Mixes of this order should, if properly applied, be fairly free from any tendency to crack. As we shall see in a later chapter, there are occasions when the cement content may be left out altogether.

Pointing

There are few operations connected with old buildings which more often fail both technically and aesthetically than that of pointing. Certain basic rules should be observed. On the technical side, joints between stones or brickwork should be raked out to a depth equal to the width of the joint and any holes filled with mortar, always leaving enough room for the pointing to go on top. Thin joints must be raked out at least 20mm. If necessary, walls should be wetted to prevent premature drying of the mortar.

The mix for pointing should not be cement-rich and should not be harder than the stonework or brickwork which it surrounds. Hard pointing against soft stone or brick prevents moisture drying through the joints, so that the walling material itself remains saturated. Then, in the event of frost, it can spall and disintegrate the stone or brick by expansion.

Pointing must also look right. It should be of a similar colour to the walling material, and slightly lighter in tone. It should also have some texture and not look slimy and smooth. To obtain texture, the mortar should be brushed over to roughen it just before it completely sets.

One of the most disagreeable sights is to see pointing buttered over the edges of the stonework. Another unattractive practice is the use of pointings such as ribbon, strap or snail-creep. These look hard, busy and vulgar. Many objectionable types of pointing have been used traditionally during the last century or so, but they are

This pointing has been properly carried out. It is kept back from the arrises of the stones, colour has been blended to tone with the masonry and it has been textured with a brush before setting. Time and the weather will do the rest.

When pointing a wall, care should be taken to avoid the mortar 'buttering over' the stones, as in this example. The arrises of the stones should stand clear by 3–5mm or more, according to the type of stonework.

Strap pointing should be avoided. It is visually damaging and helps to channel moisture into the joints. This granite wall has cambered brick arches – brick-on-end and two bricks-on-edge – with a big keystone to arch the forces outwards against the well cut and fitted abutments.

Traditional and frightful is the practice of pointing sloppily and trying to give pattern and coherence by lining in the joints, using a piece of lath or the point of the trowel. The grooves are then painted for good measure.

A very objectionable form of pointing, akin to bastard tuck pointing. It stands clear of the stones in bands which tend to gather water. These have been further emphasised with white paint.

Really obdurate ribbon pointing of a limestone wall. It is incorrect, ugly and technically unsound.

This random rubble wall has been raked out for pointing during its construction. When stonework has 'hungry' pointing it looks much the same. The masonry in this case seems rusticated to an almost manneristic degree, and it is not well enough coursed to stand the attention. Some stones have been edge bedded and a 'straight joint' has developed on the left.

no more acceptable because they have been common local practice. White lines run in on the pointing to make random or coursed rubble look more regular, and the allied practice of incising a line with the tip of the trowel, are grisly to behold. There is no need to suspend your aesthetic judgement because a method has acquired some historical precedent.

On the whole, the best pointing for stonework is one that is recessed slightly – so that it is just back from the edges of the stones, without being what is called 'hungry', or deeply recessed, which was fashionable until quite recent times (1960s to early 1970s). That has some dramatic merit but should be used with caution.

Brickwork can be pointed flush, the mortar being struck off with a trowel (Fig 31). In this same manner a bricklayer will clean off the edges when he is bedding bricks which are not to be pointed as a separate operation. An alternative is the struck and weathered joint (Fig 32). The joint can also be slightly recessed for some types of wall. Be governed by the look of the bricks and the local traditions, provided that the latter are not offensive. Always rough-brush the pointing, as for stonework, and take care with the colour.

Obtaining colour in pointing can be difficult, since cement has a way of winning over the other ingredients in the

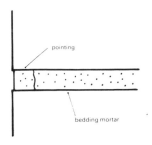

Fig 31 Brickwork may be 'flush pointed' by filling the raked-out joints with mortar and striking off the pointing level with the arrises of the bricks. In new brickwork, the bricklayer may just strike off the surplus bedding mortar with his trowel as work proceeds; the bedding mortar itself must be of the correct colour and texture.

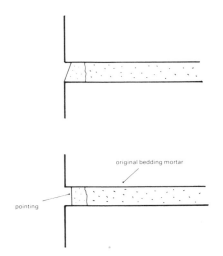

Fig 32 A struck and weathered joint *(top drawing)* is a sound method for pointing, since it sheds moisture outwards; but shallow recessed pointing *(below)* is often strictly more correct for use on old buildings.

mix, so that the result is a decidedly grey, sometimes very pale grey, network of lines over the face of the building. It depends on the colour of the sand you use. With grey sand it may be necessary to incorporate some buff colouring additive, or possibly a bit of earth pigment (very carefully judged from trial pats).

Generally speaking, add as little cement as you can – just one meagre part, to hasten the chemical action. Mortar for pointing should have as low a water:cement ratio as possible (while remaining workable) or it will tend to smear. Pure lime mortars take a very long time to harden. White cement may be obtained, which is neutral in its colour properties, giving the sand the best chance of showing its natural tone.

Technically the use of earth as a colouring pigment is frowned upon, since it can bring harmful salts into the mix. However, since the wall's own bedding material is quite likely to be clay in the first place, one wonders whether this matters. A mix which some people have found successful for pointing is cement, lime, clay subsoil (broken down as much as possible) and gritty sand, in the proportions 1:1:1:5. What must be avoided is unwashed seashore sand: this will contain large amounts of salt and will produce a white fur of efflorescence later on.

For much pointing of stonework, at any rate, a good sharp gritty sand is most suitable. It will prove less slick and dense and have a more pleasing texture. Also it is what is called less 'hungry', or in other words requires less cement and lime in the mix, since there are fewer tiny particles for these to surround and adhere to.

For lime, you make life unnecessarily complicated if you buy anything other than the easily obtained hydrated type. There are instances when unslaked lime is recommended (eg for making limewash) but it need not concern us at this stage. Nor need you worry about the clay content – whether the lime is 'fat' or not. Just go for bagged, slaked, finely powdered hydrated lime.

When using lime don't simply load it dry into the mixer, as you would for an ordinary cement and sand mortar. Get a container such as a dustbin or an old galvanised water-tank and fill the bottom with a few inches of water. Then shake the lime powder in, mixing and stirring as you go, until you have a good creamy lime putty. This can be left on site ready for use. Some people also add the sand at this stage and cover the mix with damp sacks. Those who have rarely had to mix plaster or mortar for themselves may say lime mortar should be mixed with a shovel on a board; but doing up old houses is hard enough work already without forgoing that vastly helpful piece of machinery, the cement-mixer.

Unlike mortar for stonework, bricks are usually pointed in a rather lighter colour, echoing the traditional lime mortars used in the past. But too white a mortar can be offensive. There is really only one way to get the colour right, and that is by trial and error on site. Be very painstaking and discriminating. It will make all the difference to the success of your repairs.

Before leaving the subject of pointing, which could easily fill a chapter of its own, a quick word on ashlar and on galletting. Ashlar stonework, being very precise, with fine joints, presents a problem: how do you get the pointing into such a narrow opening? The answer is to use a fine-grained hungry sand which gives a very workable mix. Make a pointing tool with a cranked handle to press the mix home (Fig 33). Finally, consider the admittedly tedious idea of sliding and pushing in the mix between two pieces of greaseproof paper: this stops the mortar getting on to the face and edges of the ashlar slabs.

Galletting is the practice of pressing little pieces of stone into the soft pointing-mortar to help stiffen and decorate the joints. From the practical point of view it is only useful where joints are rather wide and likely to slump. Aesthetically or historically it may have other merits; some people love it.

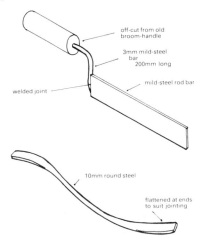

off-cut from old broom-handle

3mm mild-steel bar 200mm long

mild-steel rod bar

welded joint

10mm round steel

flattened at ends to suit jointing

Fig 33 A tool for wedging the mortar or lime putty between the fine joints of ashlar stonework can be improvised by welding together stock mild-steel sections. For general purposes, a 10mm diameter mild-steel rod can be bent to shape and flattened at the ends by heating and hammering on an anvil.

Remember too that you should ideally start pointing walls from the top and work downwards, so that when you clean off you do not mess up the work below. Non-professionals should learn to be tidy and systematic in their working methods and *always* brush up loose mortar after finishing for the day. Wash down any surfaces which could be spoiled, but not of course the actual pointing and stonework just completed.

Claddings

Regrettably, it will often be found that decently pointed but rather porous walls and joints may not be capable of resisting fierce and prolonged driving rain. Consequently it may be necessary to apply a clear silicone solution to the whole surface of an exposed wall. It brushes on and soaks into the stonework or brickwork, providing a water-repellent layer which does not prevent the wall breathing and moisture evaporating out. The only disadvantage is that these preparations are rather expensive and need renewing every few years.

Never use any kind of polyurethane solution on stone or brick walls, either inside or out – it always looks objectionable. However, there are one or two other methods of keeping out driving rain which can be effective and look good as well. Your choice will be partly governed by local traditions, and entirely influenced by the need to avoid making an old building look in any way spurious, hard, vulgar or fake. All these systems come under the heading of claddings. They are: mathematical tiling, tile-hanging, slate-hanging and weatherboarding.

The faces of mathematical tiles are made to look as if they are bricks (Fig 34), and they are laid out in a 'bond' accordingly (usually 'stretcher'). Special corner tiles are employed, and where the tiles finish against another material at corners, a batten is nailed vertically to butt against them. Although they were not invented for the purpose, mathematical tiles became exceedingly popular when real bricks were subjected to the Brick Tax in 1784 (not abolished until 1850).

Tile-hanging is straightforward enough,

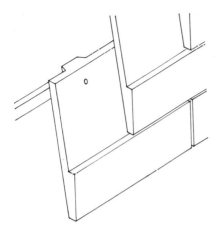

Fig 34 Mathematical tiles were used partly as an alternative to expensive and highly taxed bricks, during the late eighteenth and early nineteenth centuries. They were normally employed as a cladding for timber-framed walls, and usually imitated stretcher-bond brickwork.

but be exceedingly careful in your choice of colour and texture; there are some truly awful tiles on the market. Look for salvaged ones or make sure that new ones are mellow and of similar size to the old. It is impossible to give better advice than that you go and look at a fine, old, tile-hung wall and note its qualities before buying what you need. Tile-hanging is very much a regional tradition of the East and South-east of England.

By the end of the nineteenth century most tiles were made with 'nibs' – projecting lobes which hook over the battens or laths – but for security, one course in about three or so would be nailed as well. Tiles hung vertically must all be nailed, since they are likely to clatter in the wind if they are not really tightly fixed down. If you don't nail them, you are obliged to bed the tails in mortar instead – a tedious undertaking.

Apart from their practical water-proofing qualities, tile claddings are ideal as a disguise for new walls built in concrete blocks or other modern-looking materials. Slate-hanging too can be a particularly delightful way of cladding a wall. It was widely used in some places, for example Cornwall, during the eighteenth and nineteenth centuries. Do not, however, employ asbestos-cement slates for this purpose; they are critically visible.

Depending on the form of construction, a vertical damp-proof course of bitumen felt may be used beneath these claddings. It will not be necessary on a cavity wall, but clearly it will be required on one that has been timber-framed.

Many early nineteenth-century buildings have timber-framed walls, often starting at first-floor level, with stone or brick gable or party walls to contain flues and provide fire safety. Sometimes front and rear ground-floor walls are also of stone or brick to give a firm bearing for the first-floor joists. A great many buildings at this time, particularly in the South and South-east, were entirely timber-framed in softwood and clad in weatherboarding.

If you use weatherboarding, the best kind from the architectural point of view is 'feather-edged' (Fig 35A). It should have a sawn, not planed, finish. Other types such as shiplap (Fig 35B) have precedents in some areas, but feather-edged is the kind which looks right with most old houses.

Never employ fake stone cladding on an old building or an extension connected with it. It is often an abomination; there are some examples to be seen in the Cotswolds which, to put it mildly, do little to enhance the sense of age. Even when the material used is a form of reconstituted stone it can still look harsh, repetitive and mechanical.

Naturally, the fixings for claddings are important: use galvanised nails. At one time, copper nails were sometimes recommended, but they are prohibitively expensive. All battens should be treated with a preservative. Many builders just slosh the stuff on with a brush, but vacuum-treated timber is preferable.

Usually claddings are taken down to the level of the ground-floor window lintels, but the decision is governed by both practical and aesthetic considerations. If the wall is entirely timber-framed, the cladding will naturally go right to the ground, ending where the timber-framing ends. But where it is being used as a disguise for modern materials, it may be decided to render the ground floor and clad the upper ones. Similarly, old stone or brick

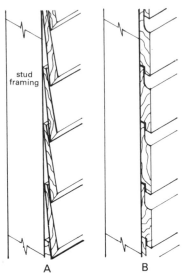

stud framing

A   B

Fig 35 The most suitable weatherboarding for an old building is the feather-edged variety (A), which should be of rough-sawn finish. Shiplap, with a planed finish (B), is also seen on older buildings; but you would make it a second choice.

walls to first-floor level may require rendering, but it is better to repair and point them if you can.

Some advice on plastering techniques and mixes is offered in Chapter 12.

Taking Measures against Penetrating Damp

When you are trying to decide how to prevent penetrating damp in an old building, the main problem is to balance the need for moisture in the walls to evaporate towards the outside against the equally pressing necessity to stop it getting too far into the walls in the first place.

Obviously all cracks and cavities must be stopped up first; but the porous nature of the walling material and faults in the bedding can channel quantities of water to the interior. What you are trying to do, in principle, is allow no more than a shallow, even penetration of moisture, while preventing any actual leaks.

A particularly vulnerable area is that above relatively impervious outer lintels. Water gathers above the lintel and tracks through the wall, finding out any loose areas of infill or bedding material. Finally it drips from the soffit of the inner lintel. The lintel decays and structural difficul-

ties then follow. Most handymen are aware of the evils which arise from the absence of a damp-proof tray between the leaves of a cavity wall, at openings. They also know the troubles which arise when damp bridges across wall-ties which have been placed the wrong way up, or have gathered blobs of waste mortar during construction.

Usually the main secret is to let old walls breathe so that moisture evaporates outwards – only applying water-repellent coatings if these do not greatly impair the process. Colourless silicones have been mentioned in this regard but there is often a good case, in exposed situations where driving rain is a real threat, for using exterior textured paints.

Such paints can brighten up a dingy-looking rendered wall, which is clearly intended to have a plaster finish, while serving a practical purpose at the same time. They can benefit rubble stone walls, and poor or ugly brickwork. All the same, as a general rule, try to expose and point any good, sound brick or stonework.

Around openings, do not depend on bitumastic and other sealing compounds. Always try to provide a mechanical joint if you can, and use paints, mortar or compounds as a second line of defence. This is a counsel of perfection – a mortar fillet may sometimes be the only way to seal around a newly fitted window.

Brick houses which post-date the Building Acts of 1774 will often have projecting reveals, behind which the window frame is fitted from the inside, forming a sound joint. A stuccoed building will provide a continuous layer of rendering at the reveals and soffit which finishes against the window frame. However, the vast majority of earlier or cruder stone buildings which are not rendered, depend solely on mortar packing and fillets to weather the abutment of frame and window jamb. This junction is a constant source of damp penetration.

From a visual point of view, any mortar in such positions should be of a colour and texture to blend with the stonework. New double-hung sashes should not be surrounded by bands of hard, smooth, grey mortar – which mortar is likely to develop hairline cracks, anyway. The rules for making mortar for pointing therefore apply where colour is concerned; but the mix may have to be richer to provide adequate adhesion. Jamb stones should first be well cleaned and keyed or roughened in the very limited area concerned. It is very important that you do not pack the underside of the sill of a new window so tight with mortar that you fill the throat or drip. This groove must be kept free of mortar. It is best to leave a clear gap between the front part of the underside of the timber sill and the slate, stone or brick one below. In some cases, particularly where windows are set in timber-framed walls, it will be necessary to introduce a lead tray which is dressed over the weatherboard, slates or tiles covering the studwork.

The use of vetical damp-proof courses to weather the gap between a window and the jambs of the opening is a modern technique and it is often difficult to carry out in old buildings. Essentially the vertical damp-proof course is designed to integrate with a cavity wall. Solid jambs of brick or stone present obvious problems.

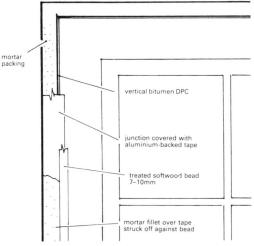

mortar packing

vertical bitumen DPC

junction covered with aluminium-backed tape

treated softwood bead 7–10mm

mortar fillet over tape struck off against bead

Fig 36 One method of weathering the vulnerable junction between a window frame and a wall (in an exposed position) is the use of aluminium tape and a treated softwood bead. The gap between the frame and the jamb is well packed with mortar and struck off flush with the outer lining. A vertical bitumen-felt dpc is placed between the mortar and the frame. The adhesive-backed aluminium tape is stuck over the mortar packing and about 12–20mm of the window lining. Its edge is then covered and secured by the bead. A conventional mortar fillet is finally applied in the normal way and struck off against the bead.

The best that can be said is that you avoid any weathering technique which encourages capillary cracks, ensure continuity of any weathering measures, ensure that new solid-frame windows have capillary grooves down their jambs or stiles and all timber is treated or is hardwood.

Measures against Rising Damp
You can react to the presence of ground moisture, in the form of rising damp, by allowing the walls to be damp and to evaporate what moisture they can, as dry weather follows wet and season follows season. If you do this, you have to either dry-line the inside of the walls, or leave the stones or bricks exposed and be prepared to live with some intermittent damp. Alternatively, you can coat the inside of the walls with a damp-proof rendering or continuous vertical damp-proof course.

If your building is made of brick, or a fairly soft but well-bedded stone, you may consider employing a specialist subcontractor to introduce a conventional damp-proof course, sawing a horizontal channel right through the wall's thickness to do so. It will work effectively, but it is expensive and many old houses are unsuited to such treatment.

You may decide to encourage a high degree of evaporation at damp-proof course level by drilling holes in the walls at roughly 300mm centres and introducing porous ceramic tubes. These are bedded in a soft and highly absorbent mortar and are set to an outward fall (Fig 37). They gather the dampness and let it both weep and evaporate towards the exterior. There are endless arguments about whether or not this system is effective. It may not always be, but reason dictates that it has its uses. Much depends on how thick the walls are, what they are made of and what the permanent conditions are inside the building. If you are prepared to tolerate some damp inside the building, ceramic tubes could prevent matters getting out of hand.

Along with the above methods, or occasionally as a remedy on its own, you may reduce the external moisture content of the subsoil around the foot of the wall. To do this you dig a trench about 300mm deeper than the internal floor level, always

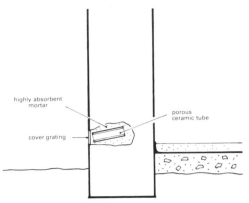

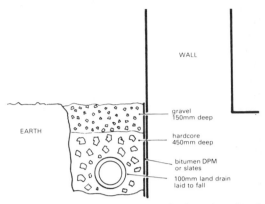

Fig 37 One method of checking some of the rising damp in old walls is the insertion of porous clay tubes which collect and evaporate moisture. It can be a useful measure in buildings where it has been decided to live with a certain amount of rising damp.

Fig 38 By laying land drains along the foot of a wall and backfilling the trench with hardcore and gravel, it is possible greatly to reduce the amount of moisture finding its way into the building from the subsoil. Avoid undermining shallow foundations.

providing the foundations are sound enough to allow it. Render the base of the wall. Then run a set of 100mm (4in) land-drain pipes along the trench so that they fall towards a specially excavated soak-away. Backfill the trench with hardcore and gravel (Fig 38).

Sometimes it is found that a spring or stream is running perilously close to the building, or even beneath it. Then it may be necessary to gather this water into a sump and divert it in a properly jointed pipe of suitable diameter. Unless it is impracticable to do so, always dig the ground level back from the walls where it is higher than the internal floor. All the same, check that the walls are not depending on a bank of earth for essential support. If you *cannot* dig away, for one reason or another, it may mean tanking the inner face of the wall. Where any degree of hydrostatic pressure is exerted, you have to give the new asphalt membrane a loading coat to resist it, and all this is work for experts (Fig 39).

'Electro-osmosis' is another treatment used to counter rising damp. This again is a specialist operation. It involves drilling holes at regular intervals around the base of the wall at what would be a suitable damp-proof course level. Into these are inserted electrodes in the form of loops in a continuous copper strip. The strip is then earthed by means of rods driven deep into the subsoil. The idea is to short-

circuit the minute electrical charges which encourage capillary action in the wall. There are both 'passive' and 'active' systems. A passive system simply earths what charges are in the wall. The active one introduces a very small charge of its own between the inside and outside of the wall. Like everything else, damp-proofing systems are the subject of fashion, experts making claims and counter-claims about their efficiency. I can neither denigrate nor commend electro-osmosis.

A favourite defence against rising damp

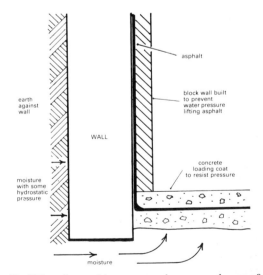

Fig 39 In cellars and basements, where some degree of hydrostatic pressure is likely from water in the surrounding subsoil, it is necessary to adopt the traditional system of *tanking*.

is the injection into the wall of a silicone water-repellent, which diffuses within the structure, forming a barrier. Like almost every other damp-proofing method which tries to fill the role of a conventional damp-proof course inside the wall itself, success is dependent on two factors. The first of these is the knowledge, skill and conscience of the contractor who carries out the work. The second is the type and condition of the wall. All these methods are best used in brick walls which have evenly distributed and compacted bedding material (mortar) and are formed of components which will themselves absorb a consistent ration of the silicone solution. This is fed into closely spaced holes by means of tubes, either from bottle reservoirs, or pumped under pressure from a large canister. Some architects swear by the silicone-injection system even for random stonework; I am slightly suspicious of it, since it is hard to see how proper continuity of absorption can be achieved, unless the walling material is free from voids and cracks.

Dry-lining inside the Building
In the last event one must admit the merits of lining or sheathing the interior of the building with a continuous vertical layer of some waterproof substance, whether it is asphalt, a bitumen compound, a waterproof plaster or corrugated lath. Naturally, with any such measures a damp-proof membrane must be laid over the floor and then turned up the walls. The weak points in these systems are around fireplaces and solid partition walls. If the latter are not dealt with, moisture will obviously get past the barrier and form patches in the corners of the rooms, which may or may not confine themselves to some minor seasonal activity. The following are the lining processes most likely to succeed.

The walls can be pointed and strapped with either vertical or horizontal vacuum-treated battens. Alternatively a stud frame is plugged and fixed to the wall surface, if space is not at a premium and it is wished to straighten up the wall area for one reason or another. Battens, which should be anything between 30 × 19mm and 50 ×

25mm in section, can be masonry-nailed, or fixed with timber plugs or a mixture of the two. A stud frame will normally be of sawn stuff measuring up to something like 100 × 50mm, according to circumstances. In most cottages you are trying to save every bit of space you can, and the smaller the battens or frames, the better. Where it is wished to replace panelling, stud framing is often essential.

Plasterboard with a foil vapour-barrier backing (Fig 40), or expanding metal lath, perhaps treated panelling, or pitch-fibre corrugated (Fig 41) is nailed to the battens or framing. However, the plasterboard can be nailed directly to the wall itself with galvanised clout nails, or by using masonry nails with washers to stop the heads pulling through. The vital things to remember are that the damp-proof course from the floor must go under the ends of the framing members or battens; and that all timber plugs, battens or framing should be treated with a preservative. Also, the damp-proof course, which is usually a polythene sheet, must not contact the damp wall surface so as to gather moisture and channel it to the wrong side of the damp-proofing system. The damp-proof lining must, of course, continue up the window and door jambs, to whatever level has been decided.

One of the most effective and authentic dry linings for old buildings is flush-beaded matchboarding. It is the perfect finish when you want something less formal than panelling but require more texture and interest than plain plaster. The boards should not be less than 140mm wide and they may be either tongued and grooved or rebated. The bead is narrow and shallow. It is a far cry from the skimpy little V-grooved boards which people usually use. Every big builders' merchant chain should stock miles of the stuff as a matter of course.

An advantage of dry-lining is that harmful salts which have accumulated in the wall, some of which actually encourage damp, can be left to their own devices. You should usually strip off old plaster and get back to the stone or brick before lining out, both to eliminate the plaster as a source of damp and to ensure a good

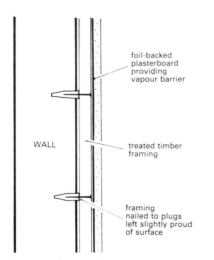

foil-backed
plasterboard
providing
vapour barrier

WALL

treated timber
framing

framing
nailed to plugs
left slightly proud
of surface

Fig 40 One of the best defences against damp is to dry-line the walls on the inside. The cheapest and easiest method is to nail foil-backed plasterboard to treated battens; but it may not suit the character of a 'rough' building.

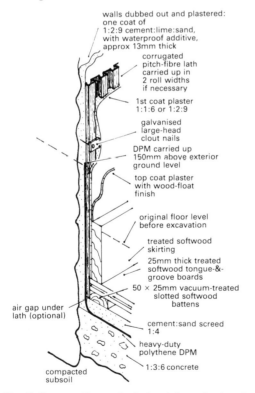

walls dubbed out and plastered:
one coat of
1:2:9 cement:lime:sand,
with waterproof additive,
approx 13mm thick

corrugated
pitch-fibre lath
carried up in
2 roll widths
if necessary

1st coat plaster
1:1:6 or 1:2:9

galvanised
large-head
clout nails

DPM carried up
150mm above exterior
ground level

top coat plaster
with wood-float
finish

original floor level
before excavation

treated softwood
skirting

25mm thick treated
softwood tongue-&-
groove boards

50 × 25mm vacuum-treated
slotted softwood
battens

air gap under
lath (optional)

cement:sand screed
1:4

heavy-duty
polythene DPM

1:3:6 concrete

compacted
subsoil

Fig 41 Damp walls may be isolated from the interior finishes of a building by lining them with a corrugated pitch-fibre lath. The system is best employed in conjunction with a damp-proof membrane in the floor, which is turned up the walls behind the lath, to a height of 150mm above the exterior ground level, if possible. The walls continue to absorb moisture from the subsoil, but the harmful effects are minimised.

fixing ground. Where there are white furry patches of efflorescence resulting from salts, the walls may be repeatedly sponged down with clean water before any plaster is applied direct. Nevertheless, trying to wash away salts is always an uncertain measure.

Remember that damp may not be evaporating from the walls on the outside because an impervious external rendering is trapping the moisture. This should be hacked off. However, some buildings are designed to be stucco-finished externally (Regency and Victorian), in which case one must live with the practical drawbacks, ensuring that the plaster is totally leak-proof, as far as that is ever possible.

Although many builders like nothing better than to plaster internal walls with a hard cement-rich mix containing a waterproof additive, I believe that this is normally counterproductive in two ways. It provides a cold, impervious surface which does not breathe, and encourages condensation; and it is also prone to crack. Aesthetically it can be far too smooth for some types of building.

The subject of plastering is more fully discussed in Chapter 12 – at present we are only considering it as it relates to damp-proofing walls.

Better than hard rendering is the use of a brush-or-trowel-applied bitumen compound, which is blinded with sharp sand and then plastered over (Fig 42). The wall surface is pointed first and dubbed out to fill any major depressions. Then the thick liquid bitumen is brushed on, and while the final coat is still tacky, sand is flicked on to the surface to provide a key for the plaster. The bitumen layer is left to harden according to the maker's instructions, then plastered over using a suitable mix.

When the contours of a cottage wall provide a pleasant visual quality, it is sometimes enough to brush a cement-sand grout on to the blinded surface of the bitumen layer. Alternatively, bitumen may be brushed on to the stones in this same manner, blinded, and rendered with a plaster which follows the overall contours of the wall.

Some firms make an asphalt-based, rather than bitumen, compound for

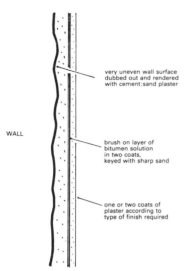

Fig 42 The principle of brush-treating damp walls with a bitumen solution in two or more coats is shown here. The last coat is keyed by blinding with sharp sand. Plastering specifications vary according to the damp-proofing product used.

WALL

very uneven wall surface dubbed out and rendered with cement:sand plaster

brush on layer of bitumen solution in two coats, keyed with sharp sand

one or two coats of plaster according to type of finish required

damp-proofing walls and recommend lightweight vermiculite mineral plaster.

Living with Damp

Before leaving the subject of damp-proofing methods, it is worth repeating that you may sometimes decide that a 'rough' building would be better served by merely pointing the inside and just lime-washing or emulsion-painting the walls. Limewash has the advantage that it breathes and absorbs condensation, but it can also flake and discolour. White emulsion paint, the more porous the better, is generally excellent for providing a clean, unaffected whitewashed look. If you decide to live with what damp the walls contain, it cannot be too strongly emphasised that the rooms should be kept warm but well ventilated and that all timber in contact with walls should be treated. It is often prudent, and visually more in keeping, to dispense with skirting boards, which are ready victims of decay.

At ground-floor level, buildings lived in without any damp-proofing measures are probably better-off with solid floors of flag-stones or brick: barn and industrial building conversions, or cottages, are suitable for this treatment. It will be argued that there are thousands of old buildings all over the country which have no damp-proofing systems of any kind, yet contain suspended timber ground floors, panelling and skirtings throughout. Fortunately, not all buildings suffer badly from damp; but there are few old ones which have no problems. This book is about dealing with difficulties that do arise. Nothing could be better than an old house which is both beautiful and serviceable, requiring the minimum fuss and intervention.

Waterproof paints or polystyrene papers applied as a means of damp-proofing are rather a waste of time; they just peel off the walls or trap moisture behind them. However, polystyrene can sometimes be useful as an anti-condensation measure, since it presents a warm surface.

With all dry-lining methods where space is left between the damp walls and the lining material, you have to decide whether it is more helpful to encourage ventilation from the room, by using such devices as hit-and-miss skirting ventilator panels, or to seal the room off completely from the trapped air-space behind the plasterboard or lathing. Remember that pressure always drives warm moist air towards the cold areas within the wall structure, upon which it will condense. This 'interstitial condensation', occurring

Black mould growths are very common in old houses and they are often caused by condensation on cold and badly ventilated wall plaster. They are relatively harmless to the structure, but affect hangings and papers.

The thrill of peeling off a truly ferocious wallpaper to reveal eighteenth-century panelling will be familiar to many house owners. The cornice of this little room in the rear wing of a town house looks late eighteenth century. The unmoulded muntins and rails of the panelling, together with the remains of canvas scrim, makes it appear that the panelling was designed as a support for wallpaper or fabric hangings.

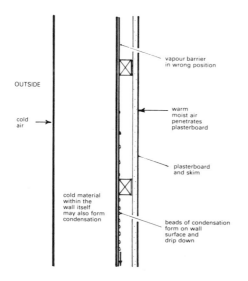

OUTSIDE

cold air →

vapour barrier in wrong position

warm moist air penetrates plasterboard

plasterboard and skim

cold material within the wall itself may also form condensation

beads of condensation form on wall surface and drip down

Fig 43 Interstitial condensation takes place if no vapour barrier is used to prevent warm moist air from the room penetrating to colder areas within the structure of the wall itself; or if the barrier is wrongly positioned.

within the actual materials of the wall itself, is a great source of damp problems. Conventionally, to prevent it, you place a vapour barrier – a foil backing on plasterboard or a plastic sheet – behind panelling, or wooden or expanding metal laths, etc. It is important that this barrier, which stops warm moist air getting into the wall, is near the surface. If it sits further back in the wall, say under the studding which supports the panelling, it will merely provide a cold surface within the wall which promotes and traps condensation, which then rots any timber it gets a hold upon (Fig 43).

Decide to ventilate the hidden air spaces only if you know that the air in the room is normally going to be dry and warm. If it is damp and warm it can be far from beneficial. Whatever you do, keep the following rules uppermost in your mind:

1 Ventilation and evaporation should be encouraged.
2 Timber should be treated.
3 Moist warm air should only be allowed to condense where it does no harm.
4 Try to keep surfaces warm by insulating them.

6

REPAIRING WALLS AND TEMPORARY SUPPORT

One of the most neglected procedures when repairing old buildings is that of providing temporary support while work is carried out. Many builders are extremely cavalier in their treatment of old walls, assuming that because they have usually been lucky they will continue to be so. For their part, handymen and amateur house restorers take alarming risks, from either impatience or ignorance.

Shoring up old walls is an expensive, time-consuming nuisance, and for that reason too many people decide that it is easier to demolish and rebuild. All over the western world, nice old buildings are being destroyed because architects, engineers, planners, developers, builders and private citizens are too penny-pinching and indifferent to do the job properly. I hope that nobody reading this book will fall for the notion that it does not matter whether you have an old mellow wall with its weathered materials and irregularities, a hard straightened-up replica or an unsuitable replacement.

Propping and Shoring Walls

Contrary to what the demolishers' structural engineers often claim, most walls *can* be supported successfully, provided that the task is undertaken cautiously and intelligently. You have to stand back and ask yourself: 'If I unbuild that wall, or replace that lintel, or make a new window in the gable end, what is likely to happen if it is not given proper temporary support?'

Take an obvious example. You need to replace a rotten timber window lintel with a new one. What will happen when the old one is removed – supposing you can get it out? The lintel may be held fast by the collapsing masonry and if you simply lever

it out with a crowbar you will further disturb the already perilous stonework.

First, therefore, place adjustable steel props about 1m back from the ends of any floor joists which the lintel is meant to support. Then fill and point the stonework above the lintel over a generous area of wall, taking special care around the lintel ends where they bear on the jambs of the opening.

Next you take further pairs of steel props and pick up the load of the masonry above the lintel by making holes through the wall and introducing stout timber needles or steel joists. You must be very careful not to disturb the old stonework by overwinding the steel props so that the needles push the already precarious

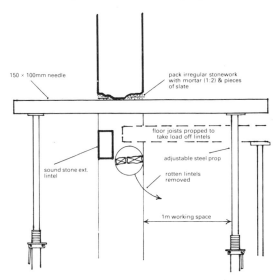

Fig 44 When removing an unsound lintel, a form of *dead shoring*, is used, consisting of a strong timber *needle* or RSJ supported by adjustable steel props. Care is taken to pack any space between the masonry and the needle so that a good bearing is obtained. Joist ends are also relieved by propping and all props are kept back from the wall to provide adequate working space.

stonework upwards. Then mix some very rich mortar (say 1:1 or 1:2) and pack it into the uneven gaps between the stones and the needles, wherever it looks necessary. Where the gaps are large, slips of slate may also be required (Fig 44).

Ensure that the steel props are so placed that you have room to work and to fit the new lintel. Not until you are satisfied that the old lintel no longer bears any appreciable load do you try to remove it. To do this it is often worth propping the lintel and taking a power saw to cut a chunk of a few inches' width out of the middle. When that has fallen clear you remove the lintel props and wriggle the decayed, embedded ends out of their sockets in the wall. What you don't want is to be forced to hammer, wrench and heave at a piece of timber which appeared to be rotten but suddenly seems to have become remarkably robust, just when it least suits you.

Sometimes it may be possible when replacing a lintel in a brick wall to dispense with propping, if the brickwork is in good condition and well bedded and bonded in sound mortar. The important thing is to make certain that the loading on the relevant section of wall is minimal. If it bears the weight of roof trusses, floor joists or beams, it is better to play safe. The joy of making new openings in brick walls is that brick is such an easy material to cut through. You can make the opening match the perpends of the brickwork and tap through alternate bricks with a steel bolster.

Temporary support for openings is a reasonably simple matter, which an amateur can undertake using a modicum of prudence and sensible technique. General shoring for walls is very much more complicated. The principle is straightforward enough and so are the processes – theoretically. Achieving them to a satisfactory standard is less easy.

This is what you do to support a wall, preventing any outward movement. If you have enough space available in front of the wall, adopt a system of 'raking shores' (Fig 45). Certain points must be remembered when shoring up in this way. The sole plate should be firmly based on a pad

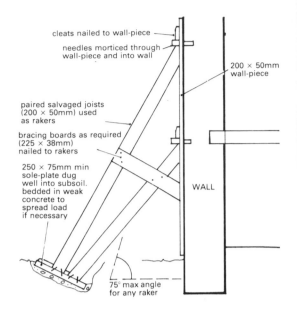

cleats nailed to wall-piece

needles morticed through wall-piece and into wall

200 × 50mm wall-piece

paired salvaged joists (200 × 50mm) used as rakers

bracing boards as required (225 × 38mm) nailed to rakers

250 × 75mm min sole-plate dug well into subsoil. bedded in weak concrete to spread load if necessary

WALL

75° max angle for any raker

Fig 45 The traditional method of giving temporary support to prevent a wall collapsing outwards is a system of *raking shores*. Theoretically, the centre line of a raker should intercept the bearing of an internal joist, so obtaining some counter-thrust.

of weak concrete or a platform of thick planks which will help to spread the load. The top shores must not be at an angle with the ground greater than 75°. The wall pieces should be cleated to the walls and well secured against upward thrust from the shore, using stout timber needles socketed into the stone or brickwork. The timbers themselves must be of adequate size to resist the pressure from the wall without bending.

While it is usually not difficult to find old pieces of timber for dead shoring and strutting openings, finding lengths of sufficient section to use as rakers is another matter. They are heavy to handle on site and must be levered up to make a snug contact with the wall pieces. All this needs strength and finesse. However, consider using salvaged floor joists and doubling them up to provide the necessary thickness. If you have any doubts, it is better to employ a scaffolding firm or other contractor to shore the front of the building using a system of steel sections.

Similarly, a flying shore, between walls, may be better left to professionals. But since you may well be involved in some more modest task involving flying shores, it is worth outlining how it is done.

Flying shores are used for propping back walls by strutting horizontally, sometimes between a firm building and an unsound one, but often between two sound structures. The obvious application is in a terrace where a building has to be demolished, thus removing the necessary shared support from the houses on either side. Equally you might strut a wall by placing flying shores obtaining support from a building on the opposite side of a lane or alley (Fig 46).

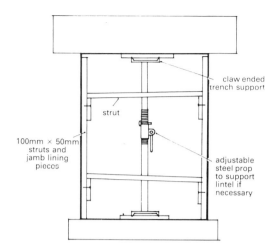

Fig 47 It may sometimes be necessary to give temporary support to lintels and window jambs – while underpinning, for example.

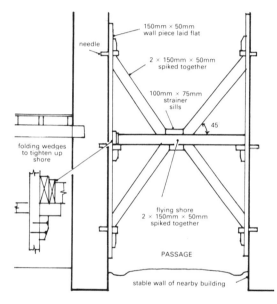

Fig 46 *Flying shores* are used to prevent outward movement of a wall by bracing it against a strong wall of another building.

When shoring is undertaken, it is necessary to strut any openings in the wall since these are points of weakness. Sometimes this needs to be done both vertically and horizontally, using pieces of $100 \times 50mm$ timber, wedged against similar pieces lining the head and jambs. Short adjustable steel reveal pins or trench supports can also be employed for strutting (Fig 47).

You may need to tie back bulging walls which are failing through lack of lateral support at arch abutments, by running wire ropes with adjustable straining screws between piers.

When arches are built they are tem-

porarily supported by curved timber assemblies called centering (Fig 48). If you are rebuilding or repairing an arch, it may be necessary to design centering to support it. This can be made from plywood for small spans, or a composite structure devised from curved timber ribs and radially placed struts can be used.

It is essential that you should at all times be able to make adjustments to shoring, so that it can be kept tight against the surfaces which it supports. This means either having adjustable steel props or using what are called folding wedges (see Fig 48), which can be tightened by tapping home with a hammer or mallet.

Whatever you are holding up, you must ensure a proper bearing face between the

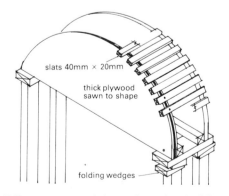

Fig 48 To support an existing arch or when building a new one, *centering* is constructed which exactly fits the soffit.

77

support and the stonework or bricks. In the case of a ragged hole in a rubble wall, you will probably have to insert bits of brick and slate and pack up with a strong mortar containing the absolute minimum of water so that it does not slump.

As you can see, raking shores are used to restrain lateral wall movements or ones which are nearly so. Dead shoring is designed to pick up the more or less vertical forces of 'dead loads'. In other words it holds up the walling material, floors and roof, while work goes on. The most typical need for dead shoring, proper, is while foundations are being underpinned or large openings made. However, in principle, any kind of vertical propping of openings is a form of dead shoring.

True dead shoring consists of stout posts placed to the inside and outside of the wall, and sufficiently far back from it to give room for working – probably about 600mm–1m from the face. The posts or dead shores themselves are bridged by robust timber needles which must obviously be strong enough not to deflect

under any load they may have to carry (Fig 49). Sometimes a short piece of rolled-steel joist or universal beam is used for this purpose. The disadvantage is that steel is more difficult to brace and secure. The feet of the dead shores rest upon heavy timber ground-sills. Adjustable steel props can be used, provided that the loading does not exceed that recommended by the manufacturers. Granite can weigh 2,600kg per cubic metre, and brick may be 1,750kg per cubic metre or more.

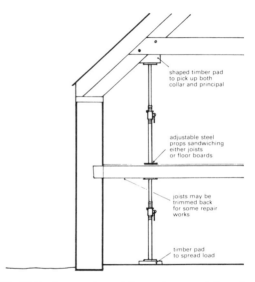

Fig 50 During repair work it may be essential to give support to floor joists, beams and roof trusses.

Inside a building it may well be necessary to remove floorboards and thread props through to higher levels. For example, a system of support may be needed to carry roof or upper-floor loads to ground level. Alternatively props may be positioned one above the other with joists and floorboards sandwiched in between (Fig 50). When any form of dead shoring is undertaken it is likely that floor joists and beams will also have to be supported, either individually or by placing a system of props and cross-timbers.

Three pieces of advice to amateurs undertaking temporary support operations are these:

1   Be absolutely painstaking in what you do and how you do it.

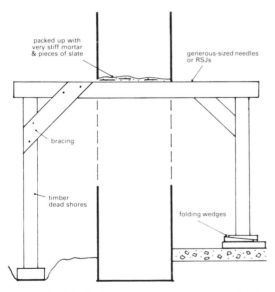

Fig 49 *Dead shoring* is used to support masonry or brick-work when forming new openings or replacing the lintels of old ones. Railway sleepers or salvaged roof and floor timbers may be used. but they must be large enough to avoid any chance of collapse or deflection when loaded. The feet of the dead shores should be on stout pads of wood or continuous timber sills to increase the area of ground upon which they bear.

2 Always choose the stouter piece of timber rather than the flimsier one. Strutting or shoring is sometimes just a precaution, like a fire extinguisher, and in that case the supports may never have any real work to do. When this is so, it doesn't much matter what size the timbers are. But the moment they are obliged to take up maximum loads, undersizing can have disastrous results.

3 Make certain that the actual process of providing temporary support does not so weaken or shake up the building that parts of it start to collapse before your carefully thought-out support structure is finally in position.

While on the subject of propping and dead shoring, we ought to consider the role of internal walls in the construction of a building. Ask yourself whether a wall is load-bearing or merely used as a partition between one space and another. It is vital that walls which carry the point loads of beams or the ends of joists, the feet of roof trusses, and so on, are not disturbed or demolished without re-routing the relevant forces. All too often an enthusiastic handyman demolishes an innocent-looking piece of wall, to find the whole ceiling collapses and the floor with it. Only then does he realise that what he had taken to be a beam was a boxed-in waste pipe, and that all the joist ends had been resting on the partition he blithely cut away.

Remember too that in some old buildings, past repair operations have already redistributed loads, by such devices as introducing stud partitions which act as hangers: these hold floors up from above, rather than propping them from below.

One method of providing temporary support for a large building, where there is no adequate land in front to set up raking shores, is to case it in with steel towers. This is a job for structural engineers; in essence, it means facing the walls with vertical poling boards and placing horizontal steel walings running between towers. The towers may be designed to resist any outward thrust, or they can be paired with others on the opposite side of the wall and then tied together with improvised lattice girders made from scaffolding sections.

Another method, of use to many people repairing old houses, is that of tying the front and back walls together, using iron rods with threaded ends. These are run through the building, preferably at floor levels, and have plates which are tightened up against the walls to restrain them. The system is used for permanent restraint of outward-thrusting walls, or in conjunction with polings and walings as a temporary measure while other work is going on (see Fig 57). This type of tying is obviously beneficial where horizontal cracks appear along a façade at first-floor level, as the upper part of the wall hinges and leans outwards because of thrust.

The problem of propping up a façade while all the internal walls and cross-members are removed is one which I hope you will not encounter. All too frequently it means that an unnecessarily wholesale destruction of the old building is taking place. However, if it *has* to be done, raking shores outside are not enough. They prevent outward movement of the wall, but something has to be devised to provide a counter-thrust from inside. Internal raking shores are clumsy but effective. Their disadvantage is that later they need to be dismantled in short sections without disturbing the new walls and floors which have taken over their role.

A better method is to plan the work so that the required counter-thrust of partitions, or tying effect of floors, is replaced as work proceeds. It always makes sense to adopt a cautious and logical step-by-step, redistribution of load, rather than subjecting the building to a single massive trauma.

If you have space, as with a building which stands in its own grounds, you can place raking shores at front and back. These are tightened up against substantial struts running through the building (Fig 51).

From all of this, you can see that giving temporary support is a highly important operation, which can be carried out in many ways to achieve different ends. The main conventional methods are outlined

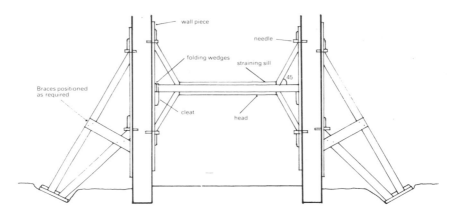

Fig 51 To support the walls of a building which has inadequate lateral tying and stiffening from floors, partitions or roof trusses, raking shores may be used against front and back walls, while flying shores brace the walls apart from within.

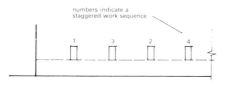

Fig 52, *top* (section). To support walls while *underpinning*, needles are inserted not more than 1.5m apart. These are carefully packed up against the brick- or stonework, using very stiff mortar and pieces of slate. At corners, where maximum loading occurs, the needles are placed at 45° so that both ends bear outside the building. The greatest caution will be required in propping at these points. Next, short lengths of footing are unbuilt, re-foundationed and rebuilt, in a staggered sequence (see lower drawing, elevation). This avoids disturbing too much wall at any one time. Success depends upon the skill with which the new footing is pinned up to the base of the old wall. Shrinkage or slump in the mortar will prove fatal. Special wedge-shaped engineering bricks are sometimes used to pack this junction.

here; some works are best left to a professional contractor, and chief among these must be underpinning, which always requires Building Regulations approval.

Underpinning and Other Options

The purpose of underpinning is either to replace an inadequate wall foundation with a better one, or to extend existing foundations downwards to reach better levels of load-bearing subsoil. The wall may be underpinned in one operation, which involves extremely well designed and constructed dead shoring, together with raking shores against the return walls at each end (Fig 52).

The method of underpinning short lengths at a time is better. In this case it may occasionally be possible to do without dead shores, provided that the wall itself is very well constructed. Brick walls sometimes come into this category. Rubble stone will almost inevitably collapse during work unless it has temporary support.

The whole process of underpinning is so labour-intensive and costly that it is the last operation you will ever carry out on an old house if it can possibly be avoided. It may prove more effective to rebuild a section of wall, putting a new foundation in at the same time. Alternatively, you may manage to redistribute the loads so that they exert less pressure on areas of foundation which are faulty. When cracks and settlement occur in connection with overstressed piers between windows, it is often the effect of the building carrying out the process of redistribution for itself. The forces exerted by the loading find a new way to the ground rather than bringing the overstressed part of the wall to a state of collapse.

There are two main drawbacks to rebuilding and foundationing pieces of wall. One is that from the aesthetic point of view it may be damaging to the building, especially if new materials are employed. The other is that further faults might arise as a result of differential movement between the very solidly founded new work and the old.

Concrete Work in Walls

The principal method of redistributing loads within walls is by using reinforced concrete, either in the form of a beam, lintel or bonder inserted into the wall and pinned up against the original stonework with pieces of slate and strong mortar, or in the form of reinforced concrete cast *in situ*. In the latter case, you fill a length of wall, which has been adequately supported, with concrete (say 1:2:4) containing mild-steel bars in the tension area near the bottom.

The idea is to bridge a weak area, or to spread the load out more evenly. A failed arch, for example, can be treated in this way. It is really a modern alternative to a relieving arch. The reinforced concrete carries the load to sound wall on either side and also ensures that the forces are

directed vertically instead of eccentrically, in the case of failed abutments (Fig 53).

Concrete or sometimes tiles, well bonded in mortar, may be used for a number of purposes within the structure of a wall. They can be used to 'stitch' across cracks and weak areas of wall where the main trouble is inadequate bond (Fig 54), or they can be used to replace rotten timber bonders. Concrete may be employed to

Fig 54 To stitch across a crack in a badly bonded wall: the wall is unbuilt to half its thickness and filled with well-bonded large clay tiles bedded in cement:sand mortar. (For clarity, the width of the crack has been exaggerated in this drawing.)

form what is called a ring-beam, to run right around the walls of a building where they have a tendency to spread outwards. Areas of reinforced concrete may also be cast *in situ* to form anchorage points for tying opposite walls of a building together by means of mild-steel tension rods (Fig 55).

Another type of repair using concrete is the formation of an internal buttress to check a wall which leans outwards. One way of making an internal buttress is to cut out a square channel in the wall, to a depth of 300–450mm. A wall beam of concrete is cast along the top of the wall, rather like a wall plate. To this the reinforcement for the buttress is linked before the concrete is poured. A trench is dug in the floor of the building in such a position, ideally, that it finishes under an existing solid internal wall. The reinforcement of the concrete in the wall is linked

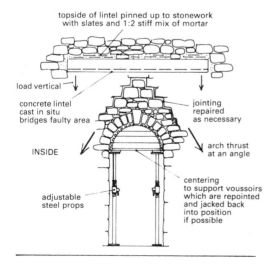

topside of lintel pinned up to stonework with slates and 1:2 stiff mix of mortar

load vertical

concrete lintel cast in situ bridges faulty area

INSIDE

adjustable steel props

jointing repaired as necessary

arch thrust at an angle

centering to support voussoirs which are repointed and jacked back into position if possible

Fig 53 It is sometimes possible to relieve the diagonal thrust on the abutments of a faulty arch by casting a reinforced concrete lintel above it, which helps to reroute the main wall load in a vertical direction.

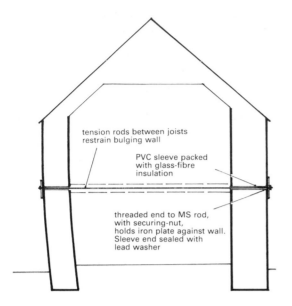

tension rods between joists
restrain bulging wall

PVC sleeve packed
with glass-fibre
insulation

threaded end to MS rod,
with securing-nut,
holds iron plate against wall.
Sleeve end sealed with
lead washer

Fig 55 A straightforward, traditional and effective way of tying in walls which lack adequate lateral restraint is to use steel tension rods spanning the building. The rods should not be more than 9m long to avoid the possibility of appreciable thermal movement. In some cases, concrete anchorage points may be cast *in situ* to tether the ends of tension rods.

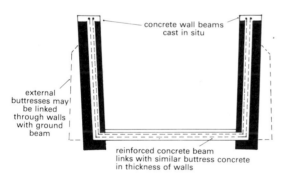

concrete wall beams
cast in situ

external
buttresses may
be linked
through walls
with ground
beam

reinforced concrete beam
links with similar buttress concrete
in thickness of walls

Fig 56 To stabilise leaning walls, internal buttresses may be devised, with reinforced concrete cast *in situ* and linked below floor level by a ground beam. The reinforcement bars must be so designed that the vertical and horizontal concrete is properly linked. An alternative method is to construct external buttresses (see dotted outlines). These are joined under the walls and floor by a ground beam on the same principle.

with reinforcement placed in the trench. The trench is also filled with structural concrete.

Clearly such measures are only effective if the continuity of the reinforcement bars is properly devised. The success of the method also depends on a sensible counterbalance of loading between the wall which needs restraint and either an internal wall or the floor itself. Naturally if the floor is to provide the balancing load, the beam will have to be sunk sufficiently deep to allow a substantial mass of concrete or stone to be loaded on to it.

Another method of restraining a wall is to link external or internal buttresses on opposite walls by means of a reinforced ground beam cast below floor level (Fig 56).

The Principles of Repairing Unstable Walls

There are two chief factors to bear in mind when designing structural restraint or support for unstable walls. You must ensure that the walling material to be supported is well enough bonded together to endure jacking up or tying back. Also, you must make certain that the wall containing the anchorage for such measures is equally robust, and sufficiently heavy and well braced to resist the new forces which are to be applied to it.

It is no use running tie-rods across a building to hook on to anchorage points in slender brick or stone piers which were only designed to transfer limited *compressive* loads to the ground. These piers may collapse completely when subjected to the horizontal pull now required. If a building is being redivided internally, as in a barn conversion for example, it may be possible to design substantial crosswalls which will either brace the external ones or provide a new source of counter-weight or anchorage points. In a stone building, the fact that such compartment walls are thick (whether of stone, brick or blockwork) will enhance the sense of age. Thick walls always look better than thin ones in such instances.

With all these methods it is essential to develop an almost physical awareness of the building's structure. Like a good sense of direction, which enables you to retain the relationship of one landmark and another in your mind's eye, you will find it extremely helpful if you can sense the way in which parts of a building are joined together and are structurally interdependent.

It is surprising how often a very straightforward, commonsense form of repair will work, provided it is properly carried out with respect for the capabilities and shortcomings of the materials involved. Clearly, in a book like this, it is impossible to give detailed instructions and specifications. In any case, these will always vary according to individual circumstances. The important step is to understand the principles.

One which should never be forgotten is that the cause of structural movement must be removed, either before or while a remedy is being carried out. If you jack a wall back into place which has been leaning because of movement in the foundations, that movement must first be stabilised. If the cause is principal rafters that are inadequately tied and spreading, then they must be tethered together at a point which will cure the trouble.

The common inclination among amateur house restorers and also lazy or incompetent builders is to patch faults up without curing them. Cracks in a gable wall containing a flue are typical candidates for this treatment. A hard mortar or filling compound is stuffed in, and that is that. But the crack soon opens up again. This may not always matter too much, but if you want to get rid of the crack for good, you must open up holes at intervals and stitch across the crack with a well-bonded arrangement of stones, tiles, bricks, blocks or concrete. At the same time you might just as well put in flue liners, provided they are to serve a hearth which does not demand the big traditional open flue, up which you can see the sky.

Sometimes it will be found that a leaning or faulty wall is stable enough if it can just be relieved of some of its load. A typical situation is when a massive external chimney-breast wall is pulling away from the gable wall to which it is attached. By unbuilding the upper part of the chimney breast you may be able to lighten the weight on the foundation subsoil while also reducing the leverage exerted by a tall, top-heavy structure.

Although it is perfectly possible, and indeed desirable, to jack or winch old walls back into an upright position, I strongly recommend that you do not undertake this kind of operation unless you know what you are doing and have adequate assistance and equipment. The method involves applying a stout timber cradle to the external face of the wall, which is tied with steel rods to anchorage points on the other side of the building (Fig 57).

The wall is provided with a new concrete foundation from the inside, to half its thickness. The internal face of the wall is wedged up against the new foundation with folding wedges. The ends of the wall are, of course, unbuilt at the quoins to allow free movement, and an *in situ*

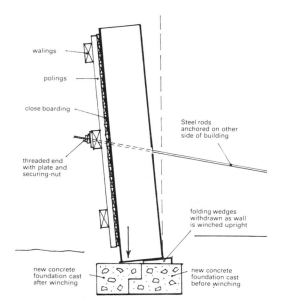

walings

polings

close boarding

threaded end with plate and securing-nut

Steel rods anchored on other side of building

folding wedges withdrawn as wall is winched upright

new concrete foundation cast after winching

new concrete foundation cast before winching

Fig 57 In rare cases it is possible to winch or screw a leaning wall back into an upright position using a stout timber cradle and cables or steel rods. In this example the wall is pushed back by tightening the securing nuts on the threaded ends of the bars; but other methods can include the use of winches and cables, or even the self-weight of the steel rods themselves. To carry out the operation, it is necessary to cast new foundations in two stages: one for the wall to settle on to when the wedges are eased out at the base; and another to pick up the other half of the load after winching. A very tricky operation.

concrete beam may be run along the top of the wall to act as a bonder and plate for roof timbers. Naturally all such timbers, and those of floors too, are cut or propped clear of the wall if necessary.

When all this has been done, the wall is wound back into an upright position, the wedges being gradually removed. Finally, the bond at the corners is made good again and the other half of the foundation, to the outside, is excavated and cast in concrete at a suitable level. This operation is not only a job of underpinning but of straightening as well. It is mentioned not because many readers will feel competent to carry it out, but rather to show that almost anything is possible with old houses if you are prepared to expend the time, money and expertise.

Small buildings, such as cottages, that are pretty but of no tremendous architectural or historical distinction, seldom merit such complicated treatment. You are more likely to leave a leaning wall alone on the basis that it may well be within the middle-third rule requirements, has stood for two hundred years and is not still moving. If it *is* moving, according to your tell-tales, then the possibility of underpinning may be considered especially if it is good stonework or brick. If it is rubble stone, then rebuilding could well be the remedy.

The Rebuilding Option

As we saw above, in some circumstances rebuilding a wall may be the most sensible answer to your problems. If you decide on this, you first prop the roof timbers and floors, and photograph the wall inside and out, after numbering the stones with whitewash. Then you carefully unbuild, methodically laying down the stones in rows. You may find that the foundations only amount to some hefty 'grounders' placed much too shallow for comfort.

You then trench and backfill with concrete to make a level foundation. Finally you rebuild the wall, bedding the stones in relatively weak mortar – perhaps 1:1:6 or 1:2:9, depending on the kind of stone you are using. Generally, as with pointing, the mortar should not be stronger than the stones themselves.

Although you probably want to rebuild the wall to look as much as possible as it did before, take the opportunity to replace any defective, broken or softened stones. Those which were wrongly bedded may be relaid correctly, if still sound. You insert a damp-proof course; and you ensure that this time the lintels are sound and have adequate bearings, if they were previously too mean. It is also worth ensuring that there are a reasonable number of 'through-stones' to tie the wall from front to back. Above all, you replace the no doubt rotten inner timber lintels, either with reinforced concrete ones or with new timbers of adequate strength. The latter should be vacuum-treated with preservative or thoroughly soaked in it. Hardwood lintels are preferable, especially if visible. Any mouldings should be copied in the new work.

While rebuilding, it is a good idea to allow recesses for pipework or wiring conduits. As you can see, there are many advantages to rebuilding; but *never forget* that a wall should still look like part of a nice old building when you have finished repairing it. On no account should it appear modernised or over-restored.

There is much argument in professional circles about the choices involved in repairing or replacing bits of old stone walls. One school of thought demands that new work should look new. Freshly cut stone is employed and allowed to weather, if it will. This, they say, is 'honest'; it is also helpful to old-buildings experts and archaeologists, who can see what has been done to the building – even at dusk, after half a dozen glasses of whisky. Personally I do not feel I am owed such debts of 'honesty'. The values involved are not moral or ethical ones, but aesthetic: there is no obligation to be 'honest' but there is a very important one to the building, and to the people who will see it every day. Let it look mellow, and if you need a new lintel, why not find a weathered piece of stone of the right kind if you can do so?

Obviously, you do not replace or tamper with ancient features which are sound, or can be repaired. It is certainly not one's intention officiously to smarten up a building, and at all times the most careful respect should be paid to detail. Most of us also have a conscience about the source from which salvaged materials are obtained: it is not helpful to damage one decent old building to repair another.

Stone Walls

When rebuilding stone walls there are one or two points to bear in mind. The first is how the stones are laid. All stone when quarried lies in what is called a natural bedding; this may be more or less horizontal or at quite a steep angle. Some Westmorland slate, for example, lies at 80°, some Cornish slate at 35°. For building purposes, stone should generally be placed in a wall so that it carries the weight at right angles to the natural bedding, which is also the principal plane of cleavage down which the stone can most easily be split (Fig 58, stone A).

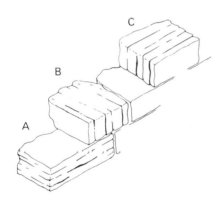

Fig 58 There are three ways to bed stones when building a wall. The correct method (A) is to place them in the natural bed in which they lay before they were quarried. To *edge bed* (B) or *face bed* (C) is aesthetically unpleasing and structurally incorrect.

If stones are positioned with the load of the wall bearing in the same direction as the laminations of the bedding, and parallel to the upright joints (perpends), it is said to be edge-bedded (Fig 58 stone B). This method is correctly employed for projections such as string courses.

Should you place the stone with the bedding plane upright and parallel to the face of the wall, it is called face-bedding (Fig 58, stone C). This is very bad practice, since it encourages the stone to spall or split away as a result of the combined

effects of moisture, frost and thermal expansion. However, stone may sometimes be laid in this way when additional structural strength is required for something like a slate lintel.

You will often have great difficulty in discerning the bedding plane of say a piece of granite, sandstone or limestone, unless your eye is very well trained. Many sandstones and limestones have no obvious plane of cleavage to suggest the bedding, whereas slate has two – one for its natural bed and one at right angles to that. Quarries often mark blocks of stone to indicate the bedding plane.

It is important that a wall is well bonded from front to back and this means introducing 'headers' or 'through-stones' at intervals to tie the inner and outer faces together.

Building stone walls requires a high degree of skill, especially when it is necessary to dress stones on site. If you try your hand at random rubble walling you will discover that despite its apparent lack of discipline, it in some ways requires more technique than laying reasonably squared stones in bonded courses.

The selection of the right stones for quoins and openings requires nice judgement and you soon realise that you need to develop at least a minimal ability to trim the stones with pitcher, chisel and punch.

Many people will be tempted to build extensions or replace lengths of wall using reconstructed stone, which is in effect a kind of concrete block made from carefully graded and coloured aggregates to simulate the real thing. Unfortunately, it is not normally very successful from the aesthetic point of view, as it tends to look monotonous, with little subtlety or variation in texture.

However, in areas where the normal building stone is a very smooth-looking freestone, the reconstructed type can be superficially convincing. Structurally, sandstone is merely aggregate in a naturally composed matrix; while concrete is also made of graded aggregates bound together by a matrix of Portland cement, itself a product of natural chalk and clay. It is easy to talk glibly about 'natural materials' and I am often tempted to do so.

However, it is sometimes hard to decide which is worse – a synthetic-looking artificial stone composed of natural aggregates or a totally *natural* one which still succeeds in looking repellent.

When choosing materials, avoid at all costs using real stone in a kind of vertical crazy paving (Fig 150, page 158). Any new stonework should match as nearly as possible the bond of existing areas, unless there is a sufficiently clear visual break between old and new to allow you to get away with something different.

Fig 59 A wedge-shaped 'save' stone, rather like a keystone, may be built above a vulnerable flat arch or lintel, to help the stonework form a kind of relieving arch, reducing the loading.

Freestone lintels over openings must be so devised that they are not overstressed. This means either using deep sections, reinforced concrete back-up lintels or employing relieving arches, or flat arches with 'saves' (Fig 59). A reasonable degree of bond should be aimed at, even in rubble walls, and joints should be kept as thin as adequate bedding and appearance allow. To try to compensate for your lack of skill by padding out with large wads of extra mortar looks terrible. In the case of 'snecked rubble' walls you make up these gaps with square 'snecks' of stone.

Try to observe traditional local methods of combining general wall areas with dressings at quoins and jambs. In some parts of the country the dressings used at openings are much more refined than in others. This is obviously so in places where good freestone is readily available. However, it seldom looks well to use around openings dressings which are obviously at variance, in terms of texture and outline, with the rest of the wall. An example would be neat limestone dressings to a wall of rough granite rubble.

Many buildings in the late nineteenth century – particularly those with utilitarian purposes – had machine-made brick dressings for arches and openings in rubble walls. Quoins too were treated in this way. The use of brick saved time, trouble and expense in squaring difficult stones like slate or granite, and was structurally useful as an alternative to stone or timber lintels. Brick can thus be correct, without doing a lot to please.

You should notice that stone dressings around windows do not normally project proud of the general wall surface, in early buildings. That tends to be a Victorian practice.

The use of exposed concrete in the repair of old walls should, on the whole, be avoided. One of the most likely reasons for employing concrete is the insertion of a new reinforced lintel over a door or window. It will certainly look 'honest', as they say, but as with many an honest person it may also be unlovely. It is better to keep structural concrete out of sight, beneath plaster or other finishes. Despite these warnings, I should not entirely rule it out for a few purposes in some kinds of building. I have seen exterior window sills cast *in situ* and left to weather, rather than being painted, and they have proved very inoffensive. Clearly concrete can never be anything but a second-rate substitute for stone, brick or timber, but it may occasionally have its place as a visible texture.

Bedding-material Repairs
One of the most persistent sources of trouble in old stone walls is decomposition of the bedding material. It is common to find that stones have been laid in a form of rough 'mortar' composed of either clay or non-cohesive soil with some clay content. Eventually this mixture dries out, crumbles and disintegrates or is washed away. The small stones and bedding material in the heart of the wall cease to serve their purpose and the situation is aggravated by rainwater seeping down and washing channels between the stones. If the wall has not been given a sound mechanical bond between the inner and outer leaves of stonework, by the generous use of through-stones, bulging of one face can occur.

One method used to repair walls in this

condition is called heart grouting; and it can be a costly and tedious business. First, the inside of the wall must be washed out with water, which can itself dislodge even more bedding material. The idea is to clean up the stones inside the wall so that the grout of cement and sand will adhere properly.

Next the walls are temporarily stopped out with clay, a few holes being left at roughly 1m staggered centres to reveal the presence of the grout at that level when it is poured into the wall. Holes are drilled, or existing ones used, to introduce the grout, of 1:1 cement and coarse sand. A clay cup is formed against a hole rather like a house-martin's nest. The grout is poured in and allowed to flow through the wall until it appears at the tell-tale holes lower down. These holes are then plugged and more of the mixture is poured in until it appears at another set of holes about 300mm further up. The grout obviously must contain enough water to be fluid, but not so much that it will not set into a reasonable solid core. This method is called gravity grouting.

Low-pressure grouting involves the same process, except that the grout is fed through a tube from a container. By raising the container or reservoir above the level at which you intend that section of grouting to finish, you introduce a static head of pressure which forces the mixture more extensively into the heart of the wall.

Low-pressure grouting may be carried out with increased amounts of static head by lengthening the feed tube and raising the height of the reservoir. A maximum height might be about 2.5–3m. However as pressure increases, the necessity to point the wall first, inside and out, becomes more apparent.

High-pressure grouting is carried out by contractors using special equipment and need not concern us here. For this method it is usual to shutter the walls with boards to resist collapse: it is not a device which will greatly appeal to non-professional house restorers. Incidentally, with *any* heart grouting method, it may be necessary to support insecure areas of stonework such as bulges.

Cob and Similar Materials

In some parts of the country – especially the South-west – many old cottages and farmhouses are built of cob. This is clay, mixed with straw, and sometimes with small stones. Cob walls are thick – anything from 2ft to 4ft – and the clay was built up in 'lifts' of about a foot at a time. It was not poured or rammed between shuttering boards, as in the case of pisé-de-terre and some of the other regional unbaked-earth methods. Almost all such forms of building involved the use of a stone or brick plinth and a good roof, often of thatch: the vital thing with mud, clay, chalk, wichert and clay-lump building is to keep the walls dry.

Because unbaked-earth walls were slow to dry out and unsuitable for load-bearing until they had done so, cottages were often constructed to first-floor joist level in stone. Cob and similar walls were sometimes plastered and normally limewashed to keep the all-destroying rain out. Corners were rounded to prevent weak and vulnerable edges being exposed to weather and damage. Lintels over openings were of timber and it may be imagined that these would soon decay if they were allowed to become impregnated with moisture absorbed from the surrounding clay.

Cob and its cousins mentioned above are fiendishly difficult materials to repair, while relatively straightforward to build. Some experiments have been made in modern cob building, but it is not a method which readily commends itself to local-authority building inspectors.

Clay-lump walls were built of roughly made pieces about the size of a 150mm thick concrete block (laid flat) and were moulded on site. If cob or clay lump has disintegrated, the only answer is to cut out the faulty material and build in clay lumps which have first been given plenty of time to shrink and dry. The whole idea is to prevent your repair work shrinking *after* it has been positioned in the wall.

The only other useful advice is to ensure that the roof has enough overhang to protect the tops of the walls and shoot rainwater well clear of the face. The walls themselves should be limewashed for practical reasons. What you do not want is a nice cob cottage covered with some grisly Tyrolean finish, or cracked and flaking hard cement render. (The merits of limewash are discussed in Chapter 12.)

Clearly, cob is not an easy material when you have to fit timber partitions and door frames. Wooden fixing plugs will have rotted and need replacing. When the old are withdrawn, it is anybody's guess whether the new (well-treated with preservative) will bed themselves securely when driven in. You can pack around them with a cement:lime:sand mortar and hope that will do the trick; the risk is that the cob is so soft that you keep having to drive your fixing plugs into new positions, gradually causing more and more wall to disintegrate as you do so.

It may sometimes be necessary to build up the jambs of openings in cob walls, using bricks or concrete blocks. This will help to retain the cob and may provide a reasonably well-plumbed fixing ground. Unless you are very clever with your plastering, however, much of the appealing contour will be lost.

Timber Buildings

The repair of timber-framed buildings requires great skill and is best left to an experienced carpenter, working under the direction of a knowledgeable architect or surveyor. But it is worth looking at the main technical and aesthetic factors involved. If you have an old timber-framed house, you may expect it to fall into one of three categories. Either it will be built on the 'cruck principle' (Fig 60); or it will be one of the many variations of the medieval, Tudor and seventeenth-century post-and-panel types of construction (Fig 61); or it will be a much more recent building with softwood framing hidden by a cladding of weatherboards, tiles, slates or plaster. The latter usually date from the late eighteenth century or nineteenth century (Fig 62).

The main point to remember with early timber buildings is that they depend for their structural stability upon tried and tested relationships between one member and another which can best be described as basic feats of engineering. Constructed

Fig 60 The curving blades of the pairs of crucks in this method of timber framing form both the principal rafters and the main wall posts. This is a *base cruck* building, with the feet of the blades resting on low plinth walls. The crucks were often paired by using the matching halves of a single tree. There are many other kinds of cruck which spring as roof timbers only from various positions within the height of an ordinary stone wall.

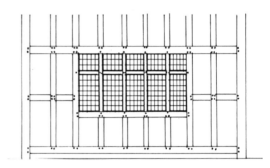

Fig 61 Many medieval, Tudor or seventeenth-century buildings were constructed on the *post-and-panel* principle, with pegged mortice-and-tenon joints. The exact pattern adopted varied at different dates and between one region and another. Panel infilling could be of wattle and daub or lath and plaster. This one, with its window of mullions and transoms and rectangular leaded panes, might be early seventeenth century.

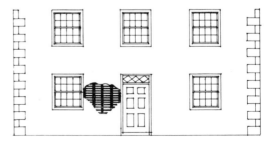

Fig 62 Rather flimsily built houses framed from imported softwoods were popular in the first half of the nineteenth century. This one is covered with lath and plaster and has simulated quoin stones. Others have slate, weatherboard or tile claddings.

from hardwood (usually oak), their walls consist of stout ground plates, with upright posts, studs and horizontal framing, which form the familiar panels of 'black and white' houses. Such houses do not rely upon the sheer mass of the walls to hold up the floors and roof, as does a Norman church or a stone cottage.

The craftsmen who built them were obliged to understand whether a piece of timber was to be subjected to compressive stresses or tensile ones. They had to use the right jointing methods, and the joints needed to be very soundly and snugly fitted. Nails were not used in framing; timbers were seldom straight or of identical section. Many timbers such as brackets, bressumers and joist ends were moulded or carved with ornament and usually most of the structural members were visible from the outside.

The framing would be worked, cut and jointed on the ground, each component being chiselled with a carpenter's mark showing which stud fitted to which mortice in the ground plate, and so on. Having been prefabricated in this way, the house would be assembled on a prepared brick or stone foundation. In some buildings the ground plates are in separate lengths, resting on foundation walls built between the main posts. More often, all the posts are tenoned into the ground plate. The most frequently employed joint was the mortice-and-tenon (Fig 63), followed by various types of notching, halving, lapping and scarfing. Most, but not all, of these joints were held together with slightly oversized tapered pegs tapped into carefully positioned holes. Despite the many sophisticated methods of jointing, there were inherent shortcomings in most timber buildings.

Foremost of these was the fact that to house a joist end into a beam, timber which contributed to its strength had to be cut away. You may therefore assume that the more massive the beam, the greater the margin for error you have when it comes to cutting housings for joists to be fitted in the traditional way – so that the top surface of both joist and beam actually support the floorboards. In the seventeenth century it became

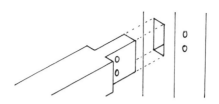

Fig 63 Variations of the pegged mortice-and-tenon were the most frequently employed carpentry joints in timber-framed houses. They also abound in every kind of internal joinery.

increasingly common for weaker imported softwood joists to be laid on edge rather than flat. This naturally provided additional strength, but for various reasons it was not a method employed in early timber-framed buildings.

In Chapter 8, we talk about jettied floors which work on the cantilever principle, thus stiffening the joists, providing extra upstairs floor space and keeping rain off lower walls. One suspects that jettied buildings were not designed only for practical reasons but were also a source of professional satisfaction to the carpenters, and objects of pride and aesthetic delight to all concerned.

The moral of all this is that you tinker about with timber-framed buildings at your peril. The results of ill-advised alterations to such houses can be seen everywhere. The introduction of new window openings, weakening the structure, is an obvious example. Walls lean out crazily where tie-beams have been cut through. Posts buckle under the additional loads imposed when intermediate studding has been removed. Trimming the opening for a new staircase may destabilise a corner post by untethering the end of a dragon beam. Because timber is a strong and flexible material, much that goes wrong with framed buildings is manifested by distortion, rather than collapse of the structure. Often, this distortion has taken place in the years directly following the building's construction and has long since ceased to present a problem. Much of the charm of such houses can be attributed to their imperfections and there is no point in straightening them up for the sake of it.

Where possible, though, you should try to remove the layers of black paint or other gunges which have been applied by previous generations. Except where framing has been tarred at an early date, it was not usual for the house to be 'black and white' in the manner we know today. That was largely a nineteenth- and twentieth-century conceit. Oak timbers never look better than when they have acquired the natural silver patina of age.

Some people claim that hardwood timbers should be treated with linseed oil, but there is little evidence that this does any particular good. New pieces of hardwood should be seasoned naturally over a considerable period – some recommend about a year per 25mm thickness – or old timber should be employed. However, the moisture content of 'new' stuff should be kept below 15 per cent and the old must be free from beetle infestation or decay.

Where infill panels, window sills and other units are to be fitted, it is worth introducing lead weathering trays, but these should be painted or felted on the underside where they rest against the timber to protect them from the tannin. Framing panels were normally filled with sprung staves, interlaced with wattle, which was then daubed with clay; or else riven laths were slotted between upright studs or posts if these were closely spaced. The latter type of construction is usually found in earlier buildings, before the acute timber shortages took hold in the second half of the sixteenth century. Although there is no hard-and-fast rule about it, the square-panelled type of framing is more often Elizabethan or seventeenth century than medieval; but regional preferences should be taken into account. The square-panelled frame was much favoured in places like Herefordshire, while upright studs and panels were common in East Anglia.

Clay daub was turned over with a shovel on site or puddled, water being added fairly generously to make it workable. Straw or hair was also added to reinforce it. The mix was thrown on to the wattle rather as you would carry out a 'spatter-dash' on a modern building. The daub was finally plastered over, or just lime-washed.

There is some controversy about the types of finish which should be given to structural timbers which have been inserted in old timber-framed buildings. However, it is generally agreed that fake distressing of timber – chisel marks, thrashing with chains, false waney edges and so on – is repugnant.

Originally, timbers were finished in one of four ways. They were either left with a pit-sawn finish or they were riven; often they were adzed and sometimes they were planed. Nobody, today, is going to convert timber with a two-man pit saw, nor bother with splitting timber when they have fast, efficient machinery to do the job of ripping down the grain. It may be that the use of an adze, which is rather like a mattock with a slightly curved flat blade (Fig 64), can be helpful to finish something like a big oak fireplace lintel or the beams and joists in a building of quality.

In the last event, the most useful tool for finishing new timber so that it has a little bit of that attractive irregularity is the hand plane. The adze, by the way, can still be purchased at some ironmongers which hold large stocks of trade tools, but those who employ it should, like the traditional craftsmen, try to achieve the most even finish the tool will allow. Otherwise the results can look faked.

An important point to bear in mind is that new framing timbers should look

Fig 64 The adze was used throughout the Middle Ages and is still the right tool for bringing hardwood beams, joists and lintels to a convincing finish. With good technique the timber can be given a surface which does not look faked, while avoiding the uniform appearance obtained by machine planing.

natural and unaffected; and above all, they must be seen to serve a structural purpose – unless you are aiming at an intentional pastiche.

The temptation when repairing a timber-framed building is to cut away more and more decayed or distorted timber, replacing it with new lengths. You should tell yourself firmly, at the very beginning of the project, that you are going to save every inch of the original you can. There is no substitute for the beauty of mellow, imperfect oak which has four hundred years of weathering. Remember that you are saving an old house, not building a new one. In some cases new pieces of moulded timber may have to be introduced to replace sections which have decayed beyond redemption. Although this point will be made again in another chapter, it cannot be said too often that machine-run mouldings are usually a disappointment when seen beside very old ones. They look too even and mechanical. One answer is to have the mouldings run very slightly oversize and finish them with hand tools.

Brickwork

The main difficulty in repairing old brick walls is to obtain bricks of the right colour, size and texture. Old bricks were hand-moulded from lumps of clay by workers standing at rough moulding-tables in the open air – sometimes on site. Brickworks were set up near the places from which suitable clay could be dug.

The process of hand-made brick manufacture was to dig the clay, puddle it with water and knead it into lumps. These were thrown into wooden moulds and the surplus cut off with a piece of wood called a 'strike'. The moulded bricks were then stacked in a herringbone pattern between layers of straw to dry.

Next they were fired in a kiln or clamp for a period of days. Much of the interest to be found in their varied colours derives from the hit-and-miss effects of firing; but the nature of the clay itself was also fundamental. By the early eighteenth century, brick clay was often puddled in a pugging mill, at more sophisticated yards. The clay was thrown into a primitive mixing machine powered by a horse which pulled round a centrally pivoted beam, similar to that used for operating corn-grinding machinery in a farm whimhouse. By the nineteenth century a variety of brick-making machines had been devised which could either mould bricks in large numbers or extrude the clay so that it could be cut to size with wires. The moulded brick is called a stock, the name arising from the use of a 'stock-board' which is spiked or pegged to the moulding table as a base for the mould itself. The thickness of the brick could be adjusted by leaving the stock-board well proud of the table or driving it further home.

It is still possible to obtain hand-made bricks, but they are very expensive and could only be contemplated for a select minority of restoration works. Mass-produced bricks cannot have the same attractive irregularities, but there are quite a number of modern bricks, both pressed and wire-cut, with good colour and texture. Wire-cut bricks do not have frogs, while hand-made stocks normally do. Some pressed bricks have frogs on both sides. The frog, if there is only one, should be laid upwards. It is a nineteenth-century innovation.

Clearly the most economic method of obtaining suitable bricks for repair work is to find old bricks and reuse them. They will then be mellow and of good colour, even if they are of relatively recent date. Most bricks have adequate crushing strength for use in restoration work, but there may be times when you are obliged to employ a very hard engineering brick to withstand, for example, the loading of steel universal beams resting upon a pier in, perhaps, a big mill-house.

When you do have to use such bricks, you can obliterate their unpleasing appearance by applying a coat or two of white emulsion paint. Sometimes it may be worth using real limewash. For work connected with flues and stoves it may be necessary to have firebricks, which have good thermal properties that prevent cracking.

The situations which can really tax your ingenuity are those which concern such features as rubbed brick arches, moulded details like cornices, drip moulds and mullions. For the latter you have no option but to pay the price and get what you want from a specialist hand-made-brick firm. If needing bricks for rubbing to specific shapes, you may be able to grind off old ones to the required taper for something like a simple arch – if you have patience and some manual dexterity. For rubbed or gauged brickwork, as it is called, craftsmen traditionally used a special soft brick called a rubber. This was made with a proportion of sand added to the clay and was fired in a less fierce and more controlled manner.

Should you be fortunate enough to have a building which has moulded brick cornices, architraves and rusticated quoin effects, you need rubbers to replace excessively worn or damaged details. In some instances you will actually be able to rub down the bricks after cutting them with a saw or a rotary disc. In others you will need to have bricks moulded to the required shapes. However, for minor replacements you should often be able to carve and rub matching details. Carving is itself another traditional way in which fine ornamental brickwork was created.

When replacing brickwork or joining new brick walls to old ones, take great care to use the correct bond, so that you both echo the spirit of the building and employ a technique which was used in that particular era. Early medieval brickwork was bonded in a random manner. The bricks usually avoided the structural weakness of having one perpend directly above another, although joints inevitably matched up here and there. By the middle of the fifteenth century 'English bond' became popular. This consisted of alternate rows of headers and stretchers. During the first third of the seventeenth century 'Flemish bond' was introduced from the Low Countries. Flemish bond is identified by its alternate headers and stretchers in every row, the headers in one course being centred below the stretchers of the one above. It continued in general use throughout the seventeenth and eighteenth centuries.

Variety of pattern was increased first by the introduction of diamond designs (diaper work) made with flared headers. These bricks were dark in tone, often almost blue or slate-grey; they were made from overburned bricks whose ends had been in contact with the smoke and flames in the kiln. Diamond decoration was often sixteenth century, and during the seventeenth century, chequer patterns were adopted.

In the eighteenth century methods of bonding became more varied. Flemish bond was continued, but adaptations, like Flemish Garden-wall bond and English Garden-wall bond, as well as English Cross bond, and a bond entirely made up of headers, were all in favour. Polychrome brickwork was still liked but could now be found in cottages and small houses as well as the more important ones.

The size of bricks can be important both in dating a building and when introducing new brickwork. The area of a brick laid flat has changed astonishingly little since the Middle Ages. It is now about 215 × 100mm (8$\frac{1}{2}$ × 4in). Saving one or two aberrations, like the 'great brick' used in the late eighteenth century- a monster, of measurements such as 11 × 5$\frac{1}{2}$ × 3$\frac{1}{4}$in – size has been very consistent. The original 'great brick' was an early medieval type, slightly larger than the Georgian one, but much thinner (perhaps 45mm). The Brick Tax regulations of 1784 to 1803 encouraged the use of monster bricks, since the tax was levied by number rather than by size. After 1803 bricks over 150cu in. in volume were taxed at a higher rate, so about that time you may find examples exactly 10 × 5 × 3in – just hitting the limit. The tax was repealed in 1850.

It is far from safe, however, to try to date buildings from brick sizes, since they varied much more than people sometimes realise. Very large bricks, for instance, were traditionally made in parts of East Anglia, quite apart from any Brick Tax implications. To some extent the thickness of the brick is an indication of age. Until well into the seventeenth century they were most often about 50mm thick; during the later seventeenth and eighteenth centuries they were still only a little more than 50mm – perhaps 57–65mm, or so. In the nineteenth century the modern thickness of about 65mm was usual. But it is impossible to lay down rules about brick measurements, and you will often find that people refer loosely to the size of a brick including its mortar joints – making a considerable difference to the dimensions.

What can be said is that much nineteenth-century brickwork, during the fever of industrial house-building, was of extremely poor quality. Not only was the mortar lacking in essential lime, but the bricks themselves might be badly made and of scrap materials. Bonding and tying of walls was often skimped and arches so roughly made that they impaired the structural stability.

Cleaning Stonework or Brickwork
The likelihood of your having to clean stonework or brickwork with your own hands is remote. If you have a country property it will probably not be dirty enough to need cleaning. If you have a town building which is sufficiently encrusted to merit treatment it will pay you to employ a specialist firm to do the job.

Most methods of cleaning require equipment not available to the average do-it-yourself house restorer, unless he or she hires it. The exception to that statement is a rudimentary form of washing with cold water, accompanied by scrubbing of particularly bad areas and, perhaps, some abrasive treatment using carborundum stones, sanding discs, and so on. Undoubtedly water-cleaning is the least injurious method in most cases, although there are some dangers to the building through saturation. If parts of the fabric which are not normally subject to moisture penetration do not dry properly after cleaning, rot may be started in timbers or rust in concealed metal fixings.

Water may be applied as a fixed spray, by hand-held guns or pressure lances, according to the circumstances. In some forms of cleaning, an abrasive grit is wet-blasted on to the stonework. Dry-blasting methods employing grit are also used with effect. Wet-blasting is generally less harsh.

Limestones are usually rather vulnerable and should not be dry-blasted except to remove particularly obdurate areas of encrusted dirt. Sulphur in the atmosphere of towns will react with limestone to form calcium sulphate which encourages the stone to flake or blister. Sulphur tends to create a thin hard coating on sandstone and this can prove very resistant to water cleaning.

Lances blowing grit at relatively low pressures, with or without water, are commonly used for cleaning old buildings; and the very delicate types, such as the pencil lance, are useful for work on carved detail and mouldings.

Dry-blasting creates pervasive dust, unless gathered by vacuum, and wet-blasting produces a mess of waste slurry. Scaffolding in towns must be sheathed in plastic sheeting, even for ordinary cold-water cleaning. The actual process of washing can cause brown staining from dirt products within the stone leeching to the surface. Sand blasting may take off surface dirt but also free other discolouring agents within the stonework, thus encouraging further stains. Blasting can be useful for cleaning up walls which have had a hard rendering removed.

As for chemical treatments, these are strictly for the specialists. The effective but highly toxic and corrosive hydrofluoric acid may be used on badly encrusted sandstones, granite or some brickwork – with all manner of safety precautions. Caustic-soda-based alkaline cleaners can be used on limestone, but may leave soluble salts behind unless the subsequent rinsing methods are very thorough.

You may, though, find yourself having to remove layers of limewash from stone or brickwork, and this can be a daunting task, depending on the porosity and general vulnerability of the walling material. If it is a tough material like granite you could blast it, but soft bricks or limestones may respond to the repeated and lengthy application of clay poultices, which lift the limewash when they are peeled from the wall. They are laid on with a plastering trowel and should contain a metal mesh reinforcement to help pull the poultice away.

People often suggest cleaning lichens and algae from their old buildings. Since these do little harm, except perhaps to encourage the occasional build-up of damp-promoting soil, they are better left alone – and they look lovely.

The golden rules when cleaning are:

1 Don't clean unless you have to for either aesthetic or practical reasons.
2 Use cold-water cleaning if possible, but don't saturate soft or badly jointed stonework to the detriment of the structure. (Do it when there is no risk of frost.)
3 Think hard and long before cleaning carved detail or mouldings and seek specialist advice on methods.
4 Leave sand-blasting and abrasive-lance-work to professionals.
5 Leave chemical treatments, particularly acid ones, to professionals.
6 Be sparing of mechanical cleaning treatments with discs or carborundum stones. A heavy hand with power tools can be disastrous to an old building.

7

ROOFS

Broadly speaking, old roofs may be classified in three main categories, which can then be subdivided into about seventy subtle variations. This thought may prompt you to sigh, but be warned – a passing interest in such matters can be habit-forming and you may eventually develop a keen regard for the difference between a *deep arch braced collar truss* and a *queen-and-crown-post double-collar tie-beam rafter roof.* I rather hope you do, because it will mean that your enthusiasm for old buildings has gone beyond the mere practicalities of making them habitable.

The first main type of roof is one supported by curving cruck blades (Fig 65) which often act as the framing for the walls as well. The second form is the *single* roof, which consists of pairs of rafters,

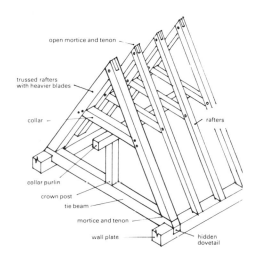

Fig 66 The early form of single rafter roof frequently had *crown posts* which carried the load of a *collar purlin.* The crown posts were jointed into the tie-beams of the pairs of heavier *trussed rafters,* constructed at intervals along the roof. The backs of the *common rafters* with their peg-jointed *collars,* lay in the same plane as those of the trussed rafters. This drawing shows the simplest kind of crown post. It has no braces. They were, however, often elaborately moulded and were an ornamented as well as a structural feature in an open-hall house.

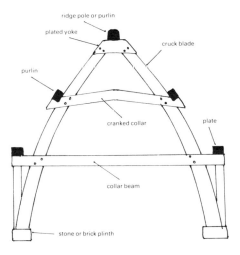

Fig 65 Roofs supported by crucks involved the use of *through purlins* and a ridge-pole or purlin, all bridging from one truss to the next. They were only limited by the lengths of the available pieces of timber. Common rafters were pegged to the backs of the purlins. This example is a 'base' cruck, with its blades springing from low stone walls.

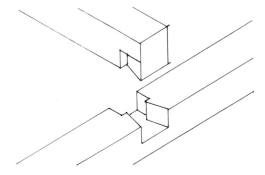

Fig 67 The halved and lapped dovetail joint in detail. It prevents the wall plate from parting company with the tie-beam, as the roof load applies an outward thrust upon the feet of the common rafters.

quite closely spaced and prevented from spreading by collars, or other tie-members, which span between the tops of the walls (Fig 66). The third roof – probably the most frequently encountered in both large and small old buildings – is the *double* arrangement of principal trusses carrying common rafters, which rest upon purlins (Fig 68). This is not to be confused with an M-shaped roof with twin gables. There are, of course, various types of flat or semi-flat lead-covered roofs constructed in different ways. These may be built up on trusses or made of stout beams with tapered firring-pieces used to provide a fall towards the gutters.

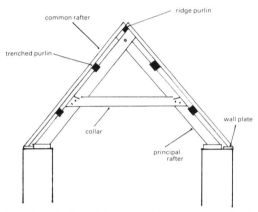

Fig 68 A *double roof* is one which is constructed of a series of pairs of principal rafters which support *purlins* and they, in turn, carry the load of *common rafters,* battens or laths, and slates or tiles. The type shown here is very frequently encountered in old buildings. The principals are 'trussed' with lapped, dovetailed-and-pegged collars, which prevent the feet of the rafters from spreading.

A mansard, so often used for eighteenth-century town houses, is merely a trussed roof with purlins raised upon a framework of beams and posts, having sides formed of common rafters. The effect is to form two slopes on each side of the roof – one very steep slope and one fairly flat one (Fig 69). The idea of the mansard is to provide good living-space in the shape of garrets.

Our concern is not so much with the architectural history of roofs as the faults and remedies associated with them. To some extent the two subjects are closely linked, since you need to know the possible age of a roof structure and the

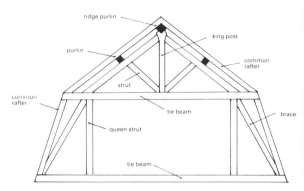

Fig 69 A mansard is constructed by forming a triangular truss upon a more or less rectangular one, so providing additional open space and headroom within the roof. Here the lower tie-beam employs queen struts to support a king-post truss, which in turn takes the load of purlins and common rafters.

sort of treatment which will be historically and aesthetically sympathetic.

Technically, you need to think of roofs in two ways: the carcassing structure, which will normally be assessed from inside the roof space; and the covering, which will be inspected from inside and out.

Faults in roofs develop from five main factors, or a combination of them: bad design, bad workmanship, bad materials, wear and tear and beetle infestation. The latter is dealt with in Chapter 11.

Bad Design

This may manifest itself in several ways. It may be that the roof was built with undersized timbers which have subsequently sagged or collapsed. Constructional joints may have been poorly thought out, so that members which are in tension have pulled apart under sustained loading. Inadequate ventilation could have encouraged decay. The lap of tiles or slates, if too mean, could have let in water, allowing laths, battens, pegs and carcassing timbers to rot.

A roof designed with inadequate pitch can sag because common rafters, purlins and trusses suffer greater stresses than their size will support. Slates or other coverings can allow moisture to 'creep' sideways into nail or peg holes via the vertical joints, should the roof have too shallow a pitch in relation to the width of the slate.

Many West Country roofs are of raised cruck construction. The curved part of the cruck blade is built into the wall masonry, at a lower level than ordinary principal rafters. The nearest collar is of the cranked type. It is mortice-and-tenoned into the cruck blade and the carpenter's marks can be seen on the face of the timbers. The second collar is cambered, while the third is a replacement.

Inadequate side or head lap to slates and tiles can allow moisture to penetrate to ferrous hanging nails, which then rust, permitting slates to slip. This is a serious matter if present throughout a roof. Moisture produces wet rot in battens and laths, which then give way. If slates have been torched or rendered from inside, this will also rot the battens, since moisture may have no opportunity to evaporate.

As we saw in Chapter 4, roofs are said to be 'nail sick' when the battens or boarding have been nailed and re-nailed so often that nails are loose and there is no sound timber left for refixing. Some people use the term 'nail sickness' to describe widespread rusting of nails in a roof. We have also seen that lack of ventilation in the roof space can build up quantities of warm damp air, ideal for promoting decay in the timbers, but condensation can also decay poor-quality slates which suffer frost attack, even in untorched or plastered slate roofs.

Slates or tiles at eaves or verges may have too little overhang to run water clear of the walls, so it is sucked or blown back into crevices between covering and timbers. Alternatively, too much projection of coverings at these points may present a broad undersurface which strong winds can catch and lift.

Lead flashings, soakers and valley gutters may be too narrow, so that water gets in where it should not. Tilting fillets, or lead rolls to slates abutting lead valleys, may have been omitted so that slates do not fit tightly. Lead may not be dressed high enough at the sides of valleys or gutters to prevent splash-back or even quite minor and temporary flooding. Melting snow can then prove a problem.

Abutments between chimney stacks and roof coverings, or between slates and

Another part of the roof pictured on the previous page provides some additional clues for dating. Here, roof braces can be seen curving up from the cruck blades. Their purpose was to check lateral movement. They suggest that the roof is probably sixteenth century. The roof is covered with big rag slates nailed directly to the common rafters rather than to battens. The purlins are deeply trenched into the backs of the crucks. This is the kind of early roof which it is essential that owners take pains to repair rather than replace.

gable parapet walls, may have been sealed without the use of lead, depending for their weather-resistance on cracked cement fillets or slate listings projecting from the brickwork. Indeed, listings were the normal practice in most earlier buildings. The extensive use of lead at abutments was an eighteenth- to nineteenth-century practice.

In better-quality Victorian houses, slates may have been nailed directly to sarking boards without the use of battens, or counterbattens, encouraging moisture to build up and discouraging ventilation. The bond of the slates or tiles themselves may have been poorly designed so that

courses start at the verge with a half-slate which will not lie down properly, instead of the more stable width-and-a-half piece. The vertical joints between slates or tiles may not be correctly staggered so that the side-lap is totally inadequate. Whether these latter points are put down to bad craftsmanship or bad design does not matter.

Old roofs, naturally, had no underfelt (which is a modern device), and although this perhaps encouraged ventilation, any moisture which found its way through faults in the covering came in. It is essential to inspect the interior of a roof in daylight, so you can search out the telltale chinks of light which inform you that a slate is slipped, missing or broken. A distinct advantage of having no sarking boards or felting is that slates or tiles may be inspected and repaired from inside the roof space.

One of the most frequent ways by which rainwater gets into walls is through roof faults. The absence of a double course or undercloak of slates or tiles at the eaves will certainly let in moisture. Gaps between the cornice or fascia board and the tiles will also let in driving rain. Here again, you will see a suspiciously wide slot of light at the eaves when you peer down between the lath and plaster of the ceiling slope and the underside of the roof covering.

Remember – of course – that you should not tread on lath and plaster in roof spaces and should carefully pick and choose which ceiling joists look capable of bearing your weight. You will find that the required view of principal rafter ends and wall plates, mentioned above, is frequently obscured by a purlin running just where it can prove most awkward. The feet of principal rafters and the ends of the beams are, in many cases, buried in the masonry or brickwork.

Looking up at the ridge with your torch, you will be dispirited to see that lengths are rotten and have been infested with beetle: another sign that damp is getting in, this time through faulty or badly bedded ridge tiles. Turn the torch off and, as your eyes adjust again, the same revealing chinks of light will emerge at the apex of the roof.

Inspect the chimney stacks – often there are water stains running down the brickwork, disclosing a fault in the abutments with the roof covering. The trouble may be the result of structural movement, or of bad design and workmanship.

When looking for leaks in a roof it really does pay to climb up through the hatch and investigate while it is pouring with rain. Old residual patches of staining are often deceptive, since they may be left from before some repairs were done, and areas damp to the touch can still be far short of showing you the actual passage by which rainwater is entering. Remember that the entry of water at a chimney stack or gable wall does not necessarily indicate a fault in the roof or abutment. Rain frequently enters by coursing down the inside of the flue and finding its way through weak joints in the brickwork. Soot-staining may give you a hint that this is happening.

While you are exploring the roof, if it is a really old one, keep your eyes open for signs of smoke-staining of all timbers, especially near the ridge and at the apex of the roof in the gable ends. It could be that the building was once an open-hall house with a central hearth, from which smoke was allowed to drift up and out through a louvre in one of these positions. Very many open-hall buildings had intermediate floors constructed during the sixteenth and seventeenth centuries, so disguising their antiquity.

To stop water entering by the flue you can either re-point and re-cap the chimney or, in some instances, provide the chimney with a new lining. It is virtually impossible to re-parge (or plaster) the inside of a soot-encrusted flue, since you cannot usually get access to the whole of its run, and to get the plaster to adhere is extremely difficult without extensive cleaning.

Alternative methods are discussed in Chapter 13, Flues and Fireplaces. As far as damp is concerned, your main purpose is to prevent rain entering the top of the flue, and this may be achieved by putting a chimney-pot where formerly there was none. However, there are some chimneys which should not have pots for historical and aesthetic reasons; and others demand an open chimney-top to burn properly.

Poor Workmanship and Materials
Bad workmanship is endemic in old roofs and so are poor materials. As one who spends much time pleading with developers and local authorities to save the ancient timbers of old roofs, it is disagreeable to have to admit that a great many so-called craftsmen of former times were both slipshod and untutored in their roofing techniques. Added to that, they often used the most weirdly shaped, ill-matched and inadequate principal rafters, collars, purlins and other carcassing pieces. These may have charm without being very structurally convincing. Despite such shortcomings, the roofs have survived and it is more often than not as a result of inadequate maintenance that rainwater has seeped in and finally brought about a roof's demise.

I must balance this statement by paying tribute to the equally and perhaps even more numerous roofers who lavished wonderful care and expertise in carrying out their trade. Generally, the older the roof, the better the work that went into its making – at any rate in the grander sort of building. By the eighteenth century matters had started to deteriorate, partly because imported softwood could be knocked up to form a roof structure without the high degree of skill demanded by hardwood; partly because the sheer volume of speculative development once the industrial revolution had taken hold encouraged jerry-building.

By and large, cottages and small vernacular houses were built by men who thoroughly understood the use of the limited materials at their disposal, but they would probably have been both shocked and gratified to hear that their efforts would be giving cause for concern three hundred years later. They would be mortified to realise that timbers embedded in cob and clay had rotted beyond redemption, but gratified that their efforts had survived long after the expected lifespan.

No enterprise undertaken by mortals can avoid the attentions of the perennial botcher, and his handiwork is self-evident in many old roofs. Slates are placed upside-down so that they do not lie properly; badly split and weakened laths are used to save the time involved in selection;common rafters fail to lie in the same plane so that the roof covering undulates – a fault which gives rise to ravishing visual effects and rather less pleasing practical ones. Scarfed purlins pull apart because the joints are not notched as well as lapped together; and nails rust away where pegs might have survived. The tails of slates or tiles lift in high winds because they are inadequately lapped and nailed. Cracked tiles or slates are put on because it is too much trouble to choose good ones. The absence of tilting fillets prevents a tight fit at eaves and verges; bedding mortar for ridge tiles is skimped or poorly pointed. Slates crack because they are nailed too tightly or nail heads pull through oversize holes.

Roof coverings are cracked and damaged because scaffolding used during repairs is laid directly upon them without the protection of sandbags or straw-filled sacks. Slates and tiles are laid in such a random and haphazard bond that joints through which water may seep are located too close to each other.

The slates and tiles themselves may be of low quality or be badly cut or formed. Good clay tiles, for example, should have a slight camber to make them sit down properly, one upon another. Old ones have a camber in both directions. Slates vary in their durability, the most beautiful not always being the most robust.

The widespread practice in some areas of grouting leaking slate roofs to make them weathertight can have disastrous results. The slates decompose, and laths and battens rot through lack of ventilation and moisture being channelled into hairline cracks. Finally, when most of the fixing pegs have either rotted through or dropped out, the roof may slide off in one piece, held together by layer upon layer of cement wash.

In counties where the traditional roof material is a big slate or stone-slate flag, nailed directly to the common rafters, any inadequacies in the choice and layout will leave substantial areas of roof virtually unsupported. Such roofs are exceedingly heavy, and rafters – often positioned at a fairly shallow pitch – may be inadequate to sustain them. Earlier Cotswold roofs of limestone were usually steep-pitched with diminishing courses. Roofing stones at the eaves were frequently of enormous size and these can present problems in lifting, positioning and supporting.

Scaffolding and Safety
Don't try to carry out roof repairs, other than fairly minor ones, from ladders. It always pays you to use scaffolding or framed towers. You must have a satisfactory base from which to work and upon which materials, tools and the ends of roof-boards and ladders may be rested. Use a guard rail so that you do not step off into eternity when standing back to admire your efforts; and employ toe-boards as well if there is any risk of dropping things on passers-by.

Tie scaffolding into the face of the building at intervals and ensure that stan-

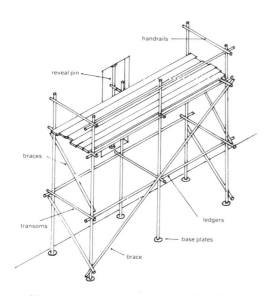

handrails

reveal pin

braces

transoms

ledgers

base plates

brace

Fig 70 A simple independent system of scaffolding is tied into a window opening by means of a reveal pin and a long transom. All connections are made with special nut-and-bolt fastened couplings – not shown in this drawing. Toe-boards around the working platform have been omitted for clarity. (The ends of the second and third boards out from the wall do not overlap the transom enough for safety.)

dards are vertical, ledgers decently horizontal, and all is properly braced (Fig 70). Never let scaffolding boards run in such a way that they will tip up by forming levers over supporting transoms or putlogs. Ensure that fittings are correctly seated and tightened and that the feet of free-standing scaffold towers are well levelled and firmly chocked-up.

Any but the shortest ladders should be lashed at the top to the scaffolding, and all ladders should bear properly on the framework and be firmly seated at ground level. One cannot emphasise too much that a sensible and cautious approach should be adopted by anybody undertaking roofing or other jobs which need scaffolding. It is second nature to experienced workers, but amateurs may be blasé once they have become accustomed to the height.

It is not good practice to rush about (which is not to imply that work must proceed at a slothful pace). Do not scorn to tie yourself on if you are working on a difficult slope of slated roof. Check with care that ladders hooked over ridges have a proper seating and cannot slip and that knots and lashings are well tied. Notice that experienced workers, when climbing ladders, do not place their hands on the rungs, but run them up the sides. This is to avoid having their knuckles stamped on by the boots of the person above.

When paying-out rope over a jinny wheel or lowering heavy objects to the ground, never be tempted to let the rope run fast through your hands. For one thing it will take the skin off your palms, and for another you can quickly lose control as the object gathers momentum.

Never forget the quite extraordinary force of the wind when you are working high up. A new ridge-board can bow in a sudden gust as you try to position it and sail off the roof, leaving you helplessly clawing at thin air. Roofing felt must be temporarily tethered with battens while work goes on. Tools and materials should not be lodged in places where wind and gravity can snatch them from you. If they can, they nearly always do, and the job of picking a couple of hundred slating nails out of the long grass is not one to be recommended.

It is worth remembering that the law requires that scaffolding must be erected by 'a competent person'. This really applies to contractors, but it could mean you, if your scaffolding is to be used by other people.

If somebody working for you or with you were to have an accident and took you to court as a result, the case could hinge on your 'competence'.

Many builders and all big contractors demand that anybody erecting scaffolding holds either a 'basic scaffolder's ticket' or an 'advanced scaffolder's ticket'. All scaffolding firms have to employ trained men or they would never get the contracts, although the law still does not specifically define who a 'competent person' is.

Please forgive the slightly nannyish tone of all this – part of the object of reading books is to avoid always having to learn the hard way.

Slates and Tiles
Returning to the question of faulty roof coverings, the options for repair are

101

limited. You can either lift, and re-nail or peg, an entire roof; or you can patch and replace here and there. The latter course has drawbacks. First of these is the difficulty of tearing slates or tiles out of the carefully devised bond in which they have been originally laid. Second is the problem of introducing the replacements. They cannot be nailed after they have been slipped up under the ones above, and if the roof has already been repaired using underfelt it is impossible to get at them for fixing from inside the roof space.

Faulty slates can be broken free of their nails by using a tool called a ripper. It is pushed beneath the wedged-up slates and has a hooked cutting edge which, with luck, severs the nail without pulling away the batten as well (Fig 71). The replacement slate is snipped to size, if necessary, with a pair of slate cutters and worked up into position.

Fig 71 A ripper is used to remove damaged slates from a roof during repair work. The blade is slid under the tail of the slate and the hooked end is engaged with the slating nail. A sharp pull on the handle then rips the slate free – if you are lucky.

There are several ways of securing the slate. You can bend a strip of lead or copper into a double-ended hook called a tingle so that it hangs on a lath at the top and is firmed around the tail of the slate to keep it from slipping. Another method, achieved from *inside,* is to glue a block of wood or polystyrene to the underside of the head of the slate so that it rests against the top edge of a slating lath or batten. Other methods involve using galvanised wire or pegs. On a scantle roof where the slates have originally been tail-bedded in mortar, you can secure the slate with another dab of the same.

To some extent all this applies to tile repairs, the difference being that tiles may have nibs to hook over laths or battens. Whichever you are using, it is very important that you patch and replace with the right materials. Nothing looks worse than livid splashes of replacement tiles of a different colour and texture on an old roof. Reuse weathered tiles of the same pattern if you can find some. In the case of slates, try to avoid patching an old Cornish Delabole scantle roof with hard grey-violet Welsh slates for example.

Slates and tiles are of varying quality and durability. Poor slates contain carbonates which can cause decomposition when regularly soaked with water, especially if the water has an acid content. Rainwater and vapour in the industrial atmosphere of big cities can be particularly damaging. It lies under the slate and attacks the back, especially around the nail holes.

Some slates are much more permeable than others and the following simple test may be carried out to check if the quality is good. Stand a slate in water so that it is immersed to half its length. Leave it for twelve hours, after which the damp line on the slate should not be more than 3mm above the water level in the bucket. A very permeable slate will have absorbed moisture inches above the water level.

Tiles may be badly formed or fired, or made from inferior clay, so that they flake, especially if subjected to frosty conditions.

Selecting a Roof Covering
The material you use to re-roof an old house or cottage is of vital importance. Few things look less appealing than a pleasant old building displaying acres of modern concrete interlocking tiles. They are the wrong texture, the wrong colour and the wrong thickness, and they have a side bond which means they cannot be arranged with the tail of one centring on the joint between the pair below.

For traditional slate roofs some people economise by using grey, grit-dusted composition tiles of small size. I do not like them and am far from convinced that they look the part. In areas like the Cotswolds there are composition, moulded imitations of the distinctive local stone slates. These are arguably more successful than most counterfeits and can look quite convincing; but you should be guided by the quality of the original building. Never use fake materials for any building of architectural distinction, even a vernacular one.

Cotswold stone roofs are particularly attractive and worthy of special attention. Because they have so much contour and irregularity the slates are laid to a steep pitch – 45° (known as 'square pitch') being a minimum. The pitch ensures that water is shed rapidly and does not 'creep'. Such roofs are extremely heavy, requiring very sound and sturdy timbering.

Even heavier are some of the roofs made from enormous riven sandstone flags. Sandstone roofs are found in the North of England, the border country of Wales and Herefordshire and around Surrey and Sussex. Northern roofs of this kind are frequently laid to a shallow pitch of as little as 30°, since the flags are so wide that 'creep' is not such a danger. The southern roofs tend to be steep, like the Cotswold ones, and composed of thinner and smaller pieces.

There are certain points to remember when dealing with stone-slate roofs. The first is their permeability. The fact that water can actually penetrate the slates themselves is one reason why they must be thicker than genuinely metamorphic slate. For this reason also, the *lap* should be greater than normal. Four inches (100mm) is considered a practical allowance. Stone slates are naturally prone to suffer from frost damage when their moisture content is high. They can split and wear so thin that they have to be replaced.

In many early buildings the oak battens are pegged to the principal rafters. In other cases the common rafters were frequently pegged to the purlins. The slates are themselves hung on single oak pegs, and these may have rotted or fallen out. Torching or overall plastering inside the roof space may have promoted decay and the latter will certainly make inspection impossible. Because of the great weight of the slates, and the steep pitch, it is vital that pegs or other fixings are robust and in good condition. Even when the slates themselves are mainly sound it may be necessary to strip the roof and relay it, replacing faulty timbers and introducing a layer of roofing felt below the battens.

Ridges and hips of old roofs like this often have specially worked stone bonnet tiles. If possible save and reuse them. Clay tiles of good colour can occasionally be employed but are less pleasing.

As with scantle slates, stone slates are laid in diminishing courses with a heavy undercloak of big pieces at the eaves. These may be bedded in mortar and fixed at a lesser pitch. The selection, matching, trimming and adjusting of stone slates requires much skill and experience. You are not advised to lay such roofs yourself, unless you first work with somebody who knows what he is doing and learn sound techniques from him. More than almost any other kind of roof, one made of stone hates to be disturbed; often it will have settled itself into an almost homogenous covering which owes less to the pegs and battens than an obscure system of mutual support.

At least one company makes a large, dark grey, composition plain tile. It is not completely unacceptable on a rough cottage but it certainly does not look like a slate since it is obviously far too thick. Nor does it look like a real tile; I should avoid it. Probably the most frequently used substitute slate is the fibre-cement copy of the regular-looking Welsh slate. Some people will disagree, but I think this a far from contemptible substitute for roofing lesser buildings, although it is very much second-best to the real thing. Some firms make textured slates – very dark in colour and cambered so they look rather more like tiles. Some products have more imitation slate texture than others. All seem to be made in relatively large sizes only.

Probably the most successful substitute for the natural slate traditionally used in any area will be found in the range of imported varieties. There are several kinds of Spanish slate and some good ones are coming in from South America. Some experts are a bit sniffy about these foreign imports but I think many of them provide an honourable alternative, even if they may sometimes be less durable and are not perhaps exactly the right colour. Shop around and see what kind of a match you can procure. I have seen Andorran slate which you could scarcely tell apart from

Cornish Delabole. Best of all would be more and cheaper British slate from our own quarries.

Asbestos-cement, of course, never weathers and mellows into the delightful and subtle blend of colours which makes old roofs so appealing. It does, however, fade and become dirty and this helps to make it more acceptable. Its lifespan is relatively short – about twenty-five years, some people say. With old buildings, we are thinking of posterity. So, resolutely shelving any ideas of a premature Armageddon, we may not consider twenty-five years enough for a building which has already survived two hundred.

In tile areas of the country, you stand a chance of obtaining new ones which are roughly in the manner of the traditional hand-made varieties of former times. It is important to select something of good colour since that smooth, bright, Edwardian red which disfigures many an old roof, as a legacy of pre- and post-war repair and replacement, is far from sympathetic. Any hand-made tile or brick is exceedingly expensive, so you will be influenced yet again by the nature of the building.

There is no question that every element of an old building matters visually. Nor is there any denying the importance of good roofing materials. But my own view, and some will think it heresy, is that too many people believe that the use of good slates or tiles for the roof of an old building is enough to guarantee the success of a restoration project. Local authorities insist on 'natural' materials for roofs, and sit back convinced that they can thus disarm any suggestion that they are allowing the buildings in their control to be unsympathetically restored. But dictating roof materials requires the absolute minimum of taste or architectural knowledge: you just keep writing 'natural slate' or 'traditional clay tiles' all over the applicants' drawings and let the other details look after themselves.

However much a fine slate or tile roof enhances the repair of an old cottage, it does little to counterbalance the harm which is done by the wrong choice of windows, doors and pointing. In many instances, the roof of an old house is far less visible to most people for most of the time than its windows (see Chapter 9). That is no reason for putting nasty roofs on old houses, but when money is really tight, and you are dealing with a modest building, you have to get your priorities right. The drawbacks of a well-chosen second-best for the roof may be vastly outweighed by the merits of really good fenestration. Anybody with public money to burn or a reasonably affluent client should be able to get *all* these things right. What requires true discrimination is to contrive an acceptable result on a very low budget. It is then that the person with a true feeling for old houses discerns the visual order of precedence.

The main points to remember about choosing roof coverings are these. The materials should be typical of those used in the region for buildings of that quality, whatever it may be. If modern substitutes are used they must be as inoffensive as possible in colour, design and texture. They should in no way draw attention to themselves. People of discretion should not stand back and say: 'They've made quite a nice job of this cottage, but what a ghastly roof!' At worst they may murmur: 'Pity about the roof, but I suppose they were short of funds. Didn't notice it at first, so it can't be too bad.' These sentiments convey the realities of an imperfect world where money dictates so many of our decisions. If you can possibly scrape together the required funds for traditional roofing tiles, slates or thatch, it will repay you in the long run.

Be very cautious about any firm you ask to estimate for re-roofing. This corner of the building industry abounds in greedy 'cowboys' who quote a lump sum and then throw the tiles on at a frightful speed with little regard for good workmanship. A quick job is not necessarily more economical as far as you are concerned. It merely serves to earn the contractor a better financial return on his time.

Bad roofing firms will tear the felting as they lay it, fail to level slating or tiling battens, use cracked tiles, and ignore rotten carcassing timbers such as purlins, common rafters and principals when they uncover them. They will skimp flashings at abutments, in valleys and upon hips; they will lose track of the bond, and nail on half-slates at verges which lift in the wind. You will be asked to pay for all kinds of extras which should be included in the price and be threatened if you argue the toss. When the cowboys have finished the job they will disappear in a cloud of dust without clearing up or removing the rubble.

Bad craftsmen and cowboys will bang slating nails in too tightly so that centre-nailed slates will bend and crack or nail heads pull through. Ridge courses of head-nailed slates may ride up at the tail because a tilting batten has been omitted. If they are in a really careless mood, they will lay slates with the rough side and spalled edges placed downwards, making the roof visually unpleasing and less water-tight. Many traditional slate roofs are laid in diminishing courses, starting with big slates at the eaves and very much smaller ones at the ridge. The margin, or exposed part, of each slate will be made slightly shallower as the slates become smaller. The layout of such roofs, often called scantle roofs in areas such as Cornwall, requires careful planning.

New random and peggie slates are small and cut to varying lengths. However, they are fairly consistent as far as width is concerned – the width is about half the length. There is some argument about whether such roofs should not only be pegged or nailed but should have the slates bedded in a weak mortar. Some-times, especially in windy coastal areas, this can have obvious advantages. You can bed the whole slate – rather a clumsy and uneconomical practice – or, more often, just bed the tail of the slate in a rough horseshoe of mortar, which is struck off with the trowel as you go (see below right).

It is very important with mortar-bedded slates to clean off the smears and drop-pings before they dry, so forming an ugly thin grout which is difficult to remove. Here and there you may find poorly matched slates have caused others to ride up at the tails, because they are of excep-tional thickness. These should be sorted out and used together, for some purpose such as forming an undercloak at the

Positioning the first slate after the doubled-up eaves course, which can be seen in the foreground. This scantle slate roof is laid 'wet' in diminishing courses towards the ridge. Lime mortar is held in the boxlike roof-hods, rested on the laths.

The corners of each slate are trimmed to help them lie flat and the peg is rested against the top edge of the thin sawn lath. The tail is bedded in a horseshoe of lime mortar, giving stability in exposed positions.

105

Punching peg holes in slates on site, using an improvised machine.

The slater uses a portable dressing-iron and a zax to trim slates for special abutments at hips, valleys, ridges and verges.

eaves, or as the final course along the ridge.

To dress slates on site, the craftsman has a wrought-iron dog or dressing-iron for trimming the edges and forming the slates into special shapes and sizes. He places the slate on the iron, striking it with a series of short sharp blows, with the zax, to trim off the waste (see below left). When slates are hand-dressed at the quarry, the same basic method is used, but the dressing-iron is bigger and mounted on a rough bench like a trestle, which the workman sits astride. Instead of a zax, which has a point on its top edge for holing slate, the quarry worker employs a tool with a long cranked handle called a whittle.

Normally, modern slates are quarry-dressed to size on a trimming-machine. This is rather like a power-operated version of the blade you find on roller-type mowing-machines. The slate is offered up to a notched size-gauge which ensures that the right amount of slate is left after trimming. A portable version of this, called a size-stick, with a spike for scoring the slate with a trimming line, is used on site.

Slates are holed on a simple punching-machine or with the spike of the zax or slating hammer. Where slates are being laid upon traditional laths, the workman will have a lath hammer, shaped on one side for driving nails and on the other as a sharp axe-head for chopping the laths to length. Laths may still be bought by bundle, but they are sawn rather than riven nowadays. When laths are used rather than battens, slates have to be peg-hung, as laths are too flimsy for nailing.

For first-class work, slating nails are of copper or some rustless alloy, but these are expensive and you cannot just go into the nearest builder's merchant and pick up a few pounds. Consequently, unless owners or architects actually specify otherwise, all builders use ordinary galvanised slating nails. Splitting and chopping up slating or tiling pegs is a good job for wet afternoons.

While today slates are normally all twice-nailed, unless they are of very small size, modern tiles are only nailed every fourth course or so. (Traditionally *slates* were often roughly trimmed off at the top corners and holed for a single nail or peg.) The other tile courses are just hung on the projecting nibs at the head. However, all the tiles at difficult or vulnerable points on the roof are nailed. Such areas include verges, eaves, ridges, hips and valleys.

The treatment of hips and valleys needs careful thought, since they are an important feature of an old roof. Both slates and tiles may be raking-cut and tilted over a lead gutter at valley intersections. Old slate roofs sometimes have secret gutters where the slates meet or nearly meet over the gutter. It looks good, but such a gutter is very prone to fill up with dust and grit and can be difficult or impossible to clean out.

A better method is to have a reasonably wide exposed valley gutter fixed as usual over a valley board which gives it a firm foundation. Some people say the gutter should be as much as 200mm wide, but half that is enough for cleaning purposes. Number 6 lead is normally employed.

The more interesting method of handling valley intersections between roofs is to sweep them or lace them. This requires considerable skill to carry out successfully and amateurs should be wary of trying it themselves, since slates and tiles are expensive and the leaks which can result from poor workmanship will be extremely troublesome to repair.

Briefly, a laced valley is one in which tile-and-a-half tiles are placed diagonally across the valley board picking up the bond from each roof slope. Swept valleys achieve a similar purpose. Spade-shaped slates called 'bottoms' are raking-cut to alternate over the valley board with pairs of rather similar slates called skews (Fig 73). Old tiled roofs do not have specially made angled valley tiles, so purists will avoid those when repairing early buildings.

For slate roofs, there are three methods of dealing with hip junctions. Some old buildings can correctly have mellow red clay bonnet, half-round or hog-back tiles. Slate-roof hips may also be weathered in this way if the building is of modest quality. The pitch, however, should be fairly steep, since very shallow slate roofs can look decidedly odd with tiled hips. Where the roof pitch is steep, it may be necessary to secure the hip tiles by fitting an iron stay-bar at the bottom to prevent them slipping.

The best way to deal with hips of slate roofs is to raking-cut the slates so that they mitre neatly together, using lead soakers at every course to weather the joints.

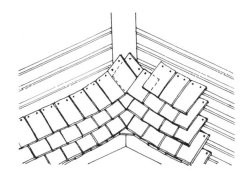

Fig 72 The internal junction between one roof slope and another can be covered by *lacing* the tiles over a valley board.

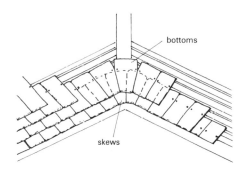

Fig 73 Where slate roof-slopes meet, skilled craftsmen may construct a *swept valley* by trimming spade-shaped *skews* and *bottoms*. The skews are paired in one course and then covered by a *bottom* in the next.

Another system, which need not look out of place, especially on later nineteenth-century roofs, is the use of lead dressed over a 50mm timber hip roll; but since the slates really have to be raking-cut, as they do for a proper mitred hip, it is doubtful whether much is to be gained. The lead, of course, lies over the top of the slates for 150mm on either side, being held with lead tacks forming a rather obtrusive junction. The lead-roll abutment is also used for dealing with ridges.

Often, your choice of method will be dictated by the necessity to match a newly repaired piece of roof with an old one, or with similar houses of the period. With tiled roofs, the less surface leadwork you can see dressed over tiles, the better.

Generally, on all old roofs, avoid the use of any sort of concrete hip or ridge tile, and this rules out the black ones. They may look as though they blend better with the colour of slate, but they are obviously not authentic. Some roofs used to have specially made ridge slates with a dowel section like a roll to run along the apex. This was slotted to receive an ordinary slate from the opposite slope.

In the Lake District, 'cut wrestler' slates were laced at the ridge – a nice conceit, but not especially durable or beautiful. Those elaborately fretted and Gothicly detailed clay ridge tiles so loved by late Victorian and Edwardian builders should not be used for Regency and Georgian buildings. Despite the spurious appearance, their origins are most respectable and genuine medieval ones are prized features of an old building.

Slate-roofed cottages did not normally have barge boards at the gable ends until the nineteenth century, if they ever had them. The verge overhang was kept to a minimum of 2–4in (50–100mm), just pointed up with mortar and weathered with capped slates. Occasionally, however, you find barge boards are useful, since they help you to square up the verge of a replacement roof which would otherwise have to be trimmed at a weird angle. In other words, you can project the roof beyond the gable wall by using an additional rafter, and devise a tapered soffit board to fill the oddly shaped space beneath the overhang. The barge board is nailed to the extra rafter and, since it will look strange on its own, you will need to employ one for the other roof-slope, to match.

All this arises from the fact that cottage-builders were not especially good at forming right angles between front and end walls. Whereas the walls themselves may display hardly any signs of being out of square, roofs will. It is a matter of judgement whether you decide to square the roof up or leave it astray.

Abutments between roof coverings and such features as chimneys or upstanding gable walls were normally weathered by mortar pointing and oversailing courses of brick or stonework. Sometimes slate or tile listings were stepped into brick or stone joints to protect the junction. These could all be attractive features and should be left in place if of architectural interest. However, new abutments should be flashed with lead, even if in some cases you restrict the work to soakers only and leave out the cover flashings. This latter course of action is only advised where some important aesthetic gain would be made.

The use of mortar bands, cast between boards, to cover the hips of roofs, is always visually unpleasant and often ineffective as well. The mortar cracks as a result of thermal movement, allowing capillary action of moisture. If it has to be repaired, only such measures as the application of bitumen paint or an adhesive foil will help. Otherwise, you have to chip off the mortar, thus damaging the mitred slates.

While on the subject of bitumen preparations, do not be tempted to use these as a cheap method of mending a leaking roof. The bitumen is normally brushed on in layers over lapped canvas scrim which serves to reinforce it and provide a key. However, it looks ugly and can only be counted a temporary measure for a roof which will have to come off in the end anyway. Once bitumen has been applied, the slates or tiles can never be reused and any future leaks are extremely difficult to trace.

If you are trying to keep the weather out of an old building which is to be properly repaired at a future date, a bitumen treatment may be an ideal way of stopping further deterioration of the roof timbers and the interior as a whole. Similarly you may resort to temporary patches of corrugated asbestos sheeting, cement grouting, asbestos-cement slates and so on. If only all the owners of empty listed buildings, which are often cynically allowed to reach a structural point of no return, would resort to these temporary devices, we should be plagued with fewer applications for demolition from ruthless developers.

The practice of grouting old slate roofs with a cement-sand slurry can only be described as a clumsy half-measure. It often promotes more trouble than it cures, preventing the roof and roof space from breathing, decomposing the slates themselves, and precluding any chance of them being properly repaired or re-laid. All the same, it is better than letting the house fall down, and in my view it never looks as ugly as a badly chosen substitute roof covering, such as concrete interlocking tiles.

In some parts of the country, you may find purpose-made stone ridge and hip tiles. These should be saved during any re-roofing job, since they are hard to come by, pretty to look at and authentic. As a general rule, it is important to save all old materials, and that means using a great deal of finesse in their removal. How many owners have said to their builders: 'Please do your best to keep all the good slates and stack them for reuse', only to find that the builders, who are on a fixed-price contract and in a hurry, have ripped the roof apart in a morning and hurled the slates down a shute into a skip?

Old pantiles are pretty and usually to be found in the east of Britain, anywhere from Scotland to London. These handmade clay tiles are the progenitors of the modern interlocking tile, since they have a side seal which permits them to be laid with a single lap. They cannot, of course, be laid in the normal stretcher bond, but are arranged more in the manner of corrugated iron. They are sometimes individually heavy but make light roofs. Suitable for shallow pitches, they are once-nailed, at the head.

Pantiles became popular through the strong trading connections with the Low Countries during the seventeenth century. They were normally torched on the inside or laid on a bed of hair- or reed-reinforced lime mortar to help to make them more weathertight. Their design is such that they are difficult to finish at eaves, verges and ridges. Nor do they lend themselves to use on anything but the most simple roof shapes, since hips and valleys present grave problems. By all means use them where they are traditional but be wary of modern varieties; many are completely out of tune with old houses. The fit of pantiles is very important and old ones were sufficiently varied in shape to require quite careful matching, hence their vulnerability to driving rain and snow.

Corrugated iron cannot be considered as a roof covering for old dwelling houses or cottages, or for any building of distinction, but I must admit to finding it relatively harmless for small outbuildings.

Asbestos-cement corrugated weathers to a dark grey, lichen-encrusted affair which is not always unpleasing. Galvanised-steel corrugated turns to a dull neutral grey and finally, if unpainted, produces a rather attractive rust; it will make neat-minded people think I have taken leave of my senses to countenance it. Black is the best colour paint for corrugated iron. I would far rather see an old stone shed with a rusty tin roof than resplendent with some harsh, new and monotonous covering. (Of course you do not let corrugated rust if you can help it.)

Thatching

Thatching roofs is hard, prickly work requiring considerable skill, and is the province of the professional. Until the sixteenth century it was probably the most common form of roof covering in country districts. It was banned in London as early as 1212 and was widely replaced with other materials during the Great Elizabethan and Stuart Rebuild.

Its visual appeal is tremendous, especially when used in conjunction with the rather soft, billowing lines of cob buildings. It displays splendours of craftsmanship in the form of ornamental ridge-lacings and eyebrow dormers. The material used is either 'true reed', 'combed wheat reed' or ordinary 'long straw'. True reed, as the name implies, can be any one of the localised marsh-grown reeds such as the famous Norfolk variety – *Phragmites communis* – which, incidentally, is also to be found in other parts of the country.

Thatchers at work in a Devonshire village lay new wheat-reed over a layer of the old.

Hazel spars like hairpins are used to secure the new half-nitches of reed to the old thatch. Here, they are temporarily positioned to prevent the ends of the completed courses from flopping over. When rethatching, this craftsman prefers the Dutch method of securing the bundles, running round straw 'sways' like ropes along the roof to secure them. The normal type of sway is made of split hazel rods, fastened down by iron 'crooks' (hooks) driven into the rafters.

Bolted timber frames called 'biddles' are spiked into the thatch to provide footholds. Hand shears like those used for sheep, are employed for trimming. The tool with a honeycomb of holes in a wooden block attached to a handle, is a 'leggett' or 'drift' for beating the ends of the wheat-reed bundles so that they are tightly packed into the roof, with just a few inches visible on the surface. This one is a Dutch drift. The hammer-headed flat iron spikes are 'reeding pins' for holding the courses in place as work progresses.

Combed wheat reed is not a reed at all but specially grown wheat which is harvested with a reaper-and-binder. It is not forced with artificial fertiliser and is cut rather green. The straw is not permitted to enter the threshing drum and the grain and leaves are combed off with a machine. It is used in the same manner as true reed. Long straw is the product of threshing, albeit a very gentle kind of threshing. It is not assembled in an orderly manner with parallel stalks and even ends and so requires a different method of laying.

Norfolk reed can last as much as sixty years, wheat reed up to about forty, and long straw not more than about twenty, at best. The quality, the situation and the skill with which the thatching material is laid will be vital in determining durability. Thatch provides excellent insulation being cool in summer and warm in winter. It is also said to be a good buffer against noise. For cob buildings it has the advantage of being light and less likely to overload the walls.

Although, as has been said, you are unlikely to try your hand at thatching, unless it is in the form of limited exercise under the tutelage of some indulgent thatcher, it is worth just mentioning the main principles of thatching technique.

For thatched roofs the pitch is usually about 50°, sometimes more. Rafters in old thatch roofs were often rough, very crooked and more widely spaced than for other coverings. Upon these, battens were nailed horizontally at about 5–7in (125–175mm) centres.

Bundles of reeds were held down with long pieces of hazel rod called sways, these in turn being secured with iron hooks spiked into the rafters. Holding-down methods vary according to local practice and the thatching material used. Sometimes the sways are tied to the battens or laths using a big iron needle and twine However, the initial courses at the eaves are still normally held down with hooks.

Split and bent hairpins of hazel, about 2ft (600mm) long, called 'spars', are also used to secure sways to the row of bundles beneath. They are also found pinning down individual bundles – especially in long-straw thatching. Reed bundles are fixed so that they present only the ends on the roof surface (plus maybe an inch or so of stalk). They are beaten up into place with a wooden block called a leggett or drift. Long straw lies down the slope of the roof, showing the straw stalks themselves and it looks much looser and wispier – less carved to shape. You will notice that long-straw roofs have lacings of split hazel or withies between horizontal rods called 'liggers'. These are laid on the surface just below the ridge, down the verges and above the eaves. While ornamental, the purpose of liggers is to hold down this volatile material in the places where it is most vulnerable to the wind. (I have avoided using too many of the specialised thatching terms in this description, because they vary from area to area and from one type of thatching to another.)

There are still a certain number of expert thatchers in regions where it is a traditional roof covering. It is not especially cheap to have a thatched roof, but it is a joy to live with, even though you may be obliged to cover it with wire netting to keep out birds. Some people treat the thatch with a fire-retardant solution – partly because insurance companies are not very keen about it as a fire risk. It is certainly worth seeing that your chimneys are sound and brushing them out regularly. But I wonder how many owners of thatched cottages have ever had a serious fire in the roof – maybe fewer than one would imagine.

A bonus is that you can forget about cleaning and painting gutters, because thatched roofs do not have them. The thatch projects about 2ft (600mm) and just shoots the water on to the ground.

Although the decorative qualities of the craft of thatching are rightly admired, it must be said that there are times when thatchers are a little over-enthusiastic in that respect. While *tours de force* in their own right, some of the carved, scalloped and oddly detailed designs used to finish thatched roofs can be too busy and intrusive. It depends on the type of building. A stocky little cottage in a village street may lend itself only to restrained decoration, while the thatched roof of a cottage ornée

may cry out for something of almost vulgar ingenuity.

Roof Timbers
The repair of the carcassing timbers of an old roof requires knowledge, skill and conscience. The knowledge is historical. Can that part of a medieval arch-brace be worth saving? Do these threaded purlins suggest a quite early roof; and if I have to put in any new trusses, should I follow the same methods? The skills are those of the competent tradesman who can cut snugly fitting joints, reinforce inadequate timbers with steel flitches, and let in new lengths of wood where needed. Conscience is required because the temptation to modernise and destroy old timbers in the dark and concealing fastness of the roof space is enormous.

In my own view a sound roof structure comes first, where more recent buildings are concerned. It is doubtful whether hidden trusses in some eighteenth- and nineteenth-century buildings merit slavish matching of methods and materials. Often you will find that one or more roofs have been added at different periods of a building's history. Early seventeenth-century principal rafters joined by cambered collars will stand abandoned beneath later eighteenth-century ones. These, in turn, carry nineteenth-century common rafters and slates or tiles. Don't take away the early roof trusses just to be tidy; they are part of the building's history. They have been left because they would have been bothersome to remove, and earlier generations of workmen had no chain-saws with which they could quickly demolish them.

That said, you will probably have to treat any old surviving timbers against beetle infestation, wet or dry rot.

Never disturb an old roof if you do not have to. The moment you begin to pick away at old timbers, laths and coverings, you will set in train a progressive dismemberment which you may live to regret. Old houses are untidy, faulty and a constant source of worry and expense. If you are unable to accept them on that basis and unprepared to live with less than technical perfection, you might well be happier with a soundly built pre-war stockbroker's dream.

Of all the many deprivations which old roof structures commonly suffer, none is more often encountered than the cutting away of collar beams. Collars were placed exactly where they could later prove the most nuisance and, although partitions might be arranged to coincide with the trusses, it still happened that the collars got in the way of the necessary door openings. One cut-away collar, or maybe two, may cause little structural disturbance, especially if attic floor joists are themselves serving as ties. But many roof timbers spread when deprived of this lateral restraint, and you may have to do something about them.

It may be that a new collar can be spiked to the principal rafters above ceiling level, but this position is frequently too high for comfort (Fig 74). Do not depend upon the restraining properties of the walls, for the many reasons mentioned in Chapter 6. You may be able to tie the feet of the rafters with mild-steel rods set between the floor joists.

Often you will find the sawn-off end of the old collar projecting from the face of the wall, where the stone or brick is built up or *ashlared* to close the gap between the wall top and the underside of the common rafters. Often in such attic rooms the

purlins will be visible, sticking out from the lath and plaster of the ceiling slope. This is a tricky place to introduce a tie, which is naturally more effective the nearer to the feet of the rafters it is located.

The joint between the blades and collars of many earlier roof trusses is the halved dovetail joint or its first cousin the lapped dovetail joint. Generally, seventeenth-century truss collars are lapped, dovetailed and pegged (Fig 75). Later buildings might have collars which were either spiked or bolted in position; but the lapped dovetail was still widely used.

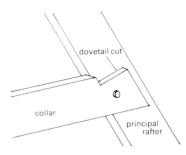

Fig 75 The *lapped* or *halved dovetail joint* was commonly used in old buildings to fix the collar to the principal rafter. The sinking is made fairly shallow (say $\frac{1}{4}$ to $\frac{1}{8}$) to avoid weakening the rafter as it might do if actually cut to half the depth. The dovetail cut restrains outward movement of the rafter.

At the apex of the roof, principal rafters which carried some form of ridge-piece would often be halved and pegged (Figs 76 & 77) or just lapped and pegged. Some early roofs, especially those of cruck construction, would have blades held together at the apex by a yoke upon which a ridge purlin could be rested (see Fig 65). In modern construction, of course, rafters are normally plumb-cut to be nailed to a ridge-board.

It will often be found that medieval roofs are of the *single* type with common rafters coupled together by simple collars and one pair in four or five trussed with tie-beams (see Fig 66). There is frequently some kind of crown post rising from the tie-beam to support a collar purlin which runs the length of the roof. In other roofs, one in every so many pairs of rafters is trussed on a substantial tie-beam and the trusses bear clasped purlins, lying *under*

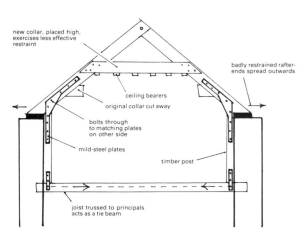

Fig 74 One way to prevent principal rafters spreading at their feet, without obstructing valuable headroom, is to truss them to the floor joists. This drawing demonstrates the general idea; but the exact method depends upon circumstances.

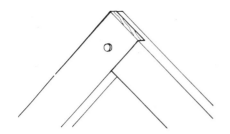

Fig 76 Principal rafters were often halved and pegged at the ridge.

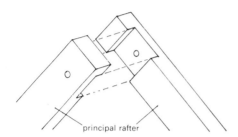

Fig 77 When the ends of the principal rafters were intended to cradle a ridge purlin, they were often *double-notched* or *halved* over each other and pegged.

the principal rafters. In this instance, the rafters may be jointed at the apex by means of pegged, open mortices-and-tenons (Fig 63).

As mentioned before, there are innumerable types of roof construction. The main thing to decide when repairing roof timbers is the type of force to which any component is being subjected. You then choose your method accordingly. The other important factor governing choice of method is the quality of the faulty timbers and how visible they are.

In any kind of hardwood roof, especially one which is open to view, like that in an open-hall house, you have to employ a good deal of finesse. The faults are often found in the most inconvenient places – at joints, or where timbers are buried in the thickness of the walls.

If you can, you match timber and jointing methods with those used before. All the same, you may be obliged in some cases to resort to steel reinforcement of the repair work – bolts, timber connectors, flitch plates, fish plates, collars, straps or stirrups. Many of these steel components will be used in addition to traditional carpentry joints. The principle of repair is to cut away and treat any rotten or

infested timber, saving all the moulded work that you can. This leaves you with either lengths of timber to replace completely or areas which must be built up with matched and seasoned stuff.

To replace parts of members which are in tension (like collars), you may peg, mortice-and-tenon a new piece into the sound old wood, to either side of the bit you have cut away. This can only be done with the thick type of collar which allows a sufficiently generous tenon or false tenon, and a decently strong area of wood each side of the mortice hole (Fig 78).

Where collars are thin (say, less than 50mm) it may be better to replace the whole collar. Collars of trusses which predate the eighteenth century were frequently cambered or cranked, and these should be preserved if possible.

When members joined with mortice-and-tenon joints have rotted or broken at the joints, there are one or two options open to you. Where aesthetic considerations are uppermost and the timber is to be matched, you may scarf in new pieces to replace the rotten ends of timbers like tie-beams and rafters (Fig 79). You may cut

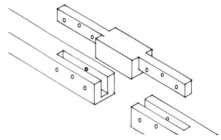

Fig 78 To splice in a replacement length, when decayed wood has been cut away, an *open mortice-and-tenon* joint can be employed. This is secured with wooden pegs or dowels.

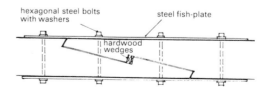

Fig 79 When connecting lengths of new timber to old, for repairs to members like rafters and purlins, a simple *scarf* joint is the traditional method. The hardwood wedges are tapped in at the centre to make the joint bear firmly at its end. Fish plates and bolts have also been used to sandwich and secure the joint.

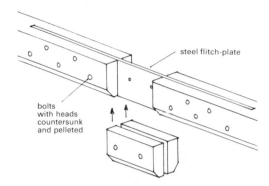

Fig 80 Steel flitch plates may be slotted into timbers such as beams, joists or collars, as a means of strengthening or introducing replacement lengths of wood.

away bad timber and slot a steel flitch plate into the sound wood and bolt it in place. The plate should be at least six times its own depth in overall length. The repair can then be covered by the original mouldings which have been previously treated and set aside. Alternatively, you may have to match the mouldings by carving or hand-running replacement lengths. You may have no mouldings but merely require some form of hardwood slip which is screwed, glued or bolted to the beam soffit, concealing the steel flitch. Slots for flitches may run the full depth of the timber or may leave some unslotted wood at either the bottom or top. Circumstances dictate the choice.

Sometimes it may prove convenient to slot a steel flitch plate into the thickness of the sound timber, and cover the part of the steel which bridges the gap left by rotten wood you have cut away. In this

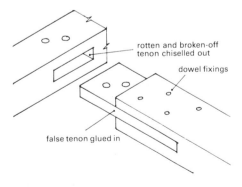

Fig 81 Decayed tenons may be replaced by slotting the end of the timber and introducing a new tenon piece. The tenon should not be glued into the mortice.

instance the carefully selected timber replacement sections are bolted through the plate, the bolt heads being countersunk and pelleted (Fig 80).

Pellets can be cut from the same timber, using a special coring bit attached to your electric drill. Do not try to cover the bolt heads with plastic wood, or glue and sawdust mixtures, if you can avoid it, since such repairs will show. Hardwood pellets can be made to present a surface which more or less matches the grain. You may even be able to replace the original few millimetres of surface wood, having specially removed it before drilling the bolt-hole.

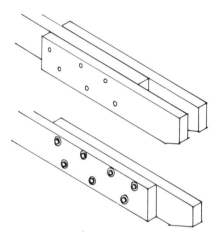

Fig 82 To replace rafter ends which have rotted and been cut away, a new length can be bolted on one side *(lower drawing)* or, to avoid twisting, the ends may be doubled up *(upper drawing)*.

When repairing softwood tie-members, rafters or purlins, which have no visual importance, you may either scarf them, or fish-plate them from both sides using staggered bolts. It is best either to treat any steel repair units with generous quantities of rustproof paint, or to choose some made of rust-resistant material: this may sometimes be galvanised steel, possibly stainless steel, or a suitable alloy. It is important to remember that hardwoods such as oak, if damp, release acids which corrode steel.

Since the mortice-and-tenon joint was so commonly used in traditional construction it is worth thinking about how to replace those which have rotted. If the tenon only has decayed it may be enough to treat all the timber and recover the original mortice hole, merely replacing the tenon itself. A false tenon is fitted and re-pegged. Do not *glue* and peg or bolt *both* ends of a false tenon, since you should allow some flexibility in repaired joints so that seasonal or thermal movement can take place (Fig 81).

The methods used for repairing end-bearings of members of roof trusses, such as tie-beams, will be dealt with in Chapter 8, which considers floors.

It is very often the case that rafter ends have decayed. The method of adding new lengths to replace the parts which have been cut away at the feet is, once again, the scarf joint. However, it will usually be unnecessary with softwood rafters seated upon a conventional wall plate. Scarfing is really a more likely choice for the union between the foot of an important and visible principal rafter which is housed into a tie-beam.

Softwood rafter ends can be contrived by doubling up with lengths of the same cross-section (Fig 82). If there is any possibility of twisting you may spike or bolt a new length of timber on *both* sides of the old rafter and thus seat a *pair* of rafter feet on the wall plate.

This brings us to wall plates, which may be external and hardly visible, or internal and represent a major architectural feature, in something like an open hall. They are devils to replace without renewing everything else at the same time. If you do put in new sections of important and visible hardwood they should officially be jointed to the old with halved and bevelled dovetail joints (Fig 83) or with halved mortice-and-tenons (Fig 84). These elaborate joints are over-demanding for ordinary lengths of hidden softwood wall plate, and you might resort to neatly halving the ends and securing with brass screws (Fig 85). If in particularly craftsmanlike humour, you might go so far as to bevel the single halving joint, but it won't add much strength.

When repairing old oak timber-framed buildings, use the correct jointing methods wherever possible; but there is

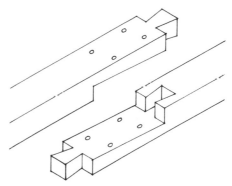

Fig 83 An elaborate but structurally very effective way of end jointing timbers like wall plates, is the *halved and bevelled dovetail*. It will not pull apart and will resist outward thrust from the rafters.

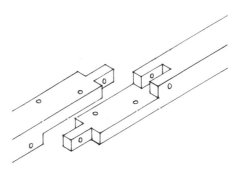

Fig 84 Similar to the joint of Fig 83, but easier to make, the *halved mortice-and-tenon* may also serve for end-to-end junctions in timbers, such as wall plates.

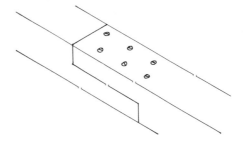

Fig 85 The normal way of joining in lengths of wall plate, where no special stresses are involved, is to *halve* and screw them together.

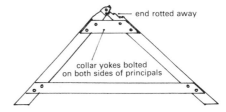

Fig 86 Where the ends of rafters have decayed, a straightforward repair can be made by cutting away the rotten timber, brush soaking with preservative, and bolting on collar yokes.

no need to be pedantic when tackling the undistinguished carpentry work of some very ordinary eighteenth- or nineteenth-century roof structure. Use judgement and conscience, as ever, and decide whether an oak king-post roof in the loft of a building can reasonably be repaired solely with devices like iron stirrups, bolts, straps and plates, or whether it calls for timber replacements.

In some cases where rafters have decayed at the apex of the roof it may be possible to avoid elaborate repairs and merely spike or bolt collars on both sides to hold them firmly in position (Fig 86). The rotten stuff can be cut away and treated, and no attempt made to replace the ends. This is a method envisaged more for principal rafters than common ones. Certainly in double roofs, the common rafters of one slope often do not meet those of the other at the ridge, but are laid in a quite haphazard manner along the purlins.

Perhaps the most frequently seen repair to the structure of a roof is the doubling-up of faulty principal rafters. People like the method because it is cheap, simple and less likely to disturb the rather frail timbers in the vicinity. All you do is introduce a new rafter next to the rotten or broken one.

The replacement can rarely be matched and positioned exactly next to its failed partner, especially if it is being fitted from inside the roof space. It usually has to be supplied with packing pieces to bring it into contact with the undersides of the purlins; or perhaps 'trenches' have to be cut out to house them.

Many old roofs present a sight which is little short of grotesque, since every few generations owners have introduced yet more botched and hasty repair work. It is not always the method which is at fault but the size of the timbers selected and the bad workmanship. A favourite roof repair is the introduction of struts to support what has sagged or to hold up new lengths of timber – purlins for example. These new compression members may them-selves be supported on bits of slate dug into the dust of a wall top. They might be seated on the head plates of rotting

internal partitions or upon wilting and undersized ceiling rafters. I have even seen them rested on lengths of floorboard laid upon the lath and plaster of the ceiling itself.

To introduce struts is a perfectly sound and sensible method of support, but struts are useless if they do not fit properly and are not seated on some firm base. One form of strut often seen is that which carries the end of the ridge-board purlin or pole where it abuts the chimney. As a result of leakages at this vulnerable roof junction, the purlin end has often rotted and the strutting piece or crutch has decayed as well. It is then necessary to scarf on a new length of sound purlin and prop it with a stout timber. This is suitably notched to receive the purlin and treated with a preservative (Fig 87).

Much the same process will have to be adopted where rotten purlin ends lie buried in stone and cob across the wall gables. These can be a problem, since it is no use making the same mistake twice and putting untreated lengths of purlin to

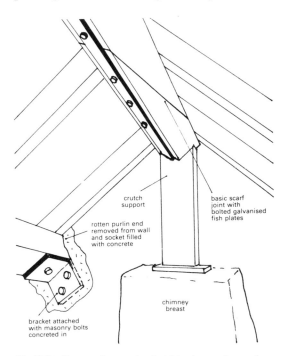

crutch
support

basic scarf
joint with
bolted galvanised
fish plates

rotten purlin end
removed from wall
and socket filled
with concrete

chimney
breast

bracket attached
with masonry bolts
concreted in

Fig 87 Purlins may be repaired within the roof space by cutting away the decayed timber and scarfing and plating in new pieces. The ends of the purlins should ideally be kept 12mm or so clear of the wall, or be butted against thick pieces of bitumen felt.

nestle against the old rotten ones, lying in the same damp bed of earth and stones. The difficulties sometimes involved in replacing purlin ends without replacing the roof covering itself may be imagined.

However, almost anything can be done if you are patient, painstaking and methodical. Circumstances vary so much from job to job that it is perhaps unwise to suggest remedies for rotted purlin ends in this position. You might do better to dig out the decayed length of timber and backfill with weak concrete, leaving the purlin end short of the gable wall. The end of the purlin could then be rested on a corbel, timber prop, steel shoe or something of that sort – inelegant, but maybe effective (Fig 87). However, you will still have to do some wedging up against the undersides of any rafters resting on the gable wall.

Another way of supporting purlin ends, where most of them have rotted, is to provide an extra truss against the inside of the gable wall to hold them up. When purlins which lie across gable walls are not replaced, you have to make certain that any verge rafters or barge boards retain some means of support. This may involve the use of strategically placed and properly treated timber plugs.

Many repairs to timbers require the use of bolts, even when care has been taken to form correctly cut joints in the wood itself. Early and traditional carpenters' joints were nearly always secured with tapered wooden pegs and these usually projected proud of the surface. The method was simple and allowed for movement, subsequent adjustment or removal. However, the job of knocking out old pegs can be more difficult than it sounds.

Where you do use bolts, you can also fit all kinds of special timber connectors in conjunction with them. Toothed ones, with holes through which the bolts pass, bite into the surface of the wood and prevent slipping (Fig 88C). They also serve to spread the effective area of the connection so that the bolts act more efficiently. Split-ring connectors are even better for those joints where you have to lap a new length of rafter over an old one which has been cut short to get rid of

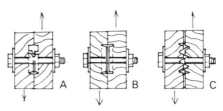

Fig 88 Timber connectors are used to increase the strength of bolted joints against failures in 'shear'. The arrows indicate the type of movement resisted. The split ring (A) is sunk into both timbers and grips tightly. Shear plates (B) are also fitted into sinkings, but grip back to back by friction. Toothed plates (C) bite into the timbers at the interface, as the bolt through the middle is tightened. There are also shear and toothed connectors for joining timber to metal.

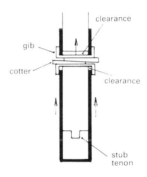

Fig 89 A slotted steel strap may be used to tighten up and secure the stub mortice-and-tenon joint between a king post and its tie-beam. Steel gibs are inserted at top and bottom of the slot, against which the wedge-shaped cotters may bear as they are driven home and tighten the joint. King 'posts' despite their name, are in tension not compression. Hence the need for a stirrup.

decayed wood (Fig 88A). The only drawback is that you must drill a special sinking in the surface of the timber exactly to fit the ring.

Where fish plates are used for sandwiching timbers together, or a flitch plate is employed as a slotted-in reinforcement, it is always good practice to fit shear plates (Fig 88B). These strengthen the bolted connections between steel and timber. Most bolt heads and nuts must have washers to prevent them damaging the wood when tightened.

Bolts themselves, if not made of stainless steel or alloy – and they can prove expensive for run-of-the-mill repair work – should be well painted to discourage rust; this also applies to any stirrups, straps, hangers, and so on. In some cases coach bolts are useful where you wish to secure metal repair sections without bolting them right through the timber.

For tightening up metal stirrups used to strap the joints of king posts to their appropriate tie-beams, the standard practice is to cut a slot through the stirrup ends and the post so that gibs and cotters can be fitted. The cotters act like folding wedges, which pull the bottom of the iron stirrup hard up against the soffit of the beam (Fig 89). This works on much the same principle as the system called 'draw boring' which is used for tightening joints with tapered timber pegs. Holes made in the two pieces of timber to be pegged together are bored slightly out of register. When the peg is driven, it forces the holes to coincide, tightening the joint.

As with everything else in repairing old buildings, the practice will often be very different from the theory. Sound techniques involving refinements of the kind I have mentioned here and there require greater practical skills, extra man-hours and, sometimes, additional materials. You must use good judgement to decide between those occasions when they are worth while and those where a simple hammer-and-nails job will do. The basic principles to remember are these:

1 Never skimp on structural repairs.
2 Never use fake-looking materials where they are being asked to serve an aesthetic purpose.
3 Repair rather than replace as much visible old timberwork as possible.
4 Remember structural flexibility and ventilation.

8

FLOORS

Nearly all neglected cottages or houses require extensive repair work to their floors. The most usual troubles in suspended floors – those with boards or other coverings rested upon spanning timbers – are decay and woodworm. Other faults can occur, of course, and we shall think about them too. We must look at the subject in two ways: structural repair and the all-important matter of choosing materials which retain the character of the building.

The structural elements, such as joists and beams, may well be covered up, so that their appearance is not particularly important. However, early buildings display (in the room below) quantities of structural timber, often of high quality, which must be dealt with in an aesthetically pleasing manner. Not only a good choice of timbers, the matching of mouldings and sympathetic finishes are required; knowledge of traditional carpentry methods is needed for jointing one element to another.

The structure of suspended floors which are open to view from the room below may encompass any number of architectural styles, according to the date and quality of the building. A fifteenth-century manor house could have massive intersecting beams with deeply cut mouldings and carved wall plates. A nineteenth-century lock-keeper's cottage might have a few very plain softwood joists and some floorboards.

Ground floors may be of solid construction, consisting of beaten earth with a thin screed of lime and ash, called a 'grip floor'. They could be made with bricks, laid on a bed of lime mortar; or of quarry tiles, flagstones, or boards nailed to battens sunk in a screed. You could find a suspended floor covering an ordinary under-floor cavity or a cellar.

The two things to distinguish in your mind are the basic structure and the covering. The latter may itself be structural or it could be just a finish. Parquet blocks were often laid, in old houses, on a structural sub-floor of ordinary boards. The system of laying them on a level bed of hard mortar is a relatively modern one. Flagstones are not normally structural, but then again you may occasionally find them spanning between supporting walls.

The boards of suspended floors are certainly structural, but they are also very important architecturally. In this book it is assumed that floorboards are going to be left exposed and polished; no account is taken of such practices as laying hardboard over floors as a base for wall-to-wall carpet. I hope that you do not intend using acres of fitted carpet in an old house. If you have allowed this blasphemous thought to cross your mind, consider that fitted carpet was unheard of before the late eighteenth century and even then it was a rare and strange device, restricted to the houses of a tiny handful of rich and sybaritic persons. Those who had carpet at all would normally have used oriental rugs, or the products of the English factories of Axminster, Kidderminster or Moorfields. Patterns were either in the Baroque or Rococo style or were imitations of Persian, Caucasian and Turkoman designs. A few might import European products. Look at the rather wooden but appealing conversation pieces by painters such as Arthur Devis, and you see that those small groups of preternaturally stiff and timid individuals are standing amidst acres of bare boards in distinctly underfurnished rooms. That was the norm in the eighteenth and seventeenth centuries. Those rich and enterprising enough to own a carpet in the sixteenth century would be just as likely to hang it on the wall or use it to cover a table. Much more could be said about carpets, but all you have to do when considering your repair work is think in terms of floorboards and other finishes which will be exposed to view.

Structural Repair

Basically, a suspended floor is made up of just two or three components. There are joists which carry boards and, quite often, beams which in turn support the joists. In buildings which have fallen into disrepair the main trouble will be decay at the ends of the joists or beams. Moisture penetrates or permeates the walls and the joist ends become damp and rot away.

Dry and wet rot are dealt with in Chapter 11. These are the two main forms of decay you will find, and arising from them may be beetle infestation. However, beetle attack is not necessarily restricted to timber which is so damp that it is decayed.

So what is the overall picture? Either the upstairs floors of your building have become so rotten that they must be stripped out altogether, or there are certain rooms or parts of rooms which need attention. Few floors in occupied buildings actually collapse as a result of overloading or because inadequate structural timbers were chosen in the first place. Nevertheless floors can collapse for those reasons, so we should discuss the dangers of overloading, spacing joists too wide apart and undersizing components.

The partial collapse of a suspended floor is frequently the result of the breakdown of a timber lintel over a door or window opening, a matter which was referred to in Chapter 5. So we shall assume that you support the floor joists with adjustable steel props and cut away any rotten timber. You replace any rotten or distorted lintels and, if required, dig out decayed bonding timbers bedded in the walls. Bonding timbers, especially if of softwood, are a menace in old buildings. They were not only used to bind brick and stonework which was thought to have insufficient cohesion, but also served as a level bearing for the ends of the joists.

This brings us to the general concept of end bearings for beams and joists. There are formulae for calculating the amount of timber which should rest on a wall to provide a satisfactory bearing but I need not trouble you with them. For the average modern joist of 50mm thickness, aim to seat no less than 75–100mm on the bearing pad or wall plate; in some instances 50mm may be enough. Remember that the joist end is exerting a crushing force on the bearing in the wall and mainly for *that reason* it is important to try to spread the load by means of a stout treated timber, a stone, or an area of strong concrete placed underneath it (Fig 90). The length of timber required to bear can be as little as 38mm on something like the flange of an RSJ (rolled steel joist).

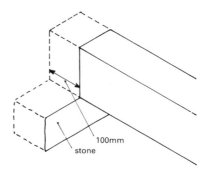

Fig 90 Floor joists should, where necessary, be provided with suitable stone, concrete or timber pads to spread the load and so avoid crushing and local settlement within the wall structure.

Where a beam is itself picking up the loads from a number of joists, it is vital that its end is generously seated at a point in the wall which is not weak in bonding or likely to buckle and distort. An obvious example of a weak bearing point is a narrow brick pier between windows. Another place where the end of a beam is frequently located is over the centre of the fireplace lintel, in a gable wall. Such beams are called spine beams. The danger with them is that they depress the timber lintel over the fireplace opening so that it deflects, crushes or even collapses. Usually collapse occurs in conjunction with other faults, such as decay. Stone lintels are frequently cracked if they are of weak composition or wrongly bedded. The continuous heating and cooling effect of the fire also tends to promote cracks in stonework. One of the difficulties of a spine beam over a fireplace is that its end may be exposed to sparks and flames within the flue, with obvious dangers. Oak, however, is extraordinarily resistant to fire and it is not uncommon to find the soffit of a timber fire lintel has been charred by successive flare-ups inside the hearth.

In buildings of hardwood timber-framed construction, the floor structure is normally more than adequate to resist the loads, provided that decay and settlement do not occur. You have to take into account the difference between the great strength of oak beams and joists which have hardened and seasoned over centuries and that of poorly kiln-dried members of modern softwood.

From an aesthetic point of view, it is again a golden rule that you do not remove any more than the minimum amount of the original floor framing and boards, especially from earlier buildings. Although the Building Regulations in their tables of spans and sizes do not take much account of the wide joist-spacing common in old buildings, this does not mean you always have to plant joists no more than 600mm apart. It may be possible in the interests of authenticity to design the floor or its repairs so that although spacings are wider, spanning boards are thicker than normal, or of superior timber. Similarly the joists themselves may be increased in section. The regulation tables B3 & 4 (Approved Documents A) only set out 'practical guidance' provisions.

This kind of deviationist behaviour is better undertaken with professional advice. Where a new beam, for instance, must carry joists, or pick up the weight of an old beam running in the opposite direction, design calculations may be required. Always remember with floors that no beam or joist, whatever its strength, is safe without adequate bearing points, and that not only goes for the ends of a beam resting on stone piers, for

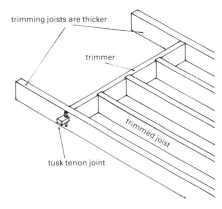

Fig 91 The conventional method of trimming floor openings for staircases and hearths involves the use of *trimming joists* and a *trimmer,* which are all made approximately 25mm thicker than the ordinary *trimmed joists.* The trimmed joists are housed into the trimmer.

example, but also for trimmed joists jointed into a trimmer (Fig 91).

While it is seldom necessary to level up the floors of early buildings – indeed it may be historically damaging to do so – repairs and replacements do have to be carried out, and weak elements reinforced. This means using good hardwood timbers, seasoned and of convincing section, together with the right sort of jointing methods.

Buildings from the late seventeenth century onwards may have softwood floor timbers, even when the boards are of something like elm. This is especially likely when floors are ceiled in with lath and plaster below. Plaster ceilings can prove a problem when it comes to strengthening and replacing individual units (see below).

Your first duty with an old hardwood floor is to notice all the finer points of detail. Are there chamfers to the edges of beams? Do they run out just short of the walls in the form of 'stops' of some traditional pattern? Are the 'stops' perhaps concealed by later wall plaster so that they could be revealed to advantage?

Beams and cross-beams, joists and posts, may all be worked with interesting mouldings. They can be very simple or splendidly elaborate. The heavy oak timber plate running along the top of the wall into which the beams are jointed may be beautifully carved with heraldic beasts or running motifs such as vine-leaf trails. If important features have decayed you may need to cut in replacement lengths which not only continue the decorative detail, but are also jointed so that they can still bear the necessary loads. In some floors, you may decide that there was an inherent weakness in the type of joint used. Then you must show ingenuity. Steel flitch plates, as described in Chapter 7, may be slotted in from top or bottom to bear the weight, being concealed from view by lengths of timber with stuck or planted mouldings.

There are several joints which were commonly used for housing joists into beams or plates. Some are more sophisticated than others. That shown in Fig 92 is fairly elaborate without being entirely efficient. As you can see, it involves cutting a mortice-and-tenon, but only half the depth of the joist bears on the beam, albeit the important tension half.

The joint in Fig 93 is often seen where joists join beams. It is very like the modern tusk tenon used between trimming joist and trimmer joist at staircase openings. However, although it does not pass right through the beam to be held by a wooden key (as in Fig 91), it is mechanically effective. The common housed joint so often cut to link trimmed joists with a trimmer is economical in so far as it only cuts to waste a small amount of trimmer in the upper zone of compression; but the trimmed joist itself has no bearing in its bottom half

A typical stopped chamfer on a beam in a seventeenth-century house. There are numerous variations. This one is probably from the second half of the century, but without other evidence they can be unreliable as a means of dating

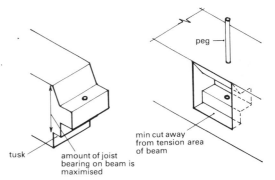

Fig 93 An excellent joist-to-beam connection is made by this version of the *tusk tenon,* which is peg-fixed within the thickness of the beam, instead of being carried through it and wedged on the opposite side

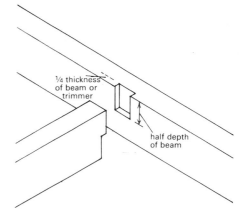

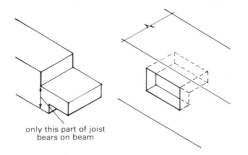

only this part of joist bears on beam

Fig 92 This joint was frequently used in old houses to connect joists to beams

Fig 94 There are three main *housing* joints for connecting bearing ends of timbers into the sides of other ones: the *half housed* type shown here; a similar joint with the top part bevelled; and a *dovetail housed* joint

122

(Fig 94). Such joints should not be heavily loaded. The simple haunched tenon is an economical and effective joint for use when the joist is less deep than the beam (Fig 95).

Sometimes, you may need to introduce new members without disturbing the other parts of the structure, but still effect

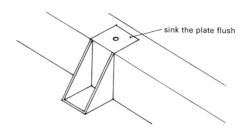

Fig 98 Mild steel *joist hangers* provide excellent bearings for joists and other members, especially if they are to be hidden by boards or plaster.

a bearing. For this purpose it was not unusual for carpenters to cut a chase, allowing the new timber to be swivelled in from the side (Fig 96) or dropped in from above (Fig 97). In concealed floor structures or occasionally for crude buildings of later vintage, steel joist hangers (Fig 98) are very effective. This is not a book on carpentry, but to mention some of these joints may remind you that, in the interests of repair rather than wholesale replacement, there are various dodges worth consideration.

In many cottages with softwood joists laid on edge which have rotted at the ends, a more or less wholesale replacement of the floor structure may be required. Especially if they are fairly late buildings (nineteenth-century perhaps) it need not be aesthetically harmful. The design fault you try to correct at the time of replacement is any sagging resulting from undersized joists, or ones which have been overloaded by being placed too far apart.

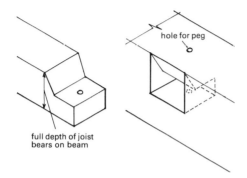

Fig 95 A simple haunched tenon may be used to advantage when the joist is less deep than the beam – as it normally was.

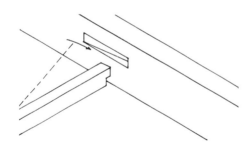

Fig 96 When a length of timber had to be fitted between existing beams, *horizontal chased mortice* was sometimes used, so that it could be worked in sideways to take up its bearing.

In open ceilings, you frequently see that rolled-steel joists or universal beams have been used to support inadequate joists – often those laid flat, or of small, almost square, dimensions. Now this may not be a bad method, provided that the joists fit into the web of the RSJ, rather than passing over it. The latter practice can look terribly contrived and unconvincing, although not without historical precedent in houses which have had earlier floors raised or replaced in the seventeenth or eighteenth centuries.

The use of structural steel is often essential, especially when you are opening up one room into another by removing lengths of load-bearing partition wall.

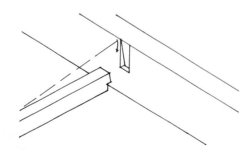

Fig 97 Another method of dropping in a timber between existing joists or beams is to use a *vertical chase mortice*.

123

However, if the ceiling is to be continuous, you must house the joist ends properly on the lower flange of the steel beam. A nineteenth-century warehouse, or something similarly lacking in historical niceties, may sometimes have exposed steelwork. Concealment is a danger area where appearance is concerned. A common fault is the casing of a steel beam with timber of unsuitable type, section or finish. Your disguise should not be so flimsy that it looks as though your joists are supported by an outsized window-box. The result will be a total lack of visual faith in the strength of the floor. I say 'visual' because the beam structure is sound enough: it just *looks* as though it cannot do the job.

Convincing disguises are most important in all repair and restoration work. Nothing is more disturbing than structural repairs hidden by cosmetic pieces of timber which look feeble and thin, without proper bearings or with grain running in the wrong direction. Ideally, one should avoid using methods involving disguise, but that is not always possible.

Take the fireplace with the beam resting above a timber lintel, mentioned earlier. If you replace the timber or stone lintel with a reinforced concrete one, you naturally feel that the concrete should be concealed. However, this concealment must itself be devised to look solid and structurally sound. It's no good applying a thin sawn board that looks as though it might be peeled off the wall or could slip down between the jamb stones (Fig 99). If

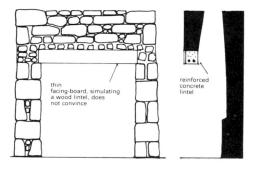

Fig 99 A fake fireplace lintel, made from a relatively thin board, seldom looks the part, especially if the ends appear to have inadequate bearings. The reinforced concrete behind it may be visible and the finish of the board itself can look very contrived.

you do not use a truly structural timber, the disguising one must be substantial enough to deceive effectively. Also, it should be finished very carefully with an adze or a hand-plane in earlier buildings. It should be thick enough to present a soffit which does not make the fireplace appear to be entirely supported by a board torn from the garden fence.

If you are not replacing whole joists, or ranges of joists, you may be obliged to stiffen a floor which sags or springs. This is no easy matter if there are plaster ceilings to preserve *in situ*. To begin with, you will probably have to lift the floorboards above to get at the space between the joists, so that some strengthening timbers can be inserted. The floorboards will, of course, run in the opposite direction to the joists, so you have to tinker with many, rather than a single board.

Juggling extra lengths of joists into place is a tricky job, which most builders find abhorrent. When you have the new joists in place they must be wedged up into position so that the boards bear upon them, or firrings must be inserted along their top edges. Bearings must be found for the ends of new joists, and these are sometimes obtained, if there are structural partitions, by cutting the new joists about 8in (200mm) over-length.

You pass the joist over the partition and then pull it back the other way to take up its bearing in the opposite wall. All too often this is not possible because the wall presents no bearing; all the joist ends were just buried in pockets in the stone as building proceeded. To get over this difficulty, cut out new pockets where needed, using a lump hammer and chisel. Brick walls present little problem, but granite, for example, may prove to be hard slogging.

You can employ steel joist-hangers, merely cutting out enough of the wall to allow you to slot the steel hanging plate over the top edge of a stone, brick or concrete wall beam. The bearing may have to be levelled up with stiff, cement-rich mortar and pieces of slate. Sometimes you can insert a joist of the same depth as its fellows, but of greater thickness, thus supplying the additional strength.

Joists which have sagged for years cannot be propped in such a way that the deflection is corrected. They have usually adopted a new shape and, when propped from below, will merely lift at the end bearings, possibly doing damage within the wall structure. This is not to say that you can never relieve sagging in this way, but you will seldom recover the deflection in the joist itself.

What you can do is stiffen timbers by bolting on new lengths or by sandwiching the old between steel fish plates. However, the more you jack the centre of a floor, the more likely you are to have to chock up, and make good cracks in the walls on either side which provide the bearings.

Excessively springy hidden joists can sometimes be stiffened by introducing one or more rows of herring-bone strutting (Fig 100). This is good thinking, since you may only have to lift a few floorboards to do the job. The strutting runs at right angles between the joists and, provided that it is efficiently fitted, so that the crossed-over strutting pieces bear tightly, it will prove an excellent work of mini-engineering. Ready-made galvanised iron herring-bone struts are also available.

Exposed floor structures can be strutted with solid lengths of timber (Fig 101), tightly wedged and nailed between joists, provided that the room below will not be spoiled by their appearance. In later and rougher buildings, a few coats of white emulsion paint will work wonders in disguising this kind of thing.

There are many ingenious methods of stiffening and reinforcing floor timbers by means of metal plates and tension cables.

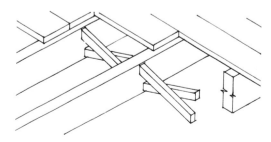

Fig 100 Herringbone strutting is a traditional and effective method of stiffening springy floors; but it should not be used anywhere it can be seen.

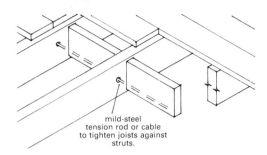

mild-steel tension rod or cable to tighten joists against struts.

Fig 101 Solid strutting seldom works well as a means of stiffening floors, unless the strut ends are perfectly fitted between the joists, which should then be wedged up from one end to make certain that the whole system is as rigid as possible. It helps if you can thread a 12–25mm mild-steel tension rod through the central axes of all the joists and tighten up with a spanner against the end one (see dotted line).

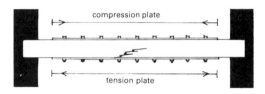

compression plate

tension plate

Fig 102 Compression and tension plates may be bolted to the top and bottom of a faulty beam as a means of repair. This drawing merely shows the principle.

The most obvious is that in Fig 102, where the beam is sandwiched between an upper compression plate and a lower one in tension. Some authorities suggest the use of a lower plate on its own. It has welded steel upstands housed into the beam soffit and is steel-wedged to bring it into tension (Fig 103).

Where steel cables are employed, the principle is to use them to provide an extremely strong form of reinforcement for the beam's own timber fibres which have been stressed in tension beyond endurance. Naturally any tension device like this must be connected at its ends with either a steel compression member or with sound compressed timber at the top of the beam (Fig 104).

Loss of strength in compression along the top of a beam or joist as a result of notching for pipes or electrical cables may need attention. The normal remedy is to insert folding wedges; or to bridge the gaps with mild-steel plates, tightly fitted and screwed down.

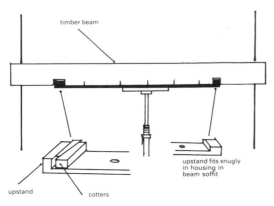

Fig 103 It is sometimes possible to prop a sagging beam and fit a wedged and screwed tension plate to the soffit, which takes up the bending stresses when the prop is removed. The SPAB suggest it in their technical pamphlet.

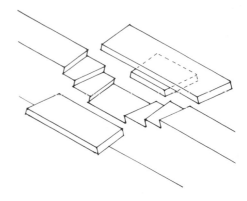

Fig 105 To replace timber in the top half of a beam so that there is a minimum loss of compressive strength: carefully matched, wedge-shaped pieces of wood are cut with dovetailed edges and are tapped home with a mallet.

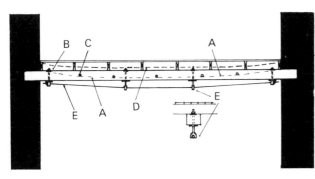

Fig 104 This drawing demonstrates (A), the principle upon which bending of a beam may be counteracted by running a firmly tethered tension cable from end to end. By anchoring the cable on the upper (compression) surface (B) and tightening it against steel bearings drilled through the beam (C), the cable may thus be kept within the floor space. To stiffen a floor *across* the joists, it may be possible to run a cable through successive joists at their mid span, in the manner shown (D). In this case the ends of the cable would be anchored to a steel compression plate fitted under one of the floorboards. If a false ceiling is to be inserted, a system of bolts and a cable could be used (E).

If it is necessary to cut away rotten timber from the top of a beam and replace it with sound wood which may be visible, the method in Fig 105 can be used. Plaster ceilings have been mentioned, in passing. Their temporary support is vital and in Chapter 12 possible methods of repair are discussed.

When replacing the *whole* of a floor bear in mind the other function which the joists may have, as well as the obvious one of holding up the floorboards. As said in Chapter 4, joists often act as tying members for the front and back walls of a building, so there are inherent dangers if you decide to run new joists the other way – parallel to the walls.

It is often a symptom of lack of lateral restraint when gaps appear between the ends or edges of floorboards where they abut front or back walls. This can, of course, happen with gable walls as well, for similar reasons. Because of the more solid construction of gable walls containing substantial chimney breasts, as they often do, movement tends to involve settlement, rather than a simple lack of tying or strutting members running from end to end of the building.

Clearly a substitute floor structure, where joists are to be run parallel to the front and back walls, should sometimes incorporate a system of ties. You can use mild-steel rods, timber or steel beams – there are various possibilities, but do not forget the principle involved.

Most joists run from front to back in buildings which are only one room deep. Buildings two rooms deep also very often have their joists running this way, for the obvious reason that the clear span is usually shorter and thus requires smaller sections of timber.

Earlier buildings with floor beams are frequently exceptions to this rule, since beams may be installed in a spinal direction, or from front to back, or in a compartment system of beams and cross-beams of the kind so often seen in late medieval houses. But the average cottage or small farmhouse of the second half of the eighteenth century or the nineteenth century can be expected to have joists laid on edge from front to back.

Joists, therefore, usually bear within the structure of exterior walls, and are thus incredibly vulnerable to damp, especially as such timbers are normally of imported softwood. The housing of joist ends is a subject for considerable care and not a little hard labour if you are dealing with a stone building. Cutting pockets for joists is a rotten job, since you shake up the walls in the process of hacking out stones and there is invariably a monumental bonding stone just where the joist end should go.

The points to remember, and we are now discussing the complete replacement of the joists, are that you must make wall pockets large enough to provide an adequate bearing, and firm enough to resist the crushing effect of the joist end (and any possible settlement in the wall below).

Obviously joist ends should be treated against decay. Sometimes, it may pay you to wrap them in plastic sheeting brought flush with the wall surface. In any event, aim to isolate them from damp, and that can include placing pieces of bitumen felt or slate underneath so they do not pick up moisture from the bearings. Ventilation is useful, provided that there is circulation

and you do not merely form a series of still cubicles of damp air.

From the point of view of fitting new joists between existing walls, you will find that they are too long to position and still have proper bearings. On the same principle as the chase-housing employed for timber beams, you can chase out the wall above one end-bearing socket (Fig 106A) or you can considerably increase the opening and depth of one socket so that the joist, when level, may be subsequently pulled back to settle on the other one (Fig 106B).

In many brick buildings and some stone ones, walls diminish in thickness at upper-floor levels, thus providing ready-made bearing ledges.

If you become discouraged with these notions, it may be worth resting the joists on a wall plate fixed to the face of the wall with rag-bolts set in concrete, or upon steel corbels. Another method is to fix a strong mild-steel angle section along the

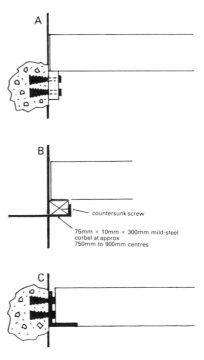

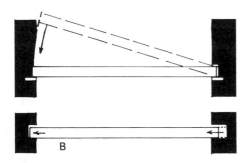

Fig 106 When new joist or beams have to be fitted between existing walls, access to bearings may be achieved by (A) chasing the wall and dropping the timber into place; or by (B) making an extra-deep pocket on one wall and pulling the joist across to rest in the one opposite.

Fig 107 Three methods of fixing joist ends without making bearings in the masonry of a wall. (A) rag-bolting a timber plate to the wall; (B) resting a timber plate on mild-steel corbels grouted into the wall; and (C) rag-bolting a mild-steel angle-iron to the wall.

wall and rest the joint ends upon the flange, using short lengths of timber packing to hold the ends upright and in position (Fig 107).

The principles of concentric and eccentric loading should also be remembered, since the rotating effect of a heavy beam resting on something like a cranked steel corbel could set up movement in the wall (Fig 108).

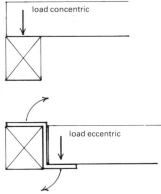

Fig 108 Where possible, it is better to load joist and beam bearings so that forces are concentrically applied (*top drawing*), rather than eccentrically *(lower drawing)*; the latter can produce problems of rotation, if considerable weight is imposed upon a flimsy wall structure.

Any work involving steel must naturally include adequate rust-protection measures and, for timber plates, a similar treatment against decay.

In a previous chapter we discussed the possibility of weakening floors or, indeed, entire timber-framed structures, by inserting new openings for such things as staircases. The standard practice for trimming stair openings (Fig 91, page 121) is well known and, in a conventional stone or brick building of reasonably sound construction, should provide few difficulties. However, there are buildings in which you must be very careful. Jettied structures depend for their stability upon the interrelationship of the floor joists and the walls. The protruding joist ends support the sill plate, beam or bressumer of the jettied wall, while the wall itself acts as a stiffening counterweight for the joists (Fig 109B). This has the merit of discouraging the sag which otherwise occurs when oak joists are placed flat and have to span rather wider spaces than they could manage unaided.

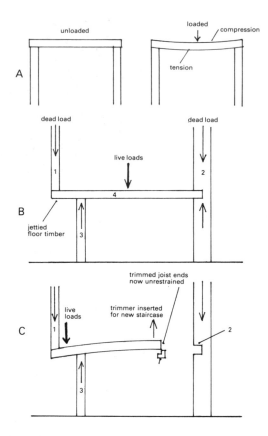

Fig 109 Jetties and their dangers. An ordinary joist which spans directly between walls has its lower surface stressed in tension under load – see (A). Jettied buildings, however, work on the cantilever principle (B). The dead load of the jetty-supported wall 1, is counterbalanced by the load at point 2 in the opposite wall. Wall 3 acts as the pivot of the cantilever, stiffening the joist so that the upper surface is stressed in tension rather than in compression, as at (A). This action greatly stiffens shallow joists and helps to support much greater live loads than would be possible without. If the restraint at point 2 (drawing C) is removed by trimming for a stair-well, or reduced by making a window opening nearby, the joist end will lift and wall 1 will settle.

You will realise that, mechanically, the success of this system depends upon a series of cantilevers which can only work if they are held down at the opposite end to the jettied wall. You can imagine what could happen if you introduced a wide window above point 2 in Fig 109C, which removed the tethering effect of the wall's dead loading. Similarly, if you open up a well in the floor for a staircase, unless you are very clever about it, you could drastically upset the balance of the construction.

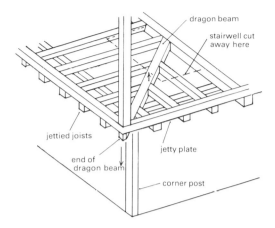

Fig 110 Any disturbance of the dragon beam and jettied joists, for such purposes as trimming a new stairwell (see broken line), can remove important restraint upon the cantilevered timbers, so that the overhanging corner of the building drops.

Floors which are integrated with cantilevered and jettied walls along two sides of a building are supported at the corner by a dragon beam (Fig 110). Here again, any disturbance of the dragon beam could remove the restraints on the beam itself and on all the joist ends connected to it – resulting in the collapse of one corner of the building from first-floor level upwards.

Another good way of weakening an otherwise sound floor is to cut bits of it away to form a fireplace hearth for an upper room. This is a much rarer occurrence in modern restorations than it used to be in the days when a heated upper chamber represented a considerable improvement in comfort for the squire or the yeoman and his lady. The modern restorer tends, alas, to think more in terms of central heating than of effective bedroom fireplaces. Clearly the dangers as far as hearths are concerned are more likely in timber-framed buildings with floors which have additional structural significance. Occasionally floor joists have been dangerously reduced in section to accommodate the laying-on of a heavy flush hearthstone, or they may merely have been overloaded by the hearth. Trimming an opening to permit a brick chimney to be inserted in the existing building can be very disruptive. All depends on the type of structure and the commonsense of the people carrying out the work.

One of the first points you notice when looking at the open-joisted floors of buildings such as old cottages is that the joists seem much too far apart. All the same, they have withstood perhaps eight generations of wear and tear without sagging unduly. As ever, the faults will probably be in the ends, which have decayed within the walls. New pieces will often have been spiked to them. Sometimes there will be plates bolted to the sides of rotten timbers; or a flat mild-steel bar may have been screwed to the joist soffit. Nine times out of ten, these measures have finally proved ineffective – usually because no action was taken to cure the decay.

If you have to replace joists wholesale, rather than repair or strengthen individual lengths, the crude but useful rule-of-thumb for assessing the size is to halve the span in feet and add 2in to the number you get. This will give a rough idea of the proper depth for a joist 2in thick. Example: if the clear span between walls is 12ft, half the span is 6ft. Add 2in to 6in, and you have a joist depth of 8in.

Your normal course, especially if your work is to be the subject of a Building Regulations application, or there are grants involved, will be to refer to the tables provided in the Regulations themselves or in one of the illustrated guides to the Regulations. Table 1 of Schedule 6 deals with floor joists and remember that the grade of timber and its resistance to fire (Section E) may also influence your calculations.

Where you need to cut away rotten end timber from hardwood beams which are to remain visible, your method of repair will normally involve the use of a steel flitch plate. This is slotted into the sound timber, and a new beam end is bolted on either side, any mouldings, chamfers or stops being continued to match the old ones. Since you cannot sit a thin, upright steel plate on a wall pad you bolt angle irons on either side so that they make a solid bearing which spreads the load (Fig 111). Other methods for steel end repairs include welded plates like that in Fig 112 and mild-steel angles, channels, plates or

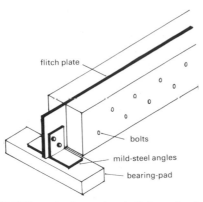

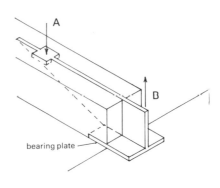

Fig 111 Where the decayed end of a beam has been cut away over its original bearing and a flitch plate has been slotted in, angle-irons may be bolted to the end of the plate to spread the load on the pad.

Fig 112 A beam end may be renewed by slotting in a triangular steel flitch plate which exploits the rotary forces at work. The flitch will bear down within the thickness of the beam at A, and the top edge has a square plate welded on to help resist this action. At B, the flitch plate will try to lift, and this movement is counteracted by the bearing plate welded underneath it, which presses against the underside of the beam. (Beam ends may also be supported with steel angles.)

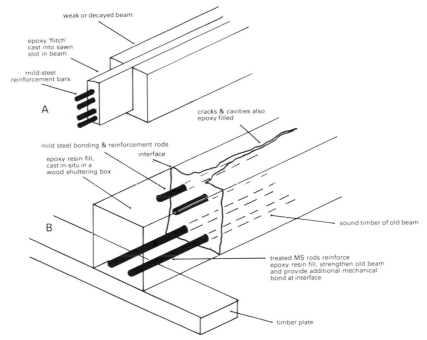

Fig 113 Some firms are now using epoxy resin techniques to fill or remake defective parts of old timbers, and to cast on new structural replacement lengths for beams and joists. The traditional method of slotting in a steel flitch plate may be substituted by casting an epoxy resin 'flitch' reinforced with mild-steel rods (A). A new end to a decayed oak beam can be cast *in situ*, with reinforcement bars bonded into drillings in the sound wood. Cracks and cavities may be filled at the same time. Glass-fibre reinforcement rods are also used in some kinds of repair. The epoxy resin may be disguised by careful toning and texturing when it must remain visible: whether this finishing process will be successful must influence the decision as to its use. An advantage of epoxy techniques is that they require minimum disturbance of the existing structure. Reinforcement of load-bearing timbers may require calculation to determine its type, size and distribution. (System A is one invented and employed by Rickards Timber Treatment Ltd. System B shows, in principle only, another method used by the same firm.)

130

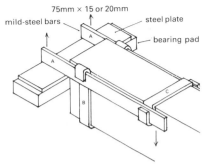

Fig 114 A method of cradling the end of a beam which has to be cut short to eliminate areas of decay is recommended by the SPAB. Flat mild-steel bars (A) act as hangers. These are restrained from upward movement by the stirrup (B), while their downward rotary movement is resisted by the top stirrup (C).

cradles. Ends may also be replaced by drilling the sound timber, inserting reinforcement rods and casting *in situ* with polyester resin (Figs 113 and 114).

Many of the repair methods for floor timbers are similar to those used for roof trusses which have been described in the previous chapter.

When dealing with floors, always look out for overloading resulting from the dead load of partitions or roof trusses. So many faults in old houses derive from ill-judged structural changes from other

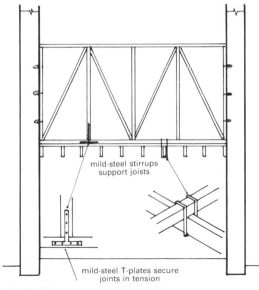

Fig 115 The general principle of giving additional support to floor joists by hanging them from a timber trussed partition above. The partition becomes, in effect, a very simple lattice girder,

eras. Partitions designed to transfer the weight of floors and roof timbers to ground level may have been removed to make bigger rooms. Where this has happened, it may be necessary to prop the partitions or to truss them, making tie-beams or sole plates which pick up bearings on solid walls (Fig 115). Universal beams or RSJs may be introduced to relieve major spans, or less ambitious methods can be used where it is simply a question of adding strength to a floor joist which supports a non-structural partition.

Suppose the floor sags a bit under a partition which only serves to divide one room from another. You may be able to jack up the floor to correct the sag before strengthening it, but as said before this is not possible if the timbers have become permanently distorted. It depends on the way the members are laid and the kind of timber used. If permanently distorted, you will probably have to accept the visual imperfections and prevent further movement.

If the partition runs the same way as the joist, you can double up the joist upon which it sits, or add joists to either side. If the floor joists are closed, with a ceiling below, you may introduce a steel plate and bolt the joist to it, but it will mean lifting the floorboards, or most of them, to get it in. You may have to chase out the top of the housings in walls and so on, as described above, to pick up bearings at either end. However, for this purpose, the bearings might be reduced to perhaps 40mm at either end, provided that they are composed of some firm material.

If the partition runs across the joists, you may have to support two or three separate joists. Occasionally, you have to prop the partition from below, using a post at some strategic point. The post can then be integrated with the layout of the room in which it stands. Cellars of old houses often contain props for the ground-floor joists or beams, and cross-walls were also built to provide additional support. The difficulty will arise when plaster ceilings have to be preserved, cornices maintained and original room shapes respected.

Sagging floors quite often contribute to the charm and sense of age in old buildings, particularly if they are of the 'rough' variety, full of exposed oak framing. Often ancient floors slope wildly from one side of a room to another. Do not try to over-restore these by levelling unless it is structurally essential. Nor, as a rule, should you level the floors for convenience and appearance by cutting tapered firring pieces and placing other boards or coverings over them.

Floorboards in early buildings, that is to say medieval ones, were sometimes laid in the same direction as the heavy, beam-like joists. The boards were very wide and the joists close together. Thus boards could be rested in rebates cut from the tops of the joists. In other old houses you will find very wide elm or similar hardwood boards jointed together with loose tongues (Fig 116). This practice, however, was also used for floorboards in softwood right into the nineteenth century. Other types of floorboards in early houses were rebated together (Fig 116); and some boards were just butted against each other and run in the same direction as the joists (Fig 116). Whatever the system used, be scrupulous about repairing and repeating what was there before if the building has any claim to architectural interest.

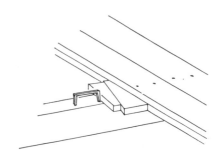

Fig 117 To cramp floorboards so that they fit tightly against each other without special equipment, spike a wrought-iron dog into the joist and lay a batten along the edge of the board. Cut wooden wedges and drive them between the dog and the batten.

When laying new floorboards, remember that it is important that they should be well cramped together, since they will tend to shrink as they lose moisture content and the building is occupied and heated. There are special cramping tools which can be fixed to the joists, though only those who are frequently called upon to lay floors will own them. However, the traditional wrought-iron dog may be banged into the top of a joist and the boards tightened up with folding wedges (Fig 117).

Boards can be secret-nailed with lost-head nails, or just punched below the surface. Secret-nailing through the angle between the tongue and the edge of the board is not really to be recommended unless you are a good neat craftsman, since you are likely to split the tongue from the board.

While on the subject of floorboards, avoid using any kind of varnish or polyurethane sealer upon them: these seldom give a satisfactory finish. If you must seal the boards, use a preparation which sinks into the surface. You might also run a little discreet colour into pale, uninteresting new boards; but you should try out different shades on sample pieces of the same timber until you find the right one. Many stains and sealers look far too red, others too black or yellow. The finish should still be many applications of wax polish, which may itself contain some colour pigment to darken new wood. A mixture of beeswax and turpentine (3:8), rendered down in a double boiler, can be

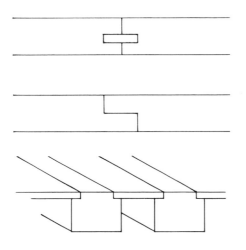

Fig 116 Three ways of laying floorboards. Loose tongues grooved into both abutting boards were often used in old houses *(top)*. Another method was to form a rebated junction *(middle drawing)*. Early floors might have boards running parallel to the joists, which had rebates in which the edges of the boards rested *(bottom drawing)*.

132

used as a truly traditional polish. But although it is authentic, having tried it I wonder if it is worth the bother, when an ordinary wax floor-polish works perfectly well. Time, dirt and polishing are the factors which give the patina of a good floor and there are no short cuts to what, at best, is a marvellous, deep rich sheen built up over maybe hundreds of years.

Always remember that your beams and joists should *not* be painted black or covered in any kind of dark gunge. Old oak timbers were not so treated, but merely aged gracefully in their natural state. There is no need to oil timbers – it does no good. Just wax polish if it seems a good idea. Some buildings may benefit from waxing structural interior timbers, while others would adopt too sophisticated an air as a result of such measures.

Never distress modern replacement timbers by hacking them about with spoke-shaves, chisels and other instruments. The effect is nearly always spurious. The best thing is to finish them, if needs be, with an adze (as mentioned before) or hand-plane them.

Go to any 'olde worlde' pub and look at the disastrous faking which has been carried out to make new timbers look old; and to make genuinely old ones look older. These blackened, waney-edged, thrashed-about abominations, far from conveying a sense of antiquity, make any sensible person imagine he or she has stepped on to the set of some Hollywood extravaganza. Public houses provide some of the best object-lessons in how not to restore and repair old buildings and you may study them at leisure for the price of a pint, taking in everything from juke-box infested William and Mary to formica-fitted early Victorian. No properly paid-up libertine could fail to come away a sadder and a wiser man after visiting the average 'listed' hotel or coaching inn.

That said, let us turn our attention to suspended floors at ground level. These are of two kinds. There is the ordinary suspended floor over a cellar which can be dealt with like its counterparts at other levels. Then there is the joisted floor which sits on brick piers, or sleeper walls. Such floors are often nineteenth- or twentieth-century. The later they are, the more likely they are to be ventilated, possibly by grills in the form of purpose-made air bricks, set below a damp-proof course if there is one.

More often than not, the number and size of the ventilation grills is inadequate and too little fresh air circulates between or through the sleeper walls which carry the timber plates and bearers or joists. The sleeper walls themselves probably have no damp-courses and take moisture up from the subsoil, causing decay. The ends of joists may touch the damp walls and also pick up damp. In earlier buildings wooden floors may have been laid on joists which sit straight on to the beaten earth or some other material such as bricks, hardcore or rubble. If such floors are not rotten, you are very lucky indeed.

Suspended timber ground floors in old buildings are practical and sometimes beautiful but they must be kept dry and well ventilated. So these are the measures you may need to employ.

Ventilators should be cleared if there are any, and if there are not, then consider making some. Even where thick rubble stone walls make it difficult to cut openings, you can perhaps introduce 150mm diameter pipes, with grills over the ends to prevent unwanted visitors. For cutting or boring ventilation openings, you need the help of a heavy-duty drill designed for the purpose; these can be easily hired.

It may be necessary, if the floor is below ground level, to adopt some of the damp-proofing measures suggested in Chapter 5. For example, if you decide to lay a mortar or concrete screed upon which to place the brick sleeper walls, you might just as well put a damp-proof membrane under it. A damp-proof course should also be placed under the wall plates which lie beneath the joists. If you decide to do without wall plates, then put the damp-proof course under the joists themselves.

It is essential that all timber should be treated against decay and that air should be able to circulate freely. This may mean placing ventilation points so that a cross-draught is encouraged. Since you want fresh air under the floorboards, it may be sensible to use some insulation board to prevent draughts and heat losses in the room above; but be careful in your choice of material, since many insulating materials readily absorb moisture. You must avoid any possibility of your floorboards lying on a raft of soggy insulation. One method would be to drape insulated roofing felt over the plates before laying the joists.

Where ventilation cannot be procured from outside, it is feasible to place hit-and-miss skirting ventilators around the room so that air can circulate into the floor space.

As always, you will have to be ruthless in cutting out decayed boards, joists and plates, and treating suspect timber connected with wall panelling and skirtings. No responsible adviser can fail to warn you of the dangers of decay; but please do not become so zealous that you wantonly rip out fine old floorboards and other features when there is no real need. Remember that as far as an atmosphere of age is concerned, your house will benefit by the retention of as much old, worn, hand-wrought, mellow and imperfect joinery as you can save.

Solid Floors and Floor Coverings

From suspended ground floors, let us move on to deal with solid ones. These take any number of forms. There are grip floors of lime and ash to be found in cottages. There are bricks laid to a pattern upon cinders or beaten earth; and flagstones and grand forms of marble and other decorative paving. You may have quarry tiles or plain clay tiles, pamments (unglazed tiles), granite slabs, sets, or even pebbles. The latter are often found in the through-passages of farmhouses, or in small manor houses where the passage divides the service rooms from the hall.

Old and interesting floors are of great importance and on no account should they be removed, unless it is absolutely unavoidable. You may find them a bit inconvenient or even cold, but their advantages greatly outweigh their disadvantages. If you cannot live with slate flagstones, granite slabs or polished boards, then perhaps living in an old house is not for you! It is inexcusable to bury old flag-

stones under new mortar screeds, or to lift and dispose of them. In many cases it is not only vandalism to remove or cover the original floor, but also a breach of the law on listed buildings. To carpet interesting old floor surfaces obviously does no lasting harm to the building, but you forfeit one of its major charms. Rugs or the bigger, non-fitted carpets will do the trick and will look much more at home in an old house.

The other type of floor which has not yet been mentioned is a timber finish lying directly upon a mortar screed. There is also the semi-suspended floor, where boards lie on battens held in floor clips or upon splayed fillets buried in the screed itself. The modern practice for laying parquet, parquet panels, wood blocks, hardwood strip and similar floors is to spread a bitumen adhesive compound straight upon the screed. The bitumen provides an additional layer of damp protection even if there is already a plastic damp-proof membrane under the screed.

Generally speaking, timber floors are warm, they look the part and are perfectly practical. The best type to choose depends upon the age and style of the building, and requires much discrimination. If I were to lay down a prescription which would be safe for anybody to follow, I should probably have to discount floor finishes which could work in a room of one old building but not in another. To give an example, I know of a 'rough' upland cottage-farmhouse which has a most successful hall floor of polished pale grey and dark grey concrete paving stones. They are laid in a chequer pattern and are pointed between the slabs with a dark grey mortar, kept slightly below the arrises of the paving. The hall probably dates from about 1700, but could be rather earlier. The adjoining part of the building, indeed its major part, is a foursquare, high-roomed cottage of about 1860.

This floor looks so like real stone paving that even well-informed visitors often fail to realise what it is made of. But the fact that here it works so well does not mean that it would necessarily do so in every similar building. Nor does it imply that *any* kind of concrete slab will serve; those textured garden-patio slabs, for example, would be disastrous in an old building. The kind of wood blocks (about 50mm thick) which we associate with village halls and schools might look splendid in the kitchen of a Georgian house or in the sitting-room of a Victorian lodge, but extremely unattractive in some other kind of building.

However, there is one sort of floor which is fairly safe on all occasions, and that is the ordinary softwood tongued-and-grooved board; or better still, a boarded floor in oak or elm. Clearly price and quality go together, and whether you should use hardwood boards depends upon the age of the building and its character. Softwood boards are perfectly correct in many eighteenth-century houses; but it may occur to you that a particular room or a particular building asks for something better. Occasionally you can get hold of second-hand floorboards, and nothing could be better if they are wide, made of oak or elm and retain a patina of age. Second-hand softwood flooring may have few aesthetic advantages but could prove cheaper.

It is worth thinking about the width of boards to use. Early boards did tend to be wide – often more than 300mm – and they were frequently quite a bit thicker than the standard 21–23mm T & G we are offered today. It is quite usual to find loose-tongued boards in seventeenth-century buildings, but there is no particular advantage in repeating this jointing method for a new floor.

Formal rooms of the late eighteenth century or nineteenth century may be fitted with hardwood strip – narrow boards not normally more than 100mm wide laid in random lengths of 1.5–2m. Equally you can lay quite narrow conventional floorboards.

On the Continent, patterned layouts of narrow hardwood boards or parquet were used much earlier than in Britain, where we normally associate parquet with Victorian houses. However, there was a time in the late seventeenth century and early eighteenth century when grander houses might be fitted with parquet, marquetry or inlaid floors, so be careful

that you do not mistake the early variety of parquet for a Victorian addition.

I have said something about polishing wood floors but there is another finish, or lack of it, which you might consider. That is the unpolished scrubbed look. It is mentioned in such books as Hannah Glasse's *The Servants Directory* (1770) and was obviously a much more common practice than might be imagined. The floor is dry-scrubbed with damp sand which is scattered over it. The sand is brushed up and sweet-smelling herbs or petals are broadcast and brushed in. Finally, these are in turn brushed up and removed. It all sounds fresh, wholesome and exactly the sort of thing to appeal to purists, or to members of the wool-and-wheatgerm fraternity. No, really, it is rather a good idea.

Clearly the intention is to avoid having damp-scrubbed floors which never dry out in our rainy climate. A few pubs still have sand- or sawdust-strewn floors – a practice which one greatly prefers to miles of raucous broadloom carpet.

If you are replacing the sub-floor of an old building, you will obviously put in a plastic damp-proof membrane and turn it up the walls. This will be covered by a protective screed of rather dry and cement-rich mortar (about 1:3).

Should you screed your own floors here are some points to remember. Battens are laid on to the sub-floor and levelled with a spirit-level placed on a straight piece of timber. These battens (or they may be pats of mortar and slate) denote the top surface of the screed, perhaps 50mm above the site concrete or whatever surface you are covering. The screed mortar is tamped down and struck off level against the battens using a straight-edged board called a feather.

It is important that you keep down the moisture content of the mix, or the screed will slump and form puddles, and water will rise to the surface as it is compacted. Too much moisture also encourages shrinkage and cracking. The screed may be left with a slight texture from the indentations of the floating rule or feather-edge board. According to the kind of finish you mean to use, it may be smoothed over with a steel plasterer's trowel when it has stiffened enough to avoid bringing cement to the surface.

Screeding floors is not as easy as it sounds and I suggest that you have this job done by a professional if you are one of those people who are practical enough but do not pride themselves on their ability to produce a fine finish. Like everything else it is a matter of experience and you may never screed enough floors to become good at it.

Remember that screeds should be damped down in warm weather while the mortar cures, especially if laid on an absorbent background such as concrete.

A screed may be less than the conventional 50mm in thickness if the background permits it. But if used to load a plastic damp-proof membrane it really needs to be at least 38mm thick. Where you are laying flagstones or tiles in mortar on a rather uneven floor surface you can get away with a fairly thin layer over the damp-proof membrane. For example, some areas may be 40mm thick, but the mortar could thin out to as little as 15mm in others, in the process of obtaining a level floor. The tiles or flagstones provide the additional load and protection for the plastic sheeting, and the only danger lies in having too wet a mix, since the moisture content can only evaporate to the surface and none will be absorbed into the sub-floor. However, there is less chance of premature drying-out, which can prevent the chemical curing process.

When laying floors consisting of paving slabs, flagstones, tiles or bricks, it is vital that you keep the joints between them thin and that mortar does not come flush with their surfaces. Nothing looks worse than rows of pale grey mortar lines between pleasantly coloured slate slabs. Keep the mortar 1–2mm below the edges of the slabs. The mortar must be coloured to tone with the flooring materials, whatever they are. Slate slabs, for example, should be either bedded upon a mortar containing a black colouring agent or pointed with a mix which contains it. It is best to do some sample pats of mortar and allow them to dry thoroughly before deciding on the mix. The idea is to obtain a tone which in no way draws attention to itself; you should never notice the mortar between the stones.

By the same token, never lay ragged pieces of stone as floors in old buildings, so that you end up with a kind of crazy paving. It will look totally out of keeping. You may use random squared stones, some with broken corners or slightly uneven sides, but make quite sure that you do not allow areas to appear in the pattern which have to be filled up with mortar. This means that you cannot skimp on the quantity of paving required. I have seen people lift slate flagstones from an old piggery and reuse them in the house to what could have been great effect: but in their desire to stretch an inadequate number of slabs to cover an extra piece of kitchen floor, they have ended up with wide, pale grey, snake-like joints, and ugly random patches of mortar infill.

Since slate is very expensive – and so are other kinds of flagstones – there is nothing to stop you using roofing slates or roofing stone for flooring. Slates are probably rather thin, and must be very firmly bedded so that no hollows remain under them, otherwise they will crack when in use. Again, be discreet in the way you point between them. The smaller the slate, stone or tile, the narrower the joint should be, within reason. Wide joints look terribly busy with small flooring units. If thin slates are laid with the ragged edge upwards, a jointing key is provided. Laid with the ragged edge downwards you obtain a crisper outline, but must take care to ensure that the edges are firmly packed and bedded.

A word here about types of finish for flagstones. Generally speaking, flagstones which have a marked cleavage, like slate, may be riven at the quarry and this presents a pleasant, rural kind of surface, less smooth and even than a sawn finish gives. Some stones are not amenable to splitting so will normally be of sawn finish anyway.

New flagstones always have sawn edges today. The edges are cut on a circular saw and the surfaces are sliced out of the block on a big water-lubricated frame saw. This system, by the way, goes for other components like sills, steps, fireplace slips and so on. Such sawn finishes are perfectly acceptable for these purposes, but never use sawn-faced rubble stone for walls. It should be hand-dressed or it will usually look hard and mechanical.

It does not normally matter that some flagstones in old houses are rather worn. That is part and parcel of having an old building. If, however, the stones have sunk unevenly, to an unacceptable degree, they may sometimes be lifted with infinite care and turned over. You have to be very patient in carrying out this task, since a very little rough treatment will crack, badly chip or break a paving stone. The danger points are the first stone you lift, when it may be difficult to get underneath it with your crowbar, and the edges of the floor, where stones may not only run under skirtings which have to be removed, but may be covered by partition walls as well.

In choosing patterns for laying-out paved halls and so on, avoid any which imply a grandeur the house does not possess. Should you be dealing with a reasonably elegant building there is scope for some fun in selecting the design for a new floor. One excellent source for such designs is Batty and Thomas Langley's *Art of Designing and Working the Ornamental Parts of Buildings*. Such enterprises prove far from cheap.

Although this is a slightly academic point, it should be remembered that when we speak of marble in English houses, we often do not mean real marble at all. English 'marbles' are hard limestones which will take a high polish. Nonetheless, they are often quite as beautiful as anything from Italy. Many marble floors in old houses were imported from Europe, so you cannot assume all floors in English buildings are hard limestone slabs.

Flagstones and paved floors may be washed, scrubbed or polished. Polish helps to protect the surface from marking but can be a nuisance if you are excessively houseproud and have a big area of floor. As far as floors are concerned, the old military maxim 'If you can't salute it, paint it' may be reinterpreted as: 'If you can't scrub it, polish it.' But it should be the

guardroom for backsliders found clutching the varnish pot.

Some of the less traditional or more recent floor finishes could have a place in your reckoning. Late Victorian and Edwardian houses may have encaustic tiles – highly patterned in little blocks of buff, black and terracotta. They are of their period and, while sometimes comic, not to be despised. One can see a place for a terrazzo floor made from well-chosen aggregates of suitable colouring. Terrazzo must be laid by specialists with proper equipment, but it could be used for a laundry, downstairs cloakroom, kitchen or bathroom. Terrazzo has impeccable antecedents, as anybody who has been to Venice can bear witness. Since nearly all our architecture from the late sixteenth century onwards owes its designs to Greece and Rome, we need feel no qualms about employing a Venetian floor in some room which can tactfully absorb it.

Cork flooring is another possible surface for a kitchen or bathroom. It may be wax-polished and, if laid from the roll (rather like linoleum), can be both practical and unassuming. Most cork, however, is bought in the form of tiles. They are not abhorrent and are more merciful to dropped wine-glasses than quarry tiles.

Tame-looking vinyl tiles may not be out of place in laundries, bathrooms, lavatories and kitchens, although one can think of nicer floor finishes. In a simple cottage or fairly modest house vinyl tiles might be all you could finally afford after spending your money on really important features like windows! Is it still possible to obtain that institutional-looking linoleum the colour of brown Windsor soup? Well polished, in the hall of a Victorian office with brass trimmings to the stair treads, it is rather comforting. It is said that there are lawyers and accountants who would sooner be surrounded by carpet and anodised aluminium, but I do not believe it. Linoleum is a descendant of the eighteenth-century oiled-canvas floor cloth, which should be mentioned in case you find some, either *in situ* or rolled up in the attic. If you do have such luck, treat it with care. It is a sought-after rarity and, if it bears a pretty pattern, you are more likely to frame some for the wall than place it underfoot.

Some eighteenth-century floors were painted with designs and these are another rarity which should be protected should you come across them. On the whole I should avoid painting floors, although some that are already painted (especially if they are white) may not warrant the bother of sanding them off. Some people take great pains to paint their floors in colours which integrate with a chosen decorative scheme and the results *can* be successful.

Old and interesting boarded floors should definitely not be sanded if they have built up a patina of wax polish. The use of the sanding machine, which, by the way, can be hired by the day, should normally be reserved for floors which have no historical merits.

Levelling or Lowering Floors

Although this has been emphasised early in the book, I should remind you again that if you have to dig out floors to procure extra headroom or to introduce damp-proof membranes, extreme caution is needed. The danger of undermining the stability of old walls must always be present in your mind. Before the advent of the new Building Regulations 1985, digging out floors was often essential to procure the 2.3m of height which building inspectors demanded for 'habitable rooms'. Now that the Regulations no longer mention required room heights it is hoped that many people will no longer feel it necessary to disturb walls and floors in this way, especially in cases where relatively low ceilings contribute much to the architectural character and atmosphere of the building. At least, it is now possible to restrict any excavation to an absolute minimum.

When repairing or restoring an old building, do not feel that floors must all be brought to a uniform level. Such actions may have the marginal advantage of allowing the full use of a tea-trolley, but they may also torture the mystery out of a building. Steps up and down from one level of a building to another relate to different stages of its historical develop-ment, and they are entertaining in their own way. True, we have all stumbled, cursing, down unexpected flights of steps in old houses with which we are unfamiliar, but we cannot blame the house for our own lack of vigilance. Building inspectors hate to find steps hidden behind doors, and many a small platform now sticks out into an otherwise inoffensive room as a result of this phobia. The trick is to make the platform look as little like a fugitive from the foyer of the Roxy Cinema as possible.

For every house converter who irons out existing changes in floor level, there is another who is busy introducing discrepancies of his own invention. These can often produce a theatrical note which calls to mind an Austrian musical comedy; you expect a banal youth in lederhosen to appear at the balustrade and yodel a mountain love-song. There are many varieties of this split-level division of spaces. They include: the quarter deck of the *Bounty*, the Rossini steps (purpose-made for budding Figaros), the slightly sinister Cromwellian gallery, the public swimming-bath, the Marigold Tea Rooms, the Anyone for Tennis? and many more. The point is that you should try not to devise split-level rooms which are subtly (or not so subtly) at variance with the spirit of the building.

Sound-insulation

To finish this chapter on floors we should consider one of the perennial complaints of those living in old houses which have been divided up into flats – the problem of noise through floors. Most of us who have lived in digs at some point in our lives remember standing on a chair to bang the ceiling with a broom-handle when the din from the flat above became too insistent. I recall a bed-sitter in London's Little Venice where the gentle girl in the room above would groan loudly and rhythmically during the dark watches of the night. It took me some time to realise that these pathetic or erotic sounds emanated from the carriage of a knitting-machine being pushed to and fro.

There are two sorts of sound against which floors or any other elements of a

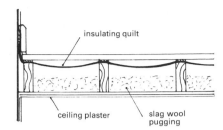

Fig 118 A simple method of providing a degree of sound insulation in a floor is to drape an insulating quilt over the tops of the joists and turn it up the wall behind the skirting. Slag wool can also be used as a pugging between the joists; and solid rubber pads may be placed between components which would otherwise be nailed or screwed directly together.

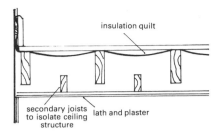

Fig 119 A more elaborate method of providing an element of sound insulation is to fix a secondary ceiling, isolated from the effects of impact sound. Extra joists are fitted below the main load-bearing ones and the ceiling plaster is attached to these. Again, an insulating quilt and rubber isolating pads may be employed.

building may be insulated. Impact sound is that which is transmitted by vibrations travelling through the structure. The noise of footsteps is the most obvious example. Then there is airborne sound, as from your neighbour's television set when it is turned up too high. Airborne sound, as the name implies, finds its way through badly fitting doors, ventilation shafts, half-open windows and, above all, through open-joisted floors which are both structurally light and full of chinks and crevices.

If you have a cottage without proper ceilings, you must accept the likelihood of being able to hear what people say in the room above and of being staggered by the din of such a simple act as pulling a chair up to a table. It is best to avoid siting children's bedrooms, for instance, over sitting-rooms in cottages of this kind. If you do, then a fitted carpet to deaden some of the childish footsteps must be conceded.

Noise like this is seldom a real problem in cottages where people organise the pattern of their lives and the occupancy of the rooms to advantage. In buildings where you share walls and floors with other tenants it can be a menace.

The first thing to understand about sound-insulation is that it has a direct relationship to the weight of the structural elements, like walls and floors. The lighter the elements, the more they will vibrate and the more sound will be transmitted. From this it follows, in theory, that your best protection is to have thick, heavy walls and floors. In many ways that principle is

not a bad one to adopt, when you can do so. (However, you must avoid leaving the kind of openings through which airborne sound can travel.)

Here we have to think about preventing sound moving through a conventional suspended floor in an old building. There are no really good answers, but useful measures can be taken in the process of *renewing* old floors. It is a good deal more difficult to deal with existing ones. The two main measures are, first, to introduce various forms of insulating material into the floor structure, such as insulation board, special quilts, felt, or lightweight slag wool pugging; second, to attempt to isolate the floor from other parts of the building such as the walls and the ceiling of the room below. By placing insulating pads at strategic points between floorboards and joists, joists and walls, floorboards and skirtings, you may reduce the transmission of sound vibrations through the structure. A ceiling can be constructed independently of the floor above so that it is fixed to its own set of separate ceiling joists, instead of to the main structural ones. This ceiling will, of course, contain insulation material, interposed with as much continuity as you can manage between the finish and the supporting joists and wall battens.

No such methods will be perfect and sound-insulation is a specialised subject. What I want to convey is the possibility of taking some effective action and the principle of deadening the floor (or, for that matter, a partition) while also isolating it.

139

The idea of isolating parts of a structure is very like that applied to damp-proofing. The measures you take are completely dependent upon the amount of continuity you can devise. When preventing the entry of damp you have to ensure that no breaks occur in the damp-proofing layers you apply and that bridges across which moisture can move are not overlooked. So it is with sound-insulation. When you have gone to great lengths to insulate a floor by placing a quilt between the boards and the joists, you should not ignore the fact that sound vibrations may still be carried up from the skirting board and into the wall.

Fig 118 shows a reasonably sensible and not too costly way of doing something towards sound-insulating a timber floor; while Fig 119 provides a much more expensive and complicated set of measures. You will see that the second drawing shows a traditional lath-and-plaster ceiling, which remains one of the best finishes for sound-insulation.

Some useful methods and general information on sound insulation may be found in the Building Regulations 1985, approved document E.

9

WINDOWS AND SHUTTERS

No feature does more to contribute to the character of most old buildings than the windows. There are some buildings which have very few windows and a great deal of roof. There are others of strictly classical flavour where colonnades, pediments and perhaps statues are the main features which catch your eye. However, in the vast majority of ordinary vernacular houses and cottages it is the fenestration which provides the principal focus of our attention.

The shape, size and spacing of the window openings are vital elements of architectural design. They spell out rhythms like passages of music. In one façade they are staccato, like the exclamations in an operatic recitative. They may march across the face of a building in a lively manner, like the allegro movement of a symphony or they may adopt a cool, reflective disposition which might be described as andante. The composition can express itself in beguiling twin minuets of white double-hung sashes enclosing a central Venetian window, like one of those elegant Haydn trios which can haunt you for the whole of a summer afternoon.

Windows may be small and secretive, defensive, serene, inviting, confident or hesitant. They are the eyes of a house and reflect its spirit. They abound in subtle and delightful details and you can ruin all of these things with one crass or careless act of modernisation.

As I said at the beginning of the book 'modernisation' is the word which should never cross your lips when talking about your plans. You do not 'modernise'. You replace or repair, or add here and there in an informed manner, but if you care anything for old buildings you treat every aspect of fenestration with loyal and educated attention. In any old town or village in the land you see the devastation which 'improvers' have inflicted upon the windows of old houses and cottages. It is a plague which remorselessly spreads from the modest artisans' terrace, to the thatched hovel; and from the eighteenth-century merchant's house to the recently converted barn.

Do not listen to the insinuating advertisements for uPVC replacement windows. Ignore the blandishments of those ignorant but well-meaning people who try to persuade you that mass-produced, neo-William-and-Mary casements, with top-opening lights, will do just as well as the real thing, and be so much cheaper.

Replacing windows is a task which may be carried out once in a lifetime and must be done properly. To achieve good results you must know something of the history of windows and their design. This book is intended to be a practical guide to restoring and repairing old houses, but we should never lose sight of the aesthetic considerations which make all the hard manual labour and sometimes punitive costs worthwhile.

Instead of giving a complete, blow-by-blow account of the development of windows in English buildings since the Norman Conquest, it seems better to think first of the types of window you are most likely to encounter in buildings which come your way. However, we shall have to outline their history, for it must be understood why certain windows are appropriate or inappropriate in different situations. On that basis, we can divide the topic into two categories: casements and double-hung (or horizontal) sliding sashes.

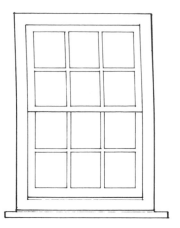

Fig 120 The main type of window used in the eighteenth and the first half of the nineteenth century was the box-framed double-hung sash with twelve or sixteen panes; a most pleasing and practical window.

Double-hung Sashes

This window (Fig 120) is the one found in most better-quality or 'smooth' buildings from the late seventeenth century through to at least 1850. Obviously there are enough exceptions to the rule to make life interesting, including the mullioned Victorian Gothic pastiches to be found in lodges, rectories, manor houses and so on. There are buildings which belong to other architectural revivals (again mainly Gothic). Despite the fairly numerous exceptions, however, the double-hung sash dominated this period, with few variations to its basic construction, although there were important adaptations of its glazing.

Sash windows are believed to have originated in Holland during the seventeenth century. The fact that the English king, Charles II, lived in exile in Holland may have a good deal to do with the fever of Dutch-influenced building in England after his return in 1660. Whatever the rival claims concerning their invention, the British used sash windows more sparingly than their Dutch neighbours. The latter crowded their limited gable frontages with big sash windows, while the former took a more spacious view of matters and inserted them in the by now fashionable Renaissance façades, with their well-proportioned and carefully designed openings.

Here it should be said that the Italians, from whom British classical architecture derived through the medium of the great pattern-book writers, including Serlio and Palladio, never embraced the double-hung sash at all. They were shutters-and-casement men. So were most other Europeans. Only Holland, North America and some of the British and Dutch colonies really learned to love a sash window.

To begin with, sashes appeared in new buildings of the grander sort, replacing the popular seventeenth-century cruciform with its rectangular leaded panes (Fig 121). The cruciform itself was an elongated version of the universal mullion and transom design, seen everywhere in England and on the Continent for hundreds of years.

The Renaissance window opening is

Fig 121 Window openings adopted the upright format dictated by Renaissance architecture in the second half of the seventeenth century. The cruciform pattern, with a central mullion and transom, was the transitional method of glazing before sash windows became almost universal in better houses. It could be of wood, stone, or even brick, and had small leaded panes and an opening casement.

taller than it is wide and is proportioned by such classical rules of thumb as the *diagonal of the square* and the *double square.* These theories can involve endless mathematical niceties which have more scholarly appeal than practical use for the average house repairer. Let us merely sketch the means of proportioning an opening or a window pane by the square of the diagonal (Fig 122) so that you can

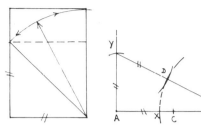

Fig 122 Window proportions. (*left*) To obtain satisfactory classical proportions for window panes and window openings, the diagonal-of-the-square may sometimes prove helpful. First draw a square of a width which allows, say, three or four panes across the sash. Then swing the diagonal in an arc to meet the extended vertical side of the square. This will give the height for the pane. (*right*) To divide a line AB according to the classical rule of the Golden Section by geometric means so that AX:XB = XB:AB (the shorter division is to the longer, as the longer is to the whole line): 1 Divide AB in half at C; 2 draw AY at right angles and equal to AC; 3 Join YB; 4 Mark YD equal to AY; 5 With radius BD, strike off X on AB. Point X will then divide AB to make the Golden Section.

use it if the need should arise. It is useful in its own way, like the Golden Section (Fig 122), to indicate some educated proportions when we are a bit nervous about what we are doing.

Architects, who are notoriously enamoured of theories which lend themselves to exercises in deft draughtsmanship, love talking about Georgian or classical proportions; but you should be warned about such neat systems. As far as vernacular buildings are concerned they are something of a chimaera.

While gentlemen who had done the Grand Tour to Italy, a few architects and other members of the very limited cognoscenti, made it their business to analyse classical proportions, most citizens had little understanding of these notions. What they knew was that the fashionable hole in the wall was higher than it was wide, that a double-hung sash provided good ventilation, looked pretty and that the Squire put them in two years ago. Builders, if they were literate and enterprising, might turn to the pages of Batty Langley's pattern books in the 1730s, but most had a stab at knocking up a classical façade or doctored an earlier one into something of more or less classical appearance.

For this reason you will find that an earnest pursuit of 'Georgian proportions',

with a textbook on how many diameters to look for in a Doric order in your hand, will prove confusing. You rapidly discover that eighteenth-century and early nineteenth-century builders broke almost as many rules as they obeyed. As in life generally, most people were happy to settle for a comfortable and practical set of standards, varying them as the occasion arose.

Lest I be accused of ignorance or over-simplification I add this rider. There are very lucidly proclaimed rules of classical architecture laid down in many a learned treatise. They provide the essential grammar for those who wish to work in that most pleasing idiom, whether they are repairing or designing something new. However, the pattern books and treatises are open to interpretation and are not in total accord when it comes to detail.

The main sources of architectural inspiration were the writings of the Roman architect Vitruvius and a rather limited number of Italian and Greek antiquities. From these few examples grew the entire 'classical architecture' industry which has given us the wonderful buildings which we now enjoy.

Because the Renaissance abounded in architects who were also artists, they were not content merely to produce endless copies of the antique orders of the Colosseum. They adapted and invented within the language of Roman design. The Romans for their part had made a similar use of Greek forms. The late-seventeenth-century English gentleman, with his Dutch box of a house, and Hobemma avenue planted last week, understood a sash window as follows.

It consisted of a box frame containing two sashes, one of which could be opened by sliding it in a channel composed of the linings, a sash stile and a central parting bead. If the window was of advanced construction, the moving sash was attached to cords running over wooden pulley-wheels in the stiles. On the ends of the cords were lead weights which balanced the weight of the sash itself.

If the window was less advanced in design, the moving sash might be kept up by pegs or metal pins pushed into holes in the box frame. Sashes were divided up into panes with glazing bars, rebated to hold the glass. Early bars were broad and flat in section, sometimes made in one piece and sometimes composed of a T section for the glass, with a bull-nosed or squat-looking ovolo or ogee fixed to it (Fig 123).

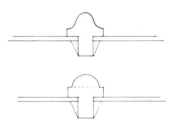

Fig 123 Two typical late seventeenth- or early eighteenth-century glazing-bar sections: broad, stubby and robust. Sometimes the curved moulding was planted on top of a plain T-shaped bar (see broken line).

The panes in the window might be anything between twelve and forty or more in number. Like most of his Georgian descendants the builder was obliged to use many relatively small panes because glass, which was a valuable commodity, had to be cut from a blown disc of perhaps only 4ft in diameter. Some glass was blown in the form of a muff-shaped cylinder which would be slit and opened out – a more expensive method. These glass panes cut from a circular disc, some of which bear the bull's-eye mark where they were broken off the metal pontil rod, are called 'crown glass'. The other sort is known as 'cylinder' or 'muff' glass. Both types display subtleties of texture and colour and pleasing imperfections which make them a joy both to look at and to look through. They catch the light in a curious and diverting way, and have that kind of variety which makes modern flat drawn or plate glass such a second-rate substitute. Indeed, this is the rub with any project which involves the fitting of lead-glazed windows in the early manner. Modern glass can look bland and rather unconvincing.

Whenever you are able, save old glass in windows, either by leaving the panes *in situ* or by reusing them. It is a counsel of perfection, since one inevitably breaks

panes in the course of removing them and, if they must then be cut to a new size, there is even further wastage. Never, though, in any type of window, employ those abominable fake panes which ape the traditional bull's eye. This practice is growing and is responsible for ruining many a reasonable new double-hung sash. More will be said later about the vile parodies of period windows.

As the eighteenth century continued, the thick flat glazing bars of the William-and-Mary and Queen Anne periods became steadily thinner and the moulding of the inside section more refined. It is extremely important to study glazing bars and use the right sort for the period in question.

The ovolo glazing bar remained the most popular form for both sashes and casements until the second half of the eighteenth century. Then bars became distinctly thinner and there was an increasing use of ogee and lamb's-tongue sections. Lamb's-tongue, ogee, and Gothic moulding were all widely used in the early nineteenth century (Fig 124). The possibility of your wishing to use cast-iron fixed-light, casements or hopper-head windows is fairly remote, unless you are repairing something like a chapel, a warehouse or a lodge.

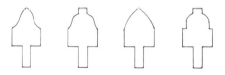

Fig 124 (*left to right*): lamb's tongue, ogee, Gothic, ovolo – the four most widely used glazing-bar sections in old buildings. These drawings are only a guide to their general outline, and should be proportioned and adapted according to circumstances.

By the second half of the eighteenth century most double-hung sashes were hung on cords and weights, both at top and bottom. We shall talk presently about their repair.

There are many architects, planners and conservationists who believe that the beautiful and traditional sash window with twelve, sixteen or more panes disappeared as soon as methods of making plate and sheet glass were developed. If the building is later than about 1840 they are ready to accept the insipid four-pane sash or its blank-looking offspring with two panes as the 'truthful' window. They should know better. Unless you are working within the disciplines of a group or terrace where clearly all the windows originally had four panes, why choose such an ugly design?

Windows with twelve and sixteen panes were made in their thousands during the third quarter of the nineteenth century and even later. The twelve-pane sash window has never really ceased to be one of the ordinary, practical options for new buildings. Why should we forgo this most sound and pleasing form just because other sorts of window have been developed since? A properly glazed sash window has the most infinitely variable capacity for ventilation. It lifts the architectural quality of even quite modest and unimaginative buildings. In almost any building containing upright window openings, built in anything approaching the classical manner, it is totally suitable.

Having said that, there are one or two vital factors to bear in mind. First of these is the proportion of the panes in relation to the shape of the window opening. The rather square-looking sixteen-pane sash was very common in the late eighteenth and early nineteenth centuries in vernacular buildings such as cottages and farmhouses, while grander buildings like Greek Revival manors and rectories would often have large upright panes – maybe twelve in number. The latter are of course correct, but rather cold and anaemic.

It is my experience that the sixteen-pane sash, in which the panes are only slightly higher than they are wide, is a wonderfully restful type of window. It has just that necessary degree of elegance, while also being comforting and commonsensical. It is a window which cries nutmeg and clean linen, pigeon pie, and Schubert serenades badly sung by candlelight.

One thing to take account of when specifying sash windows for old houses is the thickness of the sash members from front to back. It is in this respect that modern sashes, made to standard detailing by joineries, albeit to special order, differ from most older windows. The shallower section of old glazing bars relates to the equally shallow section of the framing stiles and rails.

Modern sash framing tends to be a uniform 45mm from front to back, while the Regency ones, and earlier, were often as little as 35mm. You might suppose that the odd 10mm makes little difference, but in fact it can have a pronounced effect upon the appearance of the sash-bar mouldings. Modern 45mm windows often look rather coarse, even when the type of moulding is much the same.

Another point to consider is the leading edge of the glazing bar as seen from inside. Modern glazing bars almost always finish with a neat, sharp and rather pronounced fillet. This can be unpleasing at times, although acceptable at others. If you are putting sashes in an 1850 farmhouse, although such bars may not be strictly correct, you might not worry too much. An 1820 house could be another thing altogether.

Now for sash horns. Those are the projecting ends of the sash stiles which you see protruding above the inside meeting rail and below the outside one (Fig 125). Sometimes they are moulded, sometimes bevelled, or bevelled and moulded. Always they are wrong in windows which are meant to represent the early Victorian period or before. So, for

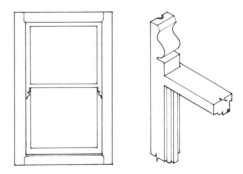

Fig 125 Sash window with horns, and (on right) detail of horn. From about 1860 onwards, sash windows usually had horns. Larger panes of glass were used without the important intermediate support of glazing bars to relieve weight on the joints between the meeting rail and stiles of the upper sash. Horns allowed for a stronger type of joint (see Figs 126 & 127). The type of window shown is often Edwardian.

windows preceding roughly 1850–60, leave them out.

The reason for employing sash horns is the type of joint between the meeting rail and the stile. By cutting a mortice-and-tenon you can procure a strong joint, better able, some would say, to withstand the heavy loading of a big single pane of glass (Fig 126). The older kind of sash had many glazing bars to help support small panes, thus exerting only moderate stresses at these joints, so allowing a simple uncluttered line for the meeting rails, which were secured with dovetails (Fig 126) or haunched tenons.

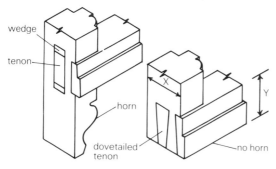

Fig 126 (*left*) The strong mortice-and-tenon used to join the meeting rail of a sash window to the stile requires a horn to resist downward movement of the tenon. Horns might be moulded, like this one, with an ogee and bevel, or be simply splayed. Sash windows before about 1860 did not have horns: the right-hand drawing shows that downward movement of the meeting rail could be resisted by forming an *open dovetailed mortice-and-tenon joint*. When copying old windows, measurements X and Y can be vital to the visual effect inside the room.

When copying old windows, take templates of the mouldings. This can be done with a steel multi-toothed adjustable template – easy to use but a trifle inaccurate. Trace the profile obtained with this instrument into your notebook immediately. Be careful not to damage vulnerable timber surfaces when you press the teeth against them. The other way of taking templates is by pressing a pad of some fast-setting material like plaster of Paris against the moulding. A simple fence mould can also be used. Other substances such as wax can be employed.

As for window furniture, the most acceptable window-catch is a brass spring fastener, or perhaps a brass screw-up fastener. Some old windows of late-eighteenth-century origin or a bit later have little inset brass spring catches which are used to wedge the sash against the framing and stop it rattling.

When you have windows made for you by a joinery (as you will) make sure that you provide them with same-size drawings of all the mouldings. Also be certain to specify the number and shape of the window-panes. The joinery may develop the exact sizes for the different members, and for the actual panes, in the workshop; but be sure they know the kind of shape you require.

It is of the first importance never to procure a window-pane which is even slightly wider than it is high. It will look terrible. Even panes which are exactly square tend to *look* as though the height measurement is less than the width. The number of panes will be governed by the shape of the window opening, to a large extent, but you can increase the width of the box linings a bit to reduce the width of the panes, especially if you can hide part of the box behind a brick or stone reveal.

If you have a very small window opening, in a cottage or similar building, consider using a double-hung sash with spring balances instead of cords and weights. This type is framed with a solid pulley-stile to which the balances are fixed. It displays only about 32mm on face instead of about 70mm. This saves roughly the thickness of two glazing bars which could be a useful economy in design terms.

There are several maverick patterns of glazing which date mainly from the early nineteenth century and might be described as transitional. There are sashes which have marginal bars (Fig 127) and some which incorporate curved work forming little Gothic arches to the upper panes; others have interlaced glazing bars in the spirit of Early English stone tracery (Fig 128). All these can be delectable in their own way and there will be occasions when they are an appropriate choice.

There are times also to ask yourself whether you should use coloured glass in some types of Victorian window. Those with marginal bars not infrequently sport

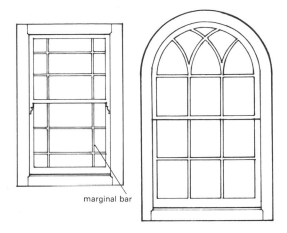

marginal bar

Fig 127 (*left*) Windows with marginal glazing bars and panes became fashionable after about 1840. This one has horns, but normally they did not.

Fig 128 (*right*) During the second half of the eighteenth century and the first half of the nineteenth century, the glazing of sash windows often reflected the Strawberry Hill or Regency Gothic fashions of architecture. This one has a semicircular head and a kind of Early English intersecting tracery.

slips the colour of wine gums, or that slightly sinister blue which one associates with medicine bottles – containing laudanum, perhaps. Corner motifs include a jaunty, frosted star. Windows of this kind were often placed at half-landings on stairs, or in glass porches and conservatories. They dispense weird shafts of light which tend to turn harmless householders into alarming creatures of the supernatural. Some people grow to love this kind of thing; but do not gratuitously introduce Victorian coloured glass into earlier houses. If such buildings already have coloured glass in conspicuous places, think seriously of removing it.

Casement Windows

The other main type of window found in old houses is the casement. This dates back much earlier in architectural history than the sash. In England it may be thought of as the first type of glazed window which could be opened. A casement is side-hung and may be opened inwards or outwards, according to the way in which the frame is rebated. In Britain, casements almost always open outwards;

in European countries usually inwards, which ensures that a façade with all its windows wide open looks as though the openings are not glazed at all. This is rather pleasing and reminds one of the fine architectural drawings by the pattern-book authors, who usually showed window openings as black voids without glazing. This device gave more emphasis to the architectural quality of the drawing, but could also look slightly dull.

As to practical comparisons between English and European casements, the first which comes to mind is that the latter are ill-adapted to resist heavy rain. Anyone who has sat in the kitchen of a French farmhouse during a summer thunderstorm knows that because the casements open inwards, water courses down the panes, on to the sill, and drips furiously back into the room. Certainly the sill can be rebated to fit the window, and a throated weather-strip applied to the bottom of the casement, but the water still tends to get in.

The outside shutters, if closed in time, may help this problem but never cure it; they are loose-fitting and stand well clear of the wall face. In Provence or Florence, why worry? The heat evaporates any moisture quickly enough. But if I were buying a building in northern Europe I should suspect that the place to look for rot might be in the joists and floorboards under the windows.

Back to England in medieval times, when the majority of houses had little or no glass in their windows. In the fourteenth century, for example, glass was very rare in domestic buildings. In the early fifteenth century there was more; but window openings were usually closed with shutters although they might also be fitted with insanitary pieces of parchment, draughty wooden or reed lattices or panels of woven withies. When glass was used it would sometimes be placed only in one or two of the lights of a mullioned window. A frequent pattern of glazing, to be seen especially in European Renaissance paintings, is that of the four-light cruciform. The top two lights, which could not be opened, might have tiny leaded diamond panes, while the two

147

major lights on each side of the mullion merely had shutters, and perhaps pieces of trellis to take the sting out of the rain. However, the glazing arrangements varied.

In grander late fifteenth-century English houses it became quite usual to have windows with glass, although they could be deemed a removable fitting which the owner would take away with him when and where he chose. Window openings, when not in churches or great halls, were normally horizontal in emphasis and divided into narrow lights with stone or timber mullions. Very many vernacular buildings still had shuttered windows with diagonally set mullions in a wooden frame, and no glass.

Where glazing did exist, only one out of a range of perhaps three or four mullioned lights might be capable of being opened. The fixed lights were composed of H-section lead cames housing a number of diamond, rectangular or polygonal panes. The cames, which were soldered at intersections, were surrounded by binder leads forming the outer margin of each light.

The leaded light would be fixed back against a rebate in the wooden or stone window frame, being bedded in glazing compound and wedged firmly in place. To stiffen the rather vulnerable window against wind pressure, horizontal or vertical iron or timber bars were placed at strategic intervals. To these the cames were tied with fine lead tapes (Fig 129).

The lights which *could* be opened were made in the same way, but were fixed to wrought-iron or wooden frames. These were side-hung with socket hinges which fitted on to gudgeon pins attached to the window jamb or mullion. This opening light was, and is, the casement. Early windows very often had quadrant-shaped iron casement stays or long hooks to hold the windows open. An iron upstand, called a 'pull', was used to draw the window closed. The casement fastener might be a form of pin-swivelled turn-buckle or one of the often quite elaborate spring catches attached to an ornamental backplate.

A mid-to-late-seventeenth-century three-light lead-glazed window. The mullions are very simple and are rebated so that the glazing lies flush with the timber. Note that the opening casement is in the middle and is hinged on gudgeon pins from a mullion rather than from the stile of the frame. This window has modern secondary glazing inside, which badly affects its appearance.

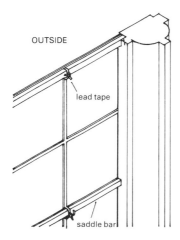

OUTSIDE

lead tape

saddle bar

Fig 129 To stiffen and retain leaded lights, they were tied back with lead tapes to iron *saddle bars*, socketed into the timber or stone jambs and mullions. This window has ovolo-moulded framing, typical of the second half of the seventeenth century.

It is of interest that wooden-mullioned windows, with glazing rebates flush with the outside, often have the opening casements hinged from the mullions themselves, rather than from the jambs of the frame. The latter is quite frequently a modern practice. The advantage of hanging from the mullion in a range of lights was that the opening ones could fold right back against the fixed lights, and

148

so were not obstructed by the wall reveals.

The diamond pane pattern tended to predominate right through until the second half of the sixteenth century, when some people began to fit windows with rectangular leaded lights. These appear to have corresponded with the transition from Gothic to Renaissance architecture and were adopted first in grander buildings. If you come across any kind of leaded light in an old building, treat it with the greatest respect. Early leaded panes are beautiful, interesting and important to a building's history. They should be repaired with care and left in place, if humanly possible. Sometimes you may be able to open up a blocked window containing a leaded light, or a plain, unglazed opening with mullions.

Older leaded windows differ in certain points from those of more recent periods. For one thing, the glass is extremely thin and obviously of crown or muff construction. Early glass also varied considerably in thickness from one part of the pane to another. Muff glass often displays a slight ripple in its surface, resulting from the process of flattening out the cylinder after it had been slit open. The leaded cames or calms which hold the panes in place will be thin, and narrow on face – delicate compared with the rather hard, obtrusive and monotonous leadwork to be found in Tudoresque houses of this century.

Cames are of H-section and now, as in times past, the panes are clasped by the lead while also being secured and waterproofed with some form of 'cement' or other bedding. An early practice was to bed the edges of the glass in tallow. One form of 'cement' used today consists of a mixture of whiting and linseed oil, with a very small quantity of yellow lead oxide to help it harden. Some turpentine will assist drying, and a little black pigment should be added to blend its colour with the lead.

The first lead panes, which were described as 'quarries', were normally of muff glass. It is suggested that it was because a diamond shape allowed for very economic use of what were small and often incomplete pieces of glass that the cames were criss-crossed in a lattice, being soldered together at intersections. Other people allege that the lattice arrangement of panes was a throwback to the reed or lath screens fitted as windbreaks, mentioned above. The lattice probably owes something to both these theories.

Very early cames were normally cast, and some people claim, with good reason, that reeds were laid in a shallow box which was then filled with molten lead. By cutting down the centres of the reeds you then obtained strips of H-shaped section when the remains of the reeds had been pulled free (Fig 130). By the seventeenth century, if not earlier, cames were being made by casting strips of lead which were then drawn through appropriately shaped dies.

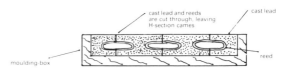

Fig 130 It is claimed that very early lead glazing cames were cast in a shallow box by placing reeds in it and pouring on the molten lead. Each reed was then cut down the middle with a sharp knife, thus procuring an H-shaped came from each pair.

Not all casements were in iron frames. Some were of hardwood and you will occasionally see these in late seventeenth-century houses. Many houses which had sash windows in most openings still retained casements in attic dormers, often with diamond panes.

Although wooden glazing bars began to appear in casement windows at about the same time that the sash window was being introduced, lead glazing continued much further into the eighteenth century than

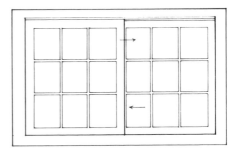

Fig 131 The horizontal-sliding sash window is ideally adapted to glazing of low, wide openings. For that reason it was frequently used when replacing small mullioned windows in the eighteenth century. Often, only one light slides open.

Molten lead is poured into an iron mould which casts three solid strips at a time. This mould is hinged at the base and is in two parts. When the handle is depressed, the back half falls away so the strips can be removed.

The 'riser' which joins the cast lead strips together is chopped off.

The lengths of lead are made into 'H' section cames of the desired size by passing them through a mill which has interchangeable mill wheels and jaws. The unfinished lead is out of sight on the other side of the machine and the operative picks up the finished came as it is fed through. It is then stretched in a vice to straighten and stiffen it.

A full-size drawing of the glazing is placed on the bench and the glass is cut to the required shapes – in this case, fairly elaborate ones. The cames are then moulded around the glass, the work being held in position with flat-sided blacksmith's nails.

When the window has been assembled, with all its glass panes in place, the junctions are soldered on both sides. The craftsman is seen holding the soldering iron in one hand and a thin rod of lead solder in the other.

Lastly, 'cement' is firmly brushed and worked into the cames to secure the glass and waterproof the window. A traditional mixture consisted of whiting, linseed oil and yellow lead oxide (litharge), thinned with turpentine and coloured with lamp black. The cement is left to set before the surplus is cleaned off and the glass is given its final polish.

The series of six pictures above show the lead glazing processes as carried out at The Cotswold Casement Company's workshops (pictures by courtesy of Higgs and Hill Building Ltd).

many people suppose. Glazing bars followed the pattern of development that has been mentioned already. On the whole, if you find a wooden casement with a fairly thin glazing bar, you can expect it to be late eighteenth or more likely nineteenth century.

In the early eighteenth century a form of the sliding sash which suited the horizontal shape of converted mullioned window openings was introduced. Some claim it for Yorkshire and others allege it to be of Cornish origin. This consisted of a double-hung sash turned on its side, without the pulleys and weights (Fig 131). It became a favourite for cottages from that time on. It is a most attractive, authentic and useful type of window which has a place in many old buildings, as an alternative to casements.

From the late seventeenth century onwards, casement windows were set in flat frames with mullions (and sometimes transoms) presenting no chamfers or other conspicuous mouldings on the outside. Inside, it was quite common for

mullions to be ovolo-moulded (Fig 132).

In earlier times, mullions were not part of a framed piece of joinery but were structural units supporting the lintel, which often had to span a considerable distance in relation to its strength. This was especially so of mullioned windows cut in freestones like limestone or sandstone. Lintels often bridged one or two narrow lights and met over a mullion.

When replacing or repairing old mullioned windows, whether they are of timber or stone, remember that the type of moulding or chamfering used is an important detail of the design. Many late Tudor and Jacobean mullions were hollow-chamfered (Fig 133) and some were plain-chamfered or splayed. Local practices varied, so you should follow the precedents set in your own neighbourhood.

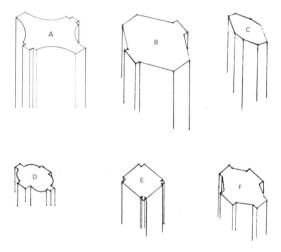

Fig 133 There are numerous types of mullions in both timber and stone which may be found in old buildings. (A) is hollow chamfered and has a glazing rebate, which may be on the outside only, or on the inside as well. (B) is splayed, and rebated for lead glazing. (C) is plain chamfered, without a rebate and is not intended for glazing. (D) is a pattern seen from the late sixteenth century, but particularly in the second half of the seventeenth century. It is ovolo moulded and rebated for glass. Another typical late seventeenth-century and early eighteenth-century section is (E). The glazing rebates are fairly minimal and bring the glass almost flush with the outside face of the mullion. Inside, pinched looking flush beads decorate the edges. Too much reliance should not be placed on the designs of (A) (B) and (C) when dating a window; while (D) and (E) are more consistent. (F) is a reserved chamfer mullion for glazing, sometimes used in the late sixteenth century.

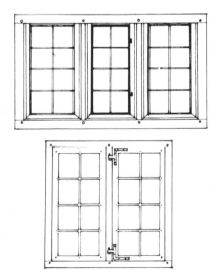

Fig 132 Oak-framed windows, divided into some fixed and some opening lights by vertical mullions, were the norm during the late medieval, Tudor and seventeenth-century periods. Both these are seen from inside. They were peg-jointed and contained lead-glazed iron- or timber-flamed casements. Variations of the timber-framed mullion window with one fixed and one opening light have never become obsolete. Early ones had timber or metal casements with lead glazing as in lower drawing. Wooden glazing-bars became common in the eighteenth century.

All this naturally applies to your treatment of other mouldings connected with windows, whether in the form of an external architrave around an early eighteenth-century sash, a drip mould or hood mould over a range of seventeenth-century mullions, or the cusps and foils of medieval Gothic work.

Repairs to stone are matters for an expert stonemason, unless you are thinking of carrying out some minor plastic repairs. Plastic repair means that you are using Roman cement or some other composition to match the rest of a mullion, transom or window jamb.

Clearly, it is quite a major operation to remove excessively worn or badly cracked stone mullions. But this may be what is required. You then need to have replacement lengths matched and built back into place. Avoid any tendency towards over-restoration; do not remove every part of a stone window that is rather worn and make all new. Just replace the bits which cannot be considered safe. This applies equally to the repair of stonework in other situations – around doors, or crenellations, string courses, cornices, copings and balustrades.

Take care that your mason – who should be a really expert stonemason, not a concrete-block-layer – understands the finish that is required. The 'labours' as they are called in dressing stone, quite apart from cutting mouldings, will have much to do with successful replacement. Stonework may be 'boasted', 'tooled', 'punched', 'picked', or treated in many other ways. Make certain that you are matching the work properly. Similarly, if you are dealing with brickwork round a window, the size and type must be matched, not to mention any rubbed detail. Highly skilled repairs must be left to experienced craftsmen, and all that I can do is alert you to the choices available, the principles involved and the overall approach.

At all costs avoid extensive repair work to stone using synthetic materials. It may be necessary to carry out small pieces of repair to carved detail and so on by making up a stiff mix of carefully matched stone-dust composition and keying it to the sound stone. But be restrained with this kind of thing, or you will produce large areas of false stone which crack away, look strange in texture and do not weather in a convincing manner. Plastic repair work can be very skilfully done by specialist firms, and I have seen men simulate Bath stone ashlar in what is really a most sophisticated plaster. It can be astonishingly realistic – or it can prove a vulgar travesty.

The principle for repair is to use a graded mix of stone dust and fine aggregate in a matrix of white cement or lime. The old stone is carefully hacked to provide a key, and holes are drilled to receive protruding non-ferrous metal dowels. These form armatures that reinforce and retain the repair mix, which is built up and modelled to continue whatever moulding or carved detail has to be replaced.

Where you can, you will prefer to introduce new stones of the correct type, if possible from the original quarry, and expertly carved with the appropriate detail; though you will probably only go to these lengths if repairing a building of considerable architectural merit. Most readers will be far more likely to need the services of a good joinery to reproduce their windows, or perhaps a skilled man who can hand-run the mouldings of oak frames, and carve simple stops and details.

Choosing Casements

Having outlined the history of the casement window, the vital question to consider is which sort to use in which building. We can start by ruling out all the bastardised patterns of casement window sold by the stock joinery firms or misguidedly ordered from specialist joineries. Dismiss any casement window containing a top opening light or lights, like those in Figs 134, 135, 136, 137. It does not matter that they have small panes; they are wrong for virtually every building earlier than about 1900, whatever it is like. (It is just possible that there may be some types of late nineteenth-century houses which had them, and if you are doing an informed replacement, well and good.)

Next come the numerous variations of

Fig 134 A typical modern casement-window arrangement with small panes and additional top-opening lights. It is *totally unsuitable* for any kind of old building; so are all similar ones.

Fig 137 Neither of these modern small-paned windows should be considered as an alternative to the double-hung sash. Despite their glazing bars, they look utterly unconvincing.

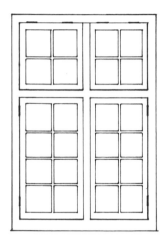

Fig 135 A bastardised version of Fig 121, this cruciform mullion and transom window, with its top-opening lights, should never be used in old buildings.

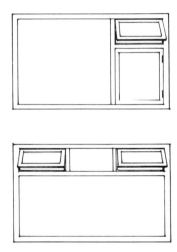

Fig 138 The numerous layouts of picture window should all be rejected when converting or restoring old buildings. The view *from* the house must not be allowed to ruin the view *of* it.

Fig 136 Both timber, aluminium and uPVC mullioned windows with leaded lights are manufactured as stock items. Some have fake lead cames applied as strip decoration. Such fakes *must be avoided*.

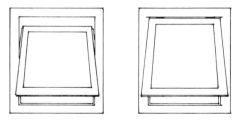

Fig 139 Many modern architects like these plain windows, which may be top, bottom or centre pivoted. They can have stained timber or aluminium frames and are considered 'honest'. Though much favoured for barn and similar conversions, they are rarely successful.

153

the picture window with blank fixed lights and opening casements, as in Fig 138. Never use them. Then there are the top-hung, centre-pivoted and side-hung windows (Fig 139) beloved of the architectural profession for use in 'honest' extensions to old buildings, barn conversions and the like. They rarely do justice to even the most 'rough' upland cottage and should be avoided like the plague in most instances. In one or two types of building there is a valid argument for their limited use. As a general rule do not be seduced into using them because they are often slightly cheaper than correct types of window; and because they allow you and your professional adviser to avoid tackling the awkward niceties of selecting the right glazing.

Do not use any uPVC or aluminium casement or any of the iron casements so popular in pre-war housing developments. Very occasionally you may come across an industrial building or, say, a Victorian lodge which has cast-iron windows with small panes; but those are exceptional circumstances which will dictate their own standards.

There is one kind of tilting light which you may need to use, and that is the twelve-pane hopper-light (Fig 140), quite often incorporated in nineteenth-century barns and stables. It is uncomfortably like the objectionable twelve-pane stock window with a top-hung opening light; but the

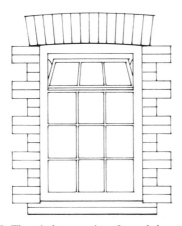

Fig 140 The timber- or iron-framed hopper-head window is often seen in nineteenth-century stables and other outbuildings. Though not particularly pleasing, it may occasionally be the 'correct' choice.

precedents for its use are quite numerous. Personally, I should choose another pattern if I could think of one which looked the part.

Now we come to the exceedingly awkward matter of lead-paned casements. From what has already been said, you can see that for many earlier buildings they are the correct form of fenestration. However, there are difficulties connected with their use.

Every reader will have seen the kind of neo-Tudor residence which abounds in nicely made leaded lights. Some of these houses cost a lot of money to build and their windows are broadly speaking of a correct pattern. Yet, despite all that, they are horribly unconvincing. It is that coy and tidy air of Tudor charm which we must avoid assiduously.

There are several ways of doing that. Ensure that the lead cames are of the right width; see that the timber frames or iron casements are correctly detailed; try to overcome the flat, bland uniformity of modern sheet glass in all those rectangular or diamond panes. Having successfully done these things, we shall still be accused of 'dishonesty', 'faking' and other horrid sins: there is a lexicon of hand-me-down phrases which you learn to recognise. They are mainly employed by those keen conservationists who have not studied architectural history quite hard enough and still have one uneasy foot in the Bauhaus. All the same, when it comes to leaded lights, you are offering these people your head on a charger unless you are extremely careful. How can you get the right sort of glass to avoid the bland look? The only way is to try to find enough pieces of old crown glass for your glazier to cut them to size. But they will never be as thin or as interesting as the old ones. Expensive reproduction 'old' glass is available, but usually looks rather fake.

Literally thousands of old cottages and smaller vernacular buildings from the sixteenth, seventeenth and early eighteenth centuries would originally have had leaded windows but now parade a motley collection of sashes and Victorian casements. The casement most frequently seen of all is that with a single (or some-

Fig 141 (*top left*) Victorianised late eighteenth-century timber casement; *(top right)* Victorianised sixteenth-century stone-mullioned window; *(bottom)* Victorian-Edwardian casement. Casement windows with horizontal glazing bars only were introduced in the Victorian period as a standard replacement for original lead glazing in earlier buildings. When used in narrow lights of stone-mullioned windows they were made of iron; but they were normally wooden when substituted for eighteenth-century casements. They are seen particularly in houses with windows of horizontal emphasis, divided into separate lights by mullions. They should generally be replaced if incorrect for the date of the building.

times two) horizontal bars and a plain dividing timber mullion (Fig 141). Some patterns have no central mullion and echo a French influence, with the casement stiles rebated, closing upon each other. The odd thing is that these Victorian casements appear to be almost universally accepted as a satisfactory window pattern for sixteenth- and seventeenth-century buildings upon which care and informed works of restoration have been lavished. The answer no doubt lies in the fact that they are inoffensive, and of long-standing derivation – people have always been accustomed to seeing them on every hand.

Be bold, and try using leaded lights if they are correct for the building. They cost money, but so do new cars, meals in restaurants, holidays abroad, fashionable clothes and all the transistorised pieces of electrical equipment with which we surround ourselves.

Here are the other sorts of casement window which you might need. There are the straightforward pairs, or ranges of six, eight-, or nine-paned wooden casements and fixed lights, so often used in late eighteenth- and early nineteenth-century vernacular buildings. These are sometimes the correct choice for warehouses and mills – especially those built of brick. A judicious mixture of casements and double-hung sashes is often acceptable, provided that you do not muddle them up on a façade which needs its fenestration to be all of a piece.

If you have a late seventeenth-century house, the evidence may call for cruciform windows, with leaded rectangular panes, of the type already discussed. They may be timber-framed, or consist of fixed lights and iron opening casements, set in stone jambs, with a mullion and transom.

A range of small, stone, or possibly brick, mullioned windows may need lead glazing. You may also have to lead-glaze the idiosyncratic windows of toll houses, lodges and other buildings of the kind which abound in the pages of Loudon's *Encyclopaedia of Farm, Cottage and Villa Architecture* (1842).

Now for the mullioned windows which may not require cames or glazing bars, or for which they are quite definitely inap-

An excellent example of transitional Renaissance windows – probably dating from the second half of the seventeenth century. The overall shape of the openings is horizontal and the medieval form of timber frame with mullions and transoms incorporates newly fashionable Palladian detailing. In effect, this is a version of the Palladian motif with a central arch and narrower side openings. Lead glazing has been used instead of wooden glazing bars and the iron casements are hung on pins spiked into the framing. The deeply overhanging moulded timber eaves of the roof are another typical seventeenth-century feature. It is clear that the building is timber framed since the windows lie flush with the face of the wall, which has been plastered and scribed to simulate stone ashlar.

Here is the kind of double-hung sash window which provides an answer to glazing horizontally shaped openings which may formerly have contained mullion windows. It has narrow fixed lights to either side of the central pair of vertical sashes. The latter may have cords and weights in the normal positions, but sometimes the cords are led across the tops of the smaller lights to boxes adjoining the window jambs.

propriate. Obviously there are very narrow mullioned windows or single, slit-like windows in early houses, cottages and barns, which call for either a plain fixed sheet of glass or a tiny casement with a minimal iron frame. The intention with these is to give the impression that they have no glass at all.

A point to remember with this empty-space look in mind is that the further back in the thickness of the wall you put the glass, the less it will reflect the light and give itself away. But there can be a balancing disadvantage to this practice, in the loss of depth to the window embrasure inside the room. Compromises will have to be made, since shallow window splays are lacking in tranquillity for rooms with thick walls. Timber-framed houses leave you no options in this matter, but I have always thought it a fault that their walls so apparently lack depth; they are conspicuously difficult to curtain for this reason, having a way of making modern curtains look intrusive and unlikely, even when made from pretty and expensive textiles. Curtains in the seventeenth century would have been of something coarse, like that linsey-woolsey union of linen and wool. The other great favourite was the linen-twill plain curtain embroidered in naïve crewelwork.

I make no apologies for mentioning curtains in the middle of a chapter on windows, because the whole secret with old houses is to see them as places in which to live. This means thinking in far broader terms than merely those of architectural history and technical repair. If you cannot think in that way, you risk killing your cottage with an excess of heartless expertise.

In the interests of easy maintenance, it has become customary for people to buy hardwood windows and doors which are left unpainted. This hellish practice is, in most cases, an act of desecration. To waste money on specially made joinery, often correctly detailed, and then wilfully make it look as raw and modern as possible is past belief. With the exception of timber-framed mullion windows with lead glazing all should be painted. Only early patterns of door should be unpainted. In some buildings it is sometimes permissible or even desirable to paint sash-boxes or frames in a different colour from the casements or the sashes themselves. Sash windows were not infrequently painted in colours like green during the eighteenth century – glazing bars and all – especially in North America. It is worth considering, but be very careful and seek advice before taking this controversial step.

Nine-pane casements may be used to advantage for asymmetrically placed windows in gable walls, on staircases, for lavatories, bathrooms and similar *ad hoc* openings in later buildings.

Pairs of twelve-, sixteen- or eighteen-pane casements can be authentic and often pleasing in cottages, mills and warehouses. Paired casements and panes, while authentic and pleasing in many instances do involve an element of unresolved duality.

Now is the time to discuss this vital point. What the theory says, and it is more a theory than a law, is this. If you juxtapose two matching architectural spaces, for example window-panes divided by a thin glazing bar, you create an unresolved duality which is disturbing to the eye. Nature appears to cry out for various kinds of holy and not-so-holy trinity – as much with windows as shouted invective. How much more resonant to declare to a tasteless entrepreneur 'You are a barbarian, a skunk and a mindless poltroon' than merely indulge in the feeble duality of 'You are a barbarian and a skunk'. The latter cries out for resolution.

But, seriously, this matter of duality is a fundamental principle which is all too often overlooked in modern architecture and, of course, in works of restoration or conversion. Unresolved pairs of spaces plague us on every side. Fig 142 shows the gable end of an extension to an old building. It speaks for itself. Unresolved dualities always invite the question 'Where is the other pane of glass?', or 'Where is the clock between the two vases on the mantelshelf?' (Fig 143). Anybody who cannot see the point of emphasising this most commonplace of architectural dilemmas and does not think it has much to do with restoring old buildings, fails to catch the spirit of the enterprise. For him or her, any good textbook on building construction (and there are many) would do. We all continually make mistakes with old buildings, but we must be alive to the implications – ready to sift ideas, right errors and discuss options.

If the notion of unresolved duality is new to you, I have put in your hands a firecracker which will continue to explode at the most inconvenient times and in the

Fig 143 The central glazing bar of the window at top left forms an unresolved duality between the vertical rows of panes on either side. It may still be the correct choice. The nine-pane casement on the right resolves the duality by having a centre row of panes. The vases on the chimney shelf again provide a duality, resolved by placing a clock in the traditional 'garniture de cheminée' arrangement.

least likely places. France, with its undoubted architectural glories, is the first which comes to mind. For there you will see, on every hand, examples of unresolved duality in their glazing. The French casement and the French door both expound duality. You will see unresolved dualities in English buildings in the form of windows divided by a central mullion. Some dualities are hard to bear, others can occasionally be endearing. Do not allow the principle to ruin your life, but I suggest that a knowledge of its dangers should be added to your armoury.

We have already disposed of the false bull's-eye pane of glass or bullion-glass pane, as it is also called. Now think about the use of obscured glass in doors and windows. In the normal way it is objectionable and usually unnecessary; rather the despised net-curtain, from behind which bourgeois busybodies allegedly squint at their blameless neighbours. Obscured glass is death to the appeal of an old window or fanlight. It should never be fitted in the front door of an old building. There are measures effective for privacy which do no damage to the sense of age of an old house. (Naturally an exception may be made for Victorian and Edwardian buildings which have traditionally employed certain kinds of frosted glass, but even these should be treated with caution.)

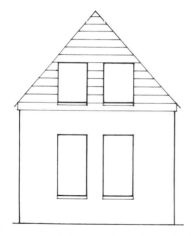

Fig 142 Unresolved duality in hateful form is demonstrated by these paired windows in a gable wall. They are a constant source of irritation, dividing the wall in half by focussing your gaze on the thin and visually meaningless centre pier.

Consider fitting a Venetian blind, with wooden slats if the room merits it; have wooden shutters, curtains, old-fashioned net on a wire; or suffer your public with dignity and indifference.

The places for obscured glass are police cells, public lavatories and pavement skylights. In the latter instance, as anybody with a city basement kitchen running under the pavement can testify, glass 'lenses' are a necessity.

No book on restoring old houses should fail to mention the ubiquitous and usually unsuitable bow window. A well-detailed bow, designed with a real understanding of genuine late eighteenth or early nineteenth-century examples, can be a valid choice. More often, the fake products which people add to their cottages, shops, restaurants and terraced houses are travesties of the most vulgar kind.

Shops in the period from about 1780 to 1830 often did have bow windows. Planners, who are fanatical about preserving the building line in a street and not obstructing pedestrians, will scarcely ever allow a bow which projects from the face of the building as it should do. Thus, modern ones are often set back into the wall opening in the most unlikely manner. They have wrongly detailed glazing bars, fake bullions and display natural wood surfaces – Georgian shop windows were

A fine example of a Venetian window, with stone facings. Many were mid- to late eighteenth century, but they started with the early eighteenth-century fashion for Palladian architecture and had Italian Renaissance origins.

almost always painted. There is nothing to stop you researching and designing a genuine-looking bow window. There are hundreds of different examples for you to copy or adapt within the grammar of the exercise. But only use a bow in the kind of building which might reasonably have had one.

The triple-light Venetian window is a possibility for a wider opening in a Georgian house. The expense will be the glazing for the rounded top of the central light (below). Eighteenth-century houses – especially pubs – often had double-hung sashes with marginal lights, which were sometimes used to fill the wide, low openings of earlier mullioned windows.

Repairing Sash Windows

So far this chapter has been given over to matters of choice and design. The reason for that is simple. A book on restoring and repairing old buildings should concentrate on those areas that need special methods which differ from standard building practice or need discussion from the historical or aesthetic point of view. Most of the practical techniques for fitting, repairing or making windows lie well within the sphere of any properly trained craftsman, and there are ample sources of reference available. However, a few remarks about window joinery might prove helpful.

Since I have insisted so fiercely on the merits of the double-hung sash, here is the way in which you can remove the sashes from the box frame, whether to allow entry of a large piece of furniture, to carry out repairs, or merely to replace the cords. A detailed drawing of its construction is shown in Fig 144.

You first remove the lower sash by prising out one or both of the staff beads which keep it in position. If the cords are not already broken, tether them by pulling enough out to knot them; or, better, wedge them against their pulley wheels with plugs of wood.

Lift the sash clear and detach the cord ends from the sash-stiles, either by pulling out the securing nails or undoing the jamming knots. If you now need to remove the upper sash, you can prise out one or

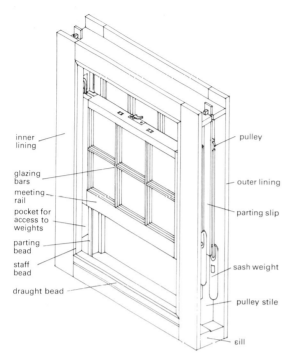

inner lining

glazing bars

meeting rail

pocket for access to weights

parting bead

staff bead

draught bead

pulley

outer lining

parting slip

sash weight

pulley stile

sill

Fig 144 The twelve- or sixteen-pane box-framed double-hung sash window is shown in detail. It is the most pleasing and practical window to use in the restoration or extension of buildings where the openings are of Renaissance proportions – higher than they are wide. The panes must not be wider than they are high. Nor must fake bullion panes be used. The frame must be painted – not stained or used in a natural timber finish, even if made of hardwood.

both parting beads with a chisel, taking care not to split or damage them. The sash may then be removed in the same way as the lower one.

When replacing the sashes, ensure the sash-cords are the correct length, or you will find that the window no longer opens and closes properly. Once removed, sashes may be planed or sanded down (if need be) to make them less tight in the frame. This must be done with discretion or you will make them too sloppy a fit, even after repainting.

To replace the cord (and you must buy genuine sash-cord), locate the little fitted shutters, called pockets, at the bottom of the pulley stiles on each side of the window. These fit absolutely flush and may be lost under layers of paint. Prise them out with a chisel.

You can now get at the lead or iron sash-weights to remove them. To run a new length of cord over the pulley wheel, you take a piece of string, tie a knot in the end and hammer around it a little lead weight made from some old flashing. This grips the string, and is called a 'mouse'. The piece of lead measures about 65 × 15mm before it is hammered into a roll small enough to pass over the pulley wheel, and drop down the sash-box, taking the string with it.

Once you have this string threaded over the pulley you use it to pull the new cord after it. The weight and sash are then refastened to the cord and the sash is replaced. Eighteenth-century sash windows have neither the strips of wood called feathers to keep the weights in the box apart, nor back linings – these prevent the weights from catching on the brickwork or masonry of the window jamb. They also prevent mortar getting behind the linings during fitting.

A common fault of old sash windows is that weights have become entangled through the absence of feathers, or protruding ledges of stone tend to snag the weights as they run up and down within the box. Pulley-wheel spindles have frequently rusted through, or come loose because the pulley-stile has decayed. Decay may most often be found in the bottom rails of sashes, where moisture tends to gather, and in the bottom part of the external linings. Sills too are very prone to decay.

Here let me give a piece of advice which some will reckon heresy. Do not think that because there is some rot in a window it must automatically be replaced; or that an expensive, skilled joiner should be called to remake and scarf in replacements for faulty lengths of timber. Decay in the outsides of windows is almost always wet rot and this only affects the damp wood. It does not spread into dry timber and brickwork like dry rot. Therefore, you may consider prolonging the life of an old sash by digging out the decayed wood during a spell of dry weather and making good with an epoxy-resin filler. This kind of filler becomes very hard and may be sculpted to regain the original shape of glazing-bar sections, sills, rails or linings. It can then

be painted over and do service for a grati-
fyingly long time if you are lucky. If you are
not, nothing much has been lost. One
product includes a fungicidal plug which
is inserted in the wood by drilling. It then
forms a protection against future damp. A
hardener is also provided for soft and flaky
timber.

Many sashes tend to pull apart at the
joints, particularly those between the
bottom rail and the stiles. There is no
harm in screwing on flat, mild-steel
mending plates to reinforce them. These
may be obtained in both straight and L-
shaped form.

It is much better to contrive a temporary repair like this, than to remove a nice
old window and replace it with some
monstrosity, pleading lack of funds to
justify your vandalism. That being said,
never use half-measures with dry rot; and
never take decay in structural timbers
such as lintels anything but seriously.

Fittings and Shutters

Window ironmongery is difficult to
choose, especially for new iron casements,
or for timber ones of early derivation. If
you have other examples of the same
period, find a blacksmith who can copy
them. That in itself may not always be easy;
but real blacksmiths know how to work
iron from the fire upon an anvil, whereas
many so-called smiths are more accurately
welders. You need a proper smith to make
copies of period ironwork.

Fastenings for casements were very
much a matter for invention on the part of
the smith and he would scroll, twist and
generally sculpture these fittings according to his talent. Search out a local
building which you believe to be of about
the same date. Make an effort to use
authentic pin-hung exterior hinges on
timber casements. They make all the
difference. (See Fig 132).

For much more recent buildings and
new timber casements where previously
there were none, you could do worse than
use the ordinary stock rat-tail fastener and
standard stay which goes with it (Fig 145).
It looks perfectly all right on a modest
nine-pane wooden casement or on such
windows as hardwood paired casements in

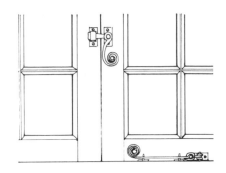

Fig 145 Although better original patterns for casement
stays and fasteners may be researched for earlier buildings, the stock rat-tail type (*top*) is a harmless choice for
small new casements in nineteenth-century houses.

the Victorian manner. However, for
earlier casements of any interest you must
have ones of a correct pattern made by a
blacksmith.

I do not believe that black fittings on
hardwood Victorian windows are always
pleasing. Brass ones may well be; but they
can cost a lot of money if you have many
windows to equip. This leads us to that
quintessentially Victorian fitting, the brass
or iron espagnolette bolt, used for French
windows and ordinary windows too. It
consists of two long bolts which run from
top to bottom of the closing stile, pivoted
at the central catch. When you turn the
handle, you close the catch, while simultaneously pushing home the bolts at top and
bottom. For iron-leaded casements,
research genuine casement stays and
catches of the period. Sometimes they
were no more than a simple hook with a
twist in the shank and an eye on the casement. These were also common on timber
windows.

Window locks are worth having for two
reasons: they deter some sneak thieves
and they give insurance companies a
feeling that you are making an effort.
Nothing will counter the experienced
burglar left quietly to his own devices,
unless you fit your old house with steel
shutters and alarms at every opening, and
introduce an effective screen in the roof as
well. When you have done all that, he can
take sledge, chisel and crowbar to your
brick wall and earnestly hammer his way
through it.

As for exterior shutters, they conjure up

such horrific visions in many people's minds that a resistance grows to the idea of ever using them. Yet there are instances when shutters may be historically correct and distinctly practical. Those we have learned to resent are the flimsy fakes which well-meaning owners regard as desirable adornment: one thinks of the whitewashed cottage with permanently fixed-back shutters and false ornamental hinges. They cannot be used and are more often than not totally out of keeping with the period of the building.

But shutters have a most useful contribution to make in the fight against driving rain and hard weather in exposed positions. Upland cottages and houses may greatly benefit from sturdy, plain, ledged and boarded shutters hung on stout iron strap hinges and gudgeon pins. They are an extremely practical buffer for those ever-vulnerable and expensive sash windows – but there is a problem. How do you close the ones on the upper floors?

As I have mentioned, continental buildings have inward-opening casements, which enable you to lean out and unhook the exterior shutters from the iron catches holding them back against the walls. This can be difficult with sash windows and almost impossible with outward-opening casements. At ground-floor level, you can just dash outside and close the shutters, dropping the bars into place across them. Upstairs, you may be able to throw up the bottom sash and lean out, but this can mean kneeling on a deep sill and ducking beneath a low meeting rail to grope perilously for the shutter as the tempest rises.

If you have casements they can effectively block any contact with the shutters. The only device could be a long shutter-pole like a boat-hook, which might enable you to complete the operation from outside. These measures may seem daunting but, as suggested, there are exposed buildings which could greatly benefit from well-designed shutters that do not clash with the architectural qualities of the building.

Shutters, often with louvre-type slats, were frequently fitted to Regency and early Victorian villas. Sometimes they slid across the windows on wheels, with iron tracks concealed by decorative timber pelmets.

If you do put shutters on an old cottage, avoid those artless little ventilation holes in the shapes of hearts, spades, clubs and diamonds. They are not without precedent, but can look twee.

Hinge-pins for fixing shutters can be a problem on rubble-stone houses made of granite – the stone you are likely to encounter in the remote, hard-weather parts of Wales, Cornwall, Yorkshire and Scotland. Hinges for shutters must be located in exactly the right positions, one plumb above the other and symmetrically placed in relation to the jambs. This will almost certainly involve drilling holes for the gudgeon-pin spikes – a far from easy task, involving the hire of a powerful and sizeable drill. For a small building you will need to drill at least twenty of these sinkings, plus the holes for the retaining catches. The shutters you can make yourself without much difficulty. Galvanised strap hinges and pins may be bought from a builders' or farmers' merchant and painted a suitable colour.

10

DOORS AND PORCHES

Although you may need to repair exterior doors and porches, they do not present many difficulties which would not occur in later buildings. The techniques are again those required by any competent joiner or builder. It is the choices of design, the materials and historical factors which must be our main concern.

Doors may be considered in two parts – the door surround and the door itself. Porches will include the closed porch, and the porch consisting of a bay with a room or rooms above. Also counted as porches are some of the fairly minimal sheltering hoods and their supports, as well as classically styled creations, some amounting to porticoes.

Porches and the Door Surround

With the architectural type of porch or surround, you must be meticulous in preserving the detail when you carry out repairs. This means copying mouldings where they are missing or need replacement. Many porches and surrounds are ruined by clumsy and uneducated restoration. Pediments are skimped, dentils or modillions scrapped because they are tedious and expensive to replace. Stone pilasters and columns are plastered over with hard mortar and painted, to avoid the trouble of proper repair. Proportions are changed, and entire elements of features such as entablatures omitted. It is very important that the orders of architecture are respected when you are obliged to work within the classical idiom. That being said, it may confuse you to find that, as we saw in Chapter 9, the builders of Georgian or seventeenth-century houses could be decidedly hit-and-miss when carrying out classical work. The variety of door surrounds involving pilasters and pediments exceeds anything dreamt of in Vitruvian philosophy.

Your general rule should be as follows. Unless a classical porch is unusually ugly and peculiar, confine your activities to repair; and repeat the details, even when you know that a column should not have a Tuscan base or that the bed mould of the cornice should not be littered with little roundels. What may be described as country classicism has its own charm, and such features are an integral part of the building's character. You may, on the other hand, remove a porch which clearly has no relationship with the building's architectural style, provided that it is sufficiently late to contribute nothing of interest to its architectural history. Here we are on slippery ground: what would qualify for removal by those standards? I do not mean the Regency latticed iron porch added to a building of 1760; nor the seventeenth-century one with Tuscan columns bearing a gabled room above, which has been stuck on to a sixteenth-century farmhouse. What you may justifiably replace is a closed Corinthian excrescence with Victorian side-glazing and marble pilasters the colour of brawn, which has been added to a fine Queen Anne frontage.

Remember, though, that you do these things at your peril if the building is listed and you have not obtained consent. Such actions are highly contentious. Removing bits of old buildings on aesthetic and architectural grounds requires a very good eye, and a fairly inclusive understanding of what could be expected at various dates on houses of a particular quality. If you do not possess the knowledge, obtain good advice.

If you should have an ugly porch of the

162

wrong period which you wish to remove, and you also possess some evidence of what was there before, perhaps an old sketch or print, you will find it much easier to convince the archaeologically minded members of the old-buildings fraternity. Many of these people do not consider buildings from an aesthetic standpoint at all, but merely as objects for study and record; they are terrified of making any judgement which could be thought subjective, and of any speculative design employing the grammar of the period which involves the imagination rather than clear-cut evidence.

One cannot over-emphasise the duty which you owe to an old building when you are replacing or repairing a porch, or adding a new one where previously there was none. You must ensure that you respect the spirit, methods, materials and architectural style of the original house. Within those guidelines, there is an astonishing amount of latitude for invention. It is not always essential to be pedantic about what you create of your own.

The pretty Regency lattice porch on the

This is the kind of restful late eighteenth- or early nineteenth-century house which can be found all over southern England and in the Midlands. It has brick walls in Flemish bond to first-floor joist level and continues upwards as a timber-framed building, with tile cladding. The charming sixteen-pane double-hung sash windows on the ground floor, with their thin glazing bars and hornless stiles, are echoed by a twelve-pane version above. This time, only four panes are allotted to the upper sash, adapting to the square format of the opening. The trellis porch has an 1825–45 look about it. It is these relatively modest vernacular houses which run such risks of inept modernisation if they do not have the protection of listed status.

rather earlier building is a case in point. You might argue that it should be removed and the original presence of the façade be reinstated with something more architectural. Certainly that might be done with effect; but the porch of 1820 can be delightful in its own right. In the end there is no substitute for good judgement coupled with understanding. Specialists will disagree with each other. Concede the possibilities, note the dangers, and enter the debate. Tired old clichés can often be right: 'Every case must be decided on its own merits.'

The practical repair of porches will frequently involve dealing with faults which arise from differential settlement between the front wall of the building and the structure of the porch itself. Porches were seldom foundationed to the same depth as the main building, and often were not properly toothed and tied to it.

Cracks appear between the two structures, flashings pull away from roofs and pediments, columns sink into the ground and walls tilt away from the house. There are four approaches to these failures. You may conclude that movement has stopped long ago and patch up the cracks. You may decide to halt the movement (which on clay could itself be seasonal), tie the porch to the building and make good. Alternatively accept the idea of some movement and provide a slip joint between the two, allowing some minor opening and shutting or rise and fall of the joint.

You can underpin the porch and improve the foundations. Finally, you can rebuild, on entirely new foundations. A lot depends upon the size and type of the porch. The structural capacity for harm of a big pillared porch, carrying a room over, is considerable. It may be necessary to prop a porch room, remove and refoundation the columns and put them back again. This will involve dead-shoring of some kind, and life being what it is the work may not stop at that. Faults will be discovered in the timbers bearing the porch room and they too may need replacing. Badly cracked columns can sometimes be strapped with non-ferrous metal, or even well-treated iron; but such repairs are always ugly and obtrusive unless they can be covered with stucco plaster – an unsuitable remedy for most such porches, which are normally too early.

You will less often find much amiss with those sturdy little stone porches with pitched slate roofs to be found on cottages in the West Country, Wales, Lancashire, Cumbria and Scotland. The very thickness of their walls gives them stability. They are seldom bonded to the building and some minor cracking at the junction with the cottage is of little consequence.

Porches of this kind are highly recommended as improvements to many 'rough' buildings. They give real shelter and often engender a visual sense of security in a building which needs it. Materials should be natural, and the porch should be faced in the manner of the cottage itself, with a rugged stone or timber lintel, a paved floor, and maybe, little slit windows in the side walls (Fig 146). Be careful with the proportions, which should avoid any suggestion of the sentry-box; leave that shape to some of the dear little timber-and-trellis porches, which not infrequently look as though a redcoat had just stepped out for a mug of ale.

I have already mentioned glass porches elsewhere in connection with windows and glazing. We need not labour the points made there. All you have to remember is that most of them date from the last quarter of the nineteenth century or even the present century. Assess the situation and act accordingly.

Fig 146 This sort of little stone porch might reasonably be added to a cottage with coursed or random rubble stone walls. To look right the walls of such porches must have substance.

What you must always have uppermost in your mind when choosing a new porch for an old cottage, or little house, is the style of the building and its general flavour. Clearly there are innumerable stock designs available from manufacturers which cannot be contemplated. Brash little wooden boxes with flat roofs and blank window panes come to mind.

Along with all of these it may be necessary to commit the façade-length hot-box conservatory to oblivion. Thousands of pleasing buildings have been ruined with them during the last hundred years or so. They are convenient and splendid for growing vines. Nothing could be more delightful than to sit in such a place on a warm summer evening drinking a gin and tonic, drowsy with the smell of green foliage, listening to the first heavy drops of rain on the glass roof. It is deeply hurtful to suggest you deprive yourself of such pleasures, but I am bound to do so, unless you have a building so late and undistinguished that it cannot matter architecturally one way or the other.

Many people encounter the difficulty of putting a new 'Georgian' porch on an old house or cottage. If it is a 'rough' vernacular building, it is usually wiser to forget it. If it is a building just sufficiently 'smooth' to make a porch or surround of that kind desirable, be very careful about what you do. We have talked of designing new features, but there are also a great number of ready-made units in fibreglass or timber – few of them in the least suitable. Some are clearly copied from genuine originals or contemporary pattern books; but even these may be wrong for your building, being out of scale, too refined or of unconvincing texture. Some manufactured door surrounds are 'Adamesque', incorporating apologetic-looking little pediments, synthetic pilaster panels, and friezes which sport dainty motifs of vaguely neo-classical mien. This kind of adornment will make a respectable old building look like an escapee from a modern housing estate.

It can make sense to give some visual emphasis to the front door of a cottage and one of the best ways is by providing a door hood. It is useful for keeping the rain off visitors' heads (and off your own while you hunt for your key). There are several types to consider, depending on the kind of building.

One category has corbels or brackets of timber or stone, supporting solid-looking pieces of moulded stone, plain slabs of slate or stone-slate (Fig 147). The same thing can be done with old tiles and a few softwood battens, or with roofing slates. A little shaped lead roof for an early nineteenth-century plastered cottage may be

Fig 147 There are hundreds of variations of the bracket-and-slab type of minimal door hood. The slab may have a correct classical cornice or it could be just a piece of squared stone. The brackets, too, can be very simple or designed in a classical manner.

held up on cast-iron brackets, or some other form of support. Weatherboards, too, are traditional enough in some parts of the country (Fig 149).

The addition of door hoods in terrace buildings which are part of a clearly defined and uniform row is often bad thinking. Most such artisans' terrace houses did not have any form of door hood, just a plain stone or brick opening for the front door, lying flush with the face of the building. You will be surprised to find how many cottages and farmhouses, too, never had any protective structure over the front door. This does not always prohibit adding one if it could prove useful, pleasing and authentic. The

Fig 148 For some smaller nineteenth-century houses, or an extension, a timber or iron porch in the Regency latticed manner, with a curved copper or lead roof, can be ideal.

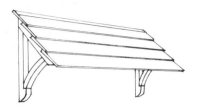

Fig 149 Cottage door hoods follow regional traditions. One often seen on late Victorian and Edwardian buildings consists of framed timber supports and a roof of weatherboards, tiles or slates.

keynotes should be simplicity and natural materials.

However, there are late Victorian houses, lodges and so on, which traditionally lend themselves to involved concoctions, with somewhat Jacobean or Gothic elaborations. One has in mind those rather clumsy and top-heavy-looking affairs with pitched roofs, barge boards, chamfers and stops, crisp quatrefoils, curved brackets and braces. They are quite fun to design and may be used in any pastiche of late nineteenth-century or Edwardian building which lends itself to a rather heavy-handed bourgeois essay in nostalgia.

Glazed-brick or concrete-block porches are anathema on an old building, as are concrete structures and grisly little protuberances covered in pebbledash or chippings. Concrete blocks which claim to simulate stonework are vile, as are sawnfaced slate blocks and slate laid like vertical crazy paving to face a wall (Fig 150). All these types of facing must be resolutely set aside when contemplating any work connected with old houses or cottages.

When dealing with exterior door openings in buildings of seventeenth-century or earlier vintage, it can be a problem to decide which doors were in use at what period, and whether those you see are in their original state. Lintels may have been removed or incorporated from ruins elsewhere. Massive jamb stones with stops and chamfers can lack similarly chamfered stones over them. The ground level may have crept upwards over the years, concealing the details at the base of the jamb. Relieving arches can be confused with blocked-up earlier openings.

What seem to be lintels turn out to be reused sills or mullions and show mouldings and sockets in all the wrong places to prove it. Squared stuff for jambs contains mysterious holes retaining stubs of wrought iron (perhaps hinge-pins); little stone quatrefoils lurk in parts of a wall where they suggest no reasonable purpose. Convincing 'mullions', with the right sort of chamfers, present sinkings for vertical iron bars, revealing their original purpose as sills or window heads. All these things are part of the history and individuality of the house. Do not gratuitously remove them and sort out every unexplained anomaly, from a desire to be tidy. All that may result is an exercise in over-restoration, and when taken away, these odds and ends of masonry serve little purpose unless you can identify their original positions.

For new door openings, and this applies equally to windows, follow local tradition. Heavy timber baulks, granite lintels, brick arches, cut voussoirs and key stones are some obvious ways of bridging an opening. Usually the jambs follow the general design and texture of the head. Never introduce arches or lintels which bear no relationship to the period.

The development of the arch is something to study. Broadly speaking, the Normans used round arches; the early medieval phase started with lancet-like pointed arches and these became progressively flatter until the four-centred Tudor

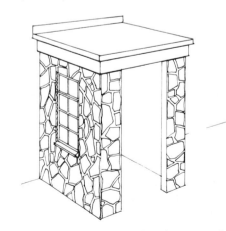

Fig 150 There are few more obnoxious ways of using stone than to apply it in the form of a vertical crazy-paving. This porch is about as ugly as it could be; yet people do add this kind of thing to old buildings.

166

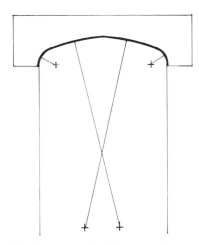

Fig 151 The four-centred arch is so named because you place the point of the pair of compasses in four positions to strike off the different parts of its shallow curve. It became fashionable around the late fifteenth century and was being used well into the seventeenth century.

type became dominant in the late fifteenth and sixteenth centuries. The seventeenth century saw the arrival of a rounded or semicircular arch, and a flat arch or lintel also in tune with Renaissance design. Rounded arches continued in use during the seventeenth century and, in conjunction with the formal orders of architecture, remained a common feature. This is the briefest of accounts and there are plenty of exceptions. The Norman shouldered arch is one such, and the four-centred arch (Fig 151) was still occasionally employed right into the early eighteenth century. In the North of England, builders of farmhouses loved to carve lintels with odd-looking up-and-down mouldings, often framing a date or the owner's initials. This practice continued throughout the seventeenth century and into the eighteenth century (Fig 152).

In stone areas, the use of bricks was at first restricted to the vicinity of seaports which could obtain them. Brickyards became more frequent in the second half of the seventeenth century and chimneys were often the first feature to depart from the stone tradition. Eventually, rubble-stone buildings, especially in places where the quality of building stone was poor, were provided with brick door heads,

window heads and jambs. Brick was cheap and tidy for these purposes, and could form a good structural arch with quite a small degree of camber. These developments tended to be nineteenth century, but were also made in the eighteenth century in some stone districts.

Not infrequently you find that almost flat brick arches have been introduced to replace rotten timber lintels of a much earlier period. Brick arches are either made up from plain, unshaped bricks, the arch being effected by filling out the joints with mortar; or alternatively of carefully cut and rubbed bricks which fitted one with another to form a curve. Arches are built with the aid of a kind of shuttering called a centering, which is propped in position while the bricks are laid upon it. A cambered arch, which has very little rise in the centre, is formed over what is called a turning-piece (Fig 153). Traditional, but botched, construction often involved building the ends of the turning-piece into the jambs and sawing out the middle, when the mortar had set. These ends could be used as fixing blocks for the door frame.

The only maverick arch is the soldier arch. This consists of bricks on end which march across the window opening and are structurally supported by a steel angle bar or a back-up concrete lintel. Usually this arch does not continue across the jambs, so it has the unnerving appearance of being totally unsupported at the ends. For that reason alone, I should never use it; it is manneristic without grandeur or panache.

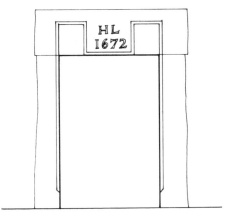

Fig 152 North of England door posts and lintel.

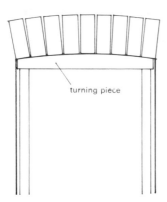

Fig 153 To support a shallow cambered or segmental arch, either for repair or rebuilding, a solid timber turning piece is propped in position.

A little-considered point about doorways is the threshold. The paving underfoot as you enter a building is worth treating with care, and steps of various kinds made from slate, granite, York stone, brick and so on come to mind. A difficulty can sometimes arise from the fact that your door will open inwards and is therefore expected to close either against or over the threshold stone. In the latter case, an open gap exists between the bottom of the door and the door sill, through which rain will drive, even when the step is slightly rebated, grooved and set to an outward fall (Fig 154).

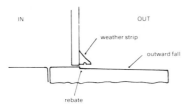

Fig 154 The normal way of weathering the bottom of a door which has no raised threshold-piece across the step is to screw on a ready-made weathering strip. This has a throat grooved underneath to catch water and drip it clear of the door.

In modern buildings a patent metal water-bar is set into the doorstep, and this often slots into a hardwood threshold-piece, against which the door closes, being rebated along the bottom rail. There are various details for this purpose and most of them tend to look rather awkward, especially in early or 'rough' houses,

where the entrance should look uncluttered. If you have a generous porch the hazard of water driving under the door is more or less avoided. If you do not, here is one method which can sometimes be made to work and can be helpful when inside floor-levels have been dug out.

The door frame, which has treated ends, finishes about 25mm clear of the external step, which is set to an outward fall in the usual way. Inside the step, below the bottom edge of the door, a narrow sump is formed. Any water dripping down the door or driving under it gathers inside the door and drains away through weep-holes to the outside (Fig 155).

This kind of threshold is probably most appropriate for flagstoned halls, where all the material underfoot can be matched for visual continuity. But one can imagine a mixture of, say, slate flagstones and a granite threshold or step. The point is to isolate any water which gathers and allow it to drain quickly away. Naturally it is vital that this damp area does not affect any timber internal fixtures, skirtings for example, or wiring.

When repairing or replacing the frames of old doors, treat them with care. Even if in a frail state as a result of wear and decay, they may still be worth saving. They may have interesting mouldings or chamfers and stops, which must not be spoiled when you are re-setting frames which have come loose. It may be necessary to have matching hardwood or softwood pieces moulded, and scarfed to the original timbers, if parts have had to be cut away.

The usual manner of fixing back door frames is to use treated softwood plugs driven into sinkings or gaps in the masonry. To these, the frame is nailed or screwed. Brickwork may permit the use of masonry nails in some instances, or perhaps of built-in pallets or driven hold-fasts. The latter are often seen in the frames of ledged doors in brick cottages.

It is obviously important that a door frame is fitted plumb, but try to avoid leaving wide gaps between jamb and post, which then need ugly and vulnerable packing. Better, when doing a good class of repair work, to plane an oversized door post so that it may be fitted snugly in a

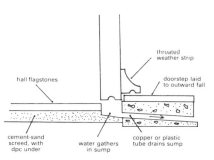

Fig 155 One method of dealing with a door which closes against the step from inside is to form a shallow waterproof sump, which catches any driven rain and allows it to drain to the outside through weep-holes. The dpc has been omitted from the drawing as its exact position (if there is one) will vary according to circumstances.

vertical position against the bricks or stones.

One of the beauties of the ledged and battened or plank door is that it does not hang on butt hinges. Hinges of that sort require very accurate fitting, and tedious adjustments have to be made to pack them or sink them if a frame or lining is out of true. Ledged doors are hung on strap hinges which are fixed to the face of the door and its frame, allowing considerable adjustment during fitting. Even door frames which are out of square, as a result either of settlement or of inaccurate construction, can accommodate plank doors. It is relatively easy to make such doors slightly oversize and shoot them to an exact fit for the existing frame.

Solid frames, whether of hardwood or softwood are required with plank doors. They should have dowelled, mortice-and-tenon joints and the dowel should be left about 1–2mm proud so that you can just see it through the paint. The rebated frame can be flush beaded to advantage. There should be *no architrave* and, while T-hinges can be employed for later doors, those before about 1850 are better hung on strap hinges.

Although panelled front doors are always fixed with butt hinges in modern practice, in the late seventeenth century and for much of the eighteenth century, they were usually hung on versions of the wrought-iron H-L hinge, nailed to the frame.

Again, for the amateur it is much simpler to use plain linings without rebates for internal doors which you fit yourself. A planted length of square-edge stop can then be nailed to meet the face of the door, ironing out minor inaccuracies. This kind of thinking will not appeal to good craftsmen, but in the world of self-help there are occasions when such methods will do the trick.

Obviously, you will increase your attention to craftsmanship in ratio to the quality of the building. On the whole, I should leave the making and fitting of framed linings (those where grounds are first fixed to the walls and panels are planted on them) to competent carpenters or joiners.

In early nineteenth-century houses, it may prove desirable to fit big double or folding doors in drawing-rooms, so that a reception room may be provided running the full depth of the building. Remember that such wide openings may remove support for an upper floor from part of a structural partition. The weight of heavy doors may also involve reinforcement of both the partition and even the floor joists beneath it.

When framing door openings in attic rooms, do not scorn the door with a clipped-off corner (Fig 156). It is a thoroughly traditional and helpful device. In order to strut a loose door frame or lining, carpenters often resort to nailing a threshold piece to the floorboards between the linings. This practice should be discouraged, as it spoils the continuity of a floor, looks ugly and gets in the way. (And it is frequently used in conjunction with fitted carpets.)

In old buildings, especially those with oak framing, doors often hang at crazy angles; the heads may slope wildly in the direction towards which the building has settled. There is no need to straighten them up; it is enough to make adjustments which allow the door to open and close with reasonable efficiency. Abandon all notions of procuring a snug, draught-proof fit. You should be sparing of straightening-up in 'rough' houses, while 'smooth' ones of style and elegance may require quite extensive remedies. Once

169

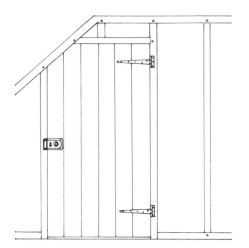

Fig 156 The device of fitting a ledged-and-battened attic door with its corner sawn off to meet the roof slope is traditional and highly practical. This door has ironmonger's black-japanned T hinges and an ordinary pressed-steel lock.

more, it is important not to over-restore. The reason for dropping door frames is often the removal of support during some badly designed alterations in the late eighteenth or nineteenth centuries; decayed timber may also play a part.

Types of Door

Medieval, Tudor and early seventeenth-century doors in vernacular buildings were almost invariably made from vertical boards nailed to ledges or to a second skin of other boards run horizontally. They were of oak or some other hardwood and were hung on strap hinges.

Those to be found of original fourteenth- or fifteenth- century origin will often have square-edged boards just butted together and tied with ledges, the intersections being covered with moulded hardwood strips. These mouldings may be very simple or relatively complicated in section.

Some early doors consisted of a system of stiles, rails and muntins, of the kind found in wall panelling, to which wide boards were fixed with clenched-over nails. The nails often had large, faceted heads. Hinges for fourteenth-, fifteenth- and sixteenth-century doors almost always ran beneath any framing members on the face of the door, and any planted

mouldings on the door post (Fig 157). This always looks a most amateur arrangement, but it is thoroughly genuine and can be imitated with impunity in some buildings.

By the seventeenth century, 'battened' doors, as this type with boards nailed to horizontal ledges are called, involved the use of tongued-and-grooved, loose-tongued or rebated techniques. Cover strips for junctions between boards became unnecessary, and the whole design of interior doors was the height of simplicity. This is lucky for those repairing and restoring old buildings who have already cringed at the cost of oak and the purpose-made ironmongery required to do a good job.

Hinges on medieval doors of important buildings tended to be more elaborate and might terminate in hooks, anchors, elaborate scrolls, spears or fleurs-de-lis designs. The average sixteenth- or seventeenth-century strap hinge is nothing more than a length of slightly tapered wrought iron, with chamfered edges, ending in a bulge, tongue or point. Decoration would be some fairly rudimentary incised lines in criss-cross and diamond patterns – anything, in fact, which allowed the smith to avoid chasing curved lines, rather than punching straight ones. You will recall a similar dislike of curves among carpenters, who restricted their marks on beams to Roman figures which could be executed with a chisel.

Once the architectural formality of the Renaissance took hold in the second half of the seventeenth century, quite ordinary houses began to have panelled doors, framed with stiles and rails. The most common pattern involved two panels divided by a lock rail. The panels were often fielded at the edges and trimmed with quite heavy bolection mouldings (Fig 158). There were slight variations, involving two big panels with a 'landscape' panel in between; and eight or ten panels can be seen in grander houses. In the same way, Jacobean buildings might have doors with arcaded panels, diamond lozenges, etc.

The details of interior-door architraves and cornices during the second half of the

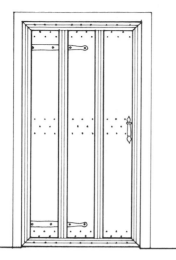

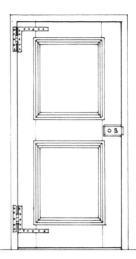

Fig 157 Early hardwood doors were made by nailing boards called ledges to the wide vertical planks on the inside face. For grander houses, moulded cover strips might be applied to the junctions between the boards, on the outside. For visual continuity, the mouldings might also be carried around top, bottom and sides. The strap hinges were usually run *beneath* the mouldings.

Fig 158 This is the most typical late seventeenth-century and early eighteenth-century panelled interior door. It has quite bold bolection mouldings around the panels, which are normally fielded off slightly at the edges. Hinges were nearly always of the H or H-L pattern, *nailed* to the frame and the face of the door. Brass or iron rim locks with rather small knobs were normally used.

seventeenth century were surprisingly uniform in design, regardless of region. They made use of heavy bolection mouldings, pulvinated friezes and a variety of Baroque touches like broken pediments or entablatures and *tête d'ange* devices. Swan-necked and scrolled pediments were also fashionable.

Exterior and interior doorways, in stone or timber, might include such universal details as cartouches, swags of fruit or flowers, masks and shells. Indeed one of the prettiest and most appealing door hoods ever invented was the big semicircular bonnet, lined with a robust plaster shell, the whole structure resting on sturdy and often ornate console brackets.

During the Queen Anne period in the early eighteenth century, door cases for front doors became less exuberant and began to adopt the simple classical arrangement of pilasters and pediments which was to continue into the 1800s. Fanlights appeared over door heads with lead, iron or wood glazing – sometimes of a refinement more easily associated with late eighteenth-century neo-classical architecture. More usual, however, were straight cornices with triangular or segmental pediments and transom lights.

Front doors themselves were almost entirely of the panelled type throughout the eighteenth century, except in small farmhouses, cottages and vernacular buildings, which continued to use the ledged door with vertical boards.

Once bolection mouldings were abandoned for interior door surrounds, simple architraves consisting of one or more flat fascia bands were used, enclosed by some fairly large moulding like an ovolo or ogee. Such details were to continue right through the eighteenth and nineteenth centuries in interiors of classical derivation. Points of difference are minimal, perhaps involving slight alterations in the size or section of mouldings. Anybody who cares to study interior joinery will find that for every hard-and-fast design trend which he thinks he discerns there will be numerous exceptions. Certainly there are motifs which may be reckoned not earlier than a particular date, when seen in the context of a door case or a chimneypiece for example, but there are many more which were used and reused in every imaginable situation.

Front-door panels throughout the eighteenth century were usually six in number for modest houses, and were either plain or fielded. Simple ovolo and ogee mouldings surrounded the panels, although one significant difference may be observed in late eighteenth-century and early nineteenth-century doors: in these, the bottom two panels would often be set flush with the frame and trimmed with some fine reeding or a line or two of flush bead moulding. The idea seems to have been to shed dirt and water more easily from this vulnerable area of the door face. Flush panels could also be made thicker, affording some extra protection against accidental damage.

All the eighteenth- and nineteenth-century revival movements engendered their own distinctive variations on classical or medieval themes. There was Strawberry Hill Gothic of the mid-eighteenth century, late eighteenth-century neo-classical, Regency Gothic, Egyptian, Greek Revival, plus Victorian Gothic, Venetian and other Italian influences. Such fashions applied not just to doors, but all the other architectural features of a building – windows, cornices, roof pitches, decorative detail, layout and proportion. If you want to try your hand at restoring old buildings you need to apply yourself to learning the main architectural trends which were pursued in the different eras.

You should know, or find out, that a typical early nineteenth-century door case in 'smoother' houses would consist of a reeded architrave, with corner blocks and roundels. A rather plain little four-panelled door with coarse-looking splayed bolection mouldings, a white china knob and a mortice lock, should suggest the second half of the nineteenth century, or later. These details are the fun of dealing with old buildings, after the trauma of eradicating dry rot and underpinning the foundations.

Always remember one thing about dating parts of a building – the most earnest specialists quite often prove to be wrong. Once innocent details are wrongly identified or tied to tempting records, the whole process degenerates into a misleading trail of dead men's footsteps.

Many experts just avoid mentioning the dates of elements which leave them feeling blank.

The selection of suitable doors and architraves for an old building is yet another occasion which calls for good judgement. Three things will govern your choice:

1 The predominant age, style and detail of other doors in the building which are of the right period.
2 The tone and texture of the building whether it is 'rough' vernacular, like a Devon longhouse, or 'smooth' like a Regency villa.
3 For inside doors, the age and architectural flavour of the rooms on either side.

If you can establish that some or most of the interior doors are original, then substitute the unsightly replacements with others of the correct pattern. Out will go all the flush doors and architraves installed during this century; out may go the Victorian mock-Gothic doors in your Elizabethan farmhouse; out will go all the cheap and nasty interlopers wished upon you by former owners. However, check that a flush door does not consist of an old one cased in modern hardboard or plywood.

Should your building contain a variety of doors which belong to different periods in its historical development, you are faced with a dilemma. The general principles are to assess the quality of what is there and to what extent a room or a wing of a building is old, attractive and architecturally coherent in its own right. For example, you would not be likely to strip out a couple of rooms full of genuine rococo plasterwork and eighteenth-century panelling, just because they were in an early seventeenth-century farmhouse. Nor would you, I hope, abolish a surviving Jacobean room in a Queen Anne manor. Whether you would feel compelled to retain an Edwardian drawing-room in a Regency rectory is another matter. Nobody is saying it is easy!

Tone and texture is, perhaps, more self-apparent. You clearly do not want nicely moulded doors and architraves with brass box locks and classically influenced architraves in a 'rough' little stone cottage with a great big down-hearth. This sort of cottage often demands simple ledged and boarded doors, with T or strap hinges, thumb latches or black-japanned ironmonger's rim latches; the latter would have plain brass knobs. It would be reasonable to ledge and also *brace* the doors, especially if the building is late nineteenth century. However, earlier ledged doors were seldom if ever braced; the broad ledges, with hand-wrought nails clenched over (Fig 159) provided enough stiffening for the purpose. Despite the general suitability of ledged doors, many cottages had modest panelled ones in the more important rooms.

The style of the rooms linked by a particular door can be of some importance, and this also applies to choice of floor finishes: they can be seen at one time in relationship to quite separate rooms, and if those rooms fail to accord in their general period detail, decoration and textures, odd results may occur.

This notion has always been well recognised and you will see eighteenth-century jib doors which match the room panelling exactly on one side, and the general door panelling of a hall on the other. But although you may have a veritable mask-of-Janus of a door, it is still sometimes left open and there is not much you can do about that. Your purpose should just be to harmonise in an informed manner.

Now let us tackle the far easier and fundamental question of what doors to shun at all costs, starting with those in outside walls. Writing in the 1980s we should first remark upon what I call the Wendy-house or doll's-house door, which incorporates its own little neo-Georgian fanlight where the top two panels should rightly be (Fig 161). Its scale and design suggest a child's plaything, but that child would have to be Violet Elizabeth Bott. Purchased from the stock joinery manufacturers' catalogues and builders' merchants everywhere, it has spread like a foul contagion throughout the land, disfiguring every imaginable kind of old building. *Do not use it.* If you already have one, go out tomorrow and replace it with

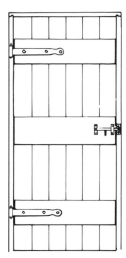

Fig 159 Softwood ledged-and-boarded doors were used in every kind of cottage and outbuilding, and in the attics of more stylish houses too. They are the main standby when restoring many vernacular buildings dating from the late seventeenth century onwards. Hung on T or strap hinges, they had sneck latches, rim locks, rim latches or hatchet latches. Note the absence of an architrave.

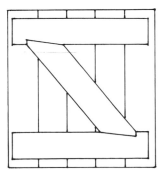

Fig 160 To stop doors sagging on the side furthest from the hinges, in the nineteenth century they were often braced diagonally by housing boards into the ledges. In rougher work the nails were still clenched over on the inside of the door.

something suitable. You will live to bless me for this advice, when you see the front of your building reinstated, once more displaying a little mellow dignity.

Other main offenders nearly all have some kind of glazing. They have panels of reeded glass; they have fretted sun-ray designs like the fronts of pre-war wireless sets, from which period they derive. There are aluminium doors which may be well enough for a modern public library or a suburban bungalow, but destroy the

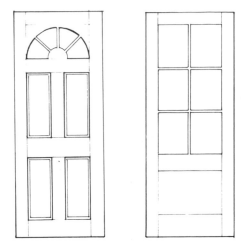

Fig 161 (*left*) This unsuitable neo-Georgian door – a Wendy-house door – with fake fanlight where the two top panels should be, has become a plague in old houses. Real Georgian fanlights are not part of the door itself but fill the space above a transom.

Fig 162 *(right)* There are many unfortunate designs for glazed doors. Those with paired panes are particularly disturbing and should be avoided.

appearance of an old terraced house, country cottage or converted barn. Some of the worst are the uPVC doors with fake glazing bars. There is the tall, narrow, panelled or battened door, with a square of four apologetic-looking frosted-glass panes squeezed in near the top. You will see half-glazed or two-thirds-glazed doors with paired panes. Like the roadhouse Johnnies they are, they seem to be whispering behind their hands 'Pardon me! I *nearly* got it right'. Many architects love *them* also (Fig 162). Watch out for the modest plank door with its varnished surface and little square pane of modern moulded bullion glass near the top. It fails to please. These are some examples, but there are many more. If you really believe in allowing old buildings to go on looking old, you will be adding your own pet hates to this list.

There are, of course, some perfectly civilised doors which have merely been abused. It is a great temptation to put panes of glass in the top two panels of a six-panel Georgian door, but try to resist it. The idea, naturally, is to give light to the hall passage. Try, as a general rule, to design a panelled door with a transom

light, which you can then glaze in some appropriate manner. Diamonds on their sides, opposing pairs of curved astragals, interlaced arches, floating rectangles and plain vertical glazing bars come to mind (Fig 163).

If you have a building earlier than about 1860, do not leave a big, plain glass fanlight over the transom just because it is there. Put in some sort of glazing-bar arrangement, otherwise it looks like an invitation for somebody to come and write Brae View Guest House in gold letters on it. Many nice stucco terraced houses of about 1830 to 1850 suffer from having blank fanlights. Incidentally, it is not effective to stick bits of wood on to fanlight glass to simulate glazing bars: they nearly always look wrong, and if the glass is broken your trouble was for nothing.

Door hoods and door surrounds should be properly detailed, with an appreciation of classical language and proportions. They should not look stiff

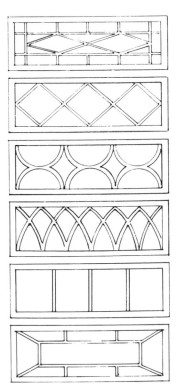

Fig 163 The glazing of transom lights or fanlights gives much scope for individuality. Here are five fairly typical designs which would be appropriate for houses of between about 1800 and 1850.

and uneasy, feeble or in any way grotesque.

One especially horrific adjunct to doors and fanlights is what passes for wrought iron, but is usually an insipid mish-mash of bent metal strips. Such iron is often worked cold, using stock mild-steel sections bolted, riveted or welded together. It is redolent of the Costa Brava and never suggests the glories of good Georgian iron, wrought or cast. It can insinuate itself into your surroundings and you can become immunised to it, so that you no longer notice it. I know one woman with considerable understanding of old houses, and excellent taste, who had an obscured-glass door of this kind filled with a big panel of 'wrought iron' left by a previous vandal; it took her nearly twenty years to get around to ripping it out.

Be careful about using wrought iron to support door hoods: there are historic precedents for doing so, and if the design and work are of a high standard it may be successful, but too frequently it looks fake and flimsy. Unless you really know what constitutes fine ironwork, with its clever variations of substance and contour, its hand-forged subtleties of detail and gracefully executed nibs and scrolls, leave it alone.

That also applies to lamp brackets, gates, well-heads and railings. However, in the case of railings, there are firms casting excellent copies of old originals which may be delightful if well chosen. You may be able to look up pre-war photographs in your local reference library to see what kind of ironwork was originally used.

Choosing Doors

Your choice of door will be governed by what is practical, and aesthetically and historically suitable. You will be selecting from the three main categories: hardwood boarded doors, possibly with stud-headed nails and other early details; panelled doors; and softwood ledged-and-battened, ledged-and-braced or framed-ledged-and-braced doors of more or less modern construction.

Of these, the most difficult to get right will be the hardwood types which follow seventeenth-century or earlier designs and methods. To be successful they require great attention to detail. If you wish to be authentic, ensure that no bolts or screws are used in the construction. Clenched-over nails with large round or faceted heads were the norm and you may possibly have to get a smith to make them.

Ledges must be generous and braces are omitted for other than very wide doors. Hinges have been previously described; but remember that they really should be nailed not screwed. Coach screws may be used, at times, but avoid any with very modern-looking heads. Sometimes ledges and boards may be dowelled together.

Boards on sixteenth- and early seventeenth-century doors were not tongued-and-grooved in the modern manner. Mainly, the edges of boards used in good-quality moulded work were once, or even twice, rebated (Fig 169). They might also be vertically moulded or have planted cover strips (Fig 168). Many doors had butt-jointed boards and these always require simple cover mouldings, perhaps 40–70mm wide.

Some of the more elaborate boarded doors had wide, concave, vertical channels – perhaps one or two per board – meeting at sharp arrises like Greek fluting. These, and features like the stuck ribs which cover vertical joints, would be worked from the solid. In all but the most exceptional buildings, such demanding and wasteful methods were avoided. Some door jambs and heads were also arched and moulded in stone or timber, and these are all features you have to study carefully and draw out to scale for your joiner or mason.

The thickness of door timbers is also important. Internal doors in the seventeenth century were often quite thin, with boards of about 25mm thickness. Early external doors in lesser domestic buildings tended to be thicker but not very much so. To obtain the timber for such joinery, visit the yards of specialist merchants who keep stocks of oak and other woods. Pay careful attention to quality, the method of sawn conversion used, and whether the timber is adequately seasoned. Timber such as oak

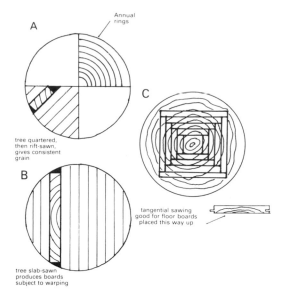

Fig 164 Radial, quarter or rift sawing (A) is wasteful but is the best method of converting a hardwood log into boards for joinery. It makes the most of any figuring. For ordinary softwood structural timber, slab sawing (B) is the most economical method. Tangential sawing (C) is advantageous when cutting pine that is not to be painted, as it emphasises the figuring of the annual rings.

Within the figure:

- Annual rings
- tree quartered, then rift-sawn, gives consistent grain
- tangential sawing good for floor boards placed this way up
- tree slab-sawn produces boards subject to warping

should be seasoned naturally in the open, covered with some type of light roof to prevent soaking from the rain. Air must be allowed to circulate freely around the timber to promote drying, softwood cross-lags being used to stack piles of boards and planks. Recently seasoned hardwoods will be left for at least a year for the sap to dry out. To speed up seasoning of timbers they can be immersed in a river for a matter of a few weeks and then cut into baulks for stacking. The type of wood and the size of the planks greatly influence the time natural dry seasoning takes. A stack of 150mm thick planks could take three years to mature.

Hardwood boards for doors should ideally be rift-sawn, meaning that they are cut radially from the log, which is first sawn into four quarters (Fig 164). The end of a rift-sawn board should show very upright annual rings, forming an angle of at least 45° with the surface. Rift-sawing is wasteful but produces boards which are far less prone to shrink or warp. Another major gain is the attractive figuring produced by the appearance of the medullary

rays on the surface. Timber for hardwood doors in early houses may be finished with an adze or be hand-planed, but should not be machined to a smooth modern-looking surface.

Ledged doors may be closed with iron thumb latches or wooden sneck latches. Sometimes a latch may be lifted with a string or leather thong passed through a hole in the door. More important doors often opened by means of a wrought-iron ring rigidly attached to an iron spindle, lifting an iron latch (Figs 165 and 166). The ring frequently had an ornamental back plate.

Many doors of early buildings were secured with heavy timber bars socketed into the jambs. Wooden bolts were also used, and iron ones too. Furniture for ledged and battened doors became more sophisticated in the seventeenth century and the fact that iron was cheaper and more readily available promoted the use of huge, heavy, metal rim locks. However, the hardwood-cased stock lock was always a favourite and was still widely employed in the nineteenth century. Before the seventeenth century locks were very much a luxury and you are most unlikely to find an original on the door of a vernacular building.

Although hinges for seventeenth-century framed and panelled interior doors were normally in the shape of an iron H or H-L (with ornamental variations of the former, like the cockshead), ledged doors were usually hung on strap hinges. You can safely employ the strap hinge for these, unless there is some strong argument for the other type.

For any 'smoother' building from the late seventeenth century to 1900 you will

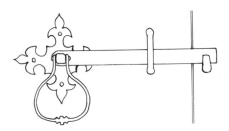

Fig 165 Doors in medieval and Elizabethan houses often had latches operated by drop-ring handles with ornamental back plates.

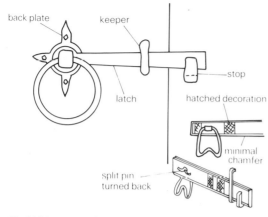

back plate   keeper
stop
latch   hatched decoration
minimal chamfer
split pin turned back

Fig 166 Seventeenth-century doors were also fitted with ring handles on through pins which pivot the latch up and down. Back plates tended to be plain. The through pin for the ring was often split and turned back to secure it on the latch side.

have to repair or fit framed and panelled doors. I have mentioned the patterns you will see in different periods. For the rather light and thin William-and-Mary or Queen Anne two-panelled door, I should use H-L hinges.

Obviously, the thinner the door, the more it demands a face-fixed hinge. Do nail the hinge – it looks much more authentic. These hinges are neither very large nor of thick metal, but should ideally be hand-wrought – without countersinking. It will help if the holes for the nails make a good fit with the shanks, and for absolute authenticity the nails should be hand-made, or at least look somewhat uneven. Converting the heads of ordinary modern roundwire nails is certainly faking but if you can make a good job of it, I see nothing against it. Clearly you will not wish to see the round and mechanical shanks of modern nails clenched on the ledge of a new oak door, so obtain wrought-iron clout nails if you are determined on total attention to detail.

Some lesser two-panelled doors had latches, as mentioned above, but the best furniture for earlier ones (late seventeenth century to early eighteenth century) is the brass or steel rim lock. This had a small round, oval or shaped knob. Later rim locks had either knobs or, more often, shield drop handles. Black-painted steel rim locks are perfectly suitable for

less important doors, although unlikely to prove very secure.

For interior doors of later cottages, whether ledged-and-braced or panelled, the standard japanned rim latch is ideal. This has a brass knob or sometimes a wooden or china one. Many Victorian and later panelled doors have mortice locks, again with brass, wooden or china knobs.

The panelled doors about which we have so far talked are mainly in softwood. Panelled doors of high quality in polished hardwoods, or late seventeenth-century and early eighteenth-century ones with gilding and graining, are rather beyond the scope of this book. Early marbled, grained or polished doors would be heavy and hang on much more substantial H-L hinges than those used for minor ones. The hinges could well be of brass. Iron hinges were normally painted over. Butt hinges were developed during the eighteenth century and there were very fine brass ones with ornamental turned-brass pins.

No panelled door of natural hardwood should be varnished. Polish is the correct finish. Grand doors with gilded mouldings should not have these details picked out in gold paint – only gold leaf will do. Many houses of the late eighteenth and early nineteenth centuries have beautiful mahogany doors, finely figured and of superb quality; nobody needs to be reminded to respect these. However, there is a distressing tendency for people to strip painted pine doors with the notion that this gives a sense of age. Pine doors should not be stripped! They were nearly always painted, and this goes for William-and-Mary and Queen Anne doors as well as those of the late eighteenth and early nineteenth centuries.

However, there were hardwood doors and panelling in those periods, which were left unpainted and might have been wax-polished. Some late seventeenth-century doors were grained, marbled or covered in leather; but these are exceptional and if you encounter them you need to seek specialist advice.

If you are having ledged or framed-and-braced doors made in the Victorian manner, for such buildings as converted

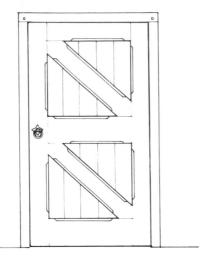

Fig 167 Victorian public buildings, lodges and rectories are full of solid framed-and-braced doors with chamfered mouldings. The style is affectionately known as 'board-school Gothic'.

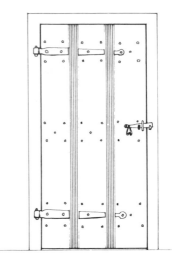

Fig 168 An appropriate door for an early seventeenth-century house of presence but no great sophistication is made from three wide planks with cover mouldings over the intersections. Ledges are nailed on the opposite side.

barns, schools, chapels and lodges, there are some little touches which help denote the period: stop-chamfer the framing members or ledges, for example (Fig 167). Door frames, too, may be stop-chamfered; and everything may be made from machine-planed softwood, with a gloss-paint finish.

For the general run of cottages, make your ledged-and-battened doors fairly sturdy with T or strap hinges according to the date. Most boards can be of 25mm tongued-and-grooved softwood. On the whole I should not be so authentic as to clench over the nails on the ledges. They will catch on things if not well punched down and may even crush or split the timber. Use standard lost-head or oval wire nails, punched home below the surface of the boards. Dovetail nailing is good practice when nails are not to be clenched over.

It may not always be convenient to fix T or strap hinges to the boarded side of ledged doors since you may wish to close the door with the boards against stop mouldings nailed to the lining. This gives you one uncluttered and civilised side to the door, while the other can be all hinge, ledge and brace. By fixing T hinges directly to the frame or lining and to the boards on the ledged side, you can fit the

door flush with the lining (Fig 159). The early practice often involved nailing the strap hinges to the boards on the plain side of the door, while clumsily sandwiching the hinge between the boards and mouldings (Fig 168). Sometimes strap hinges were cranked to procure a fit, but very often the more awkward the method used, the more likely it is to be original.

Whether you choose to leave softwood ledged doors unpainted will depend on the character of the building. Normally they were painted, but there may be a case for not painting them in some 'rough' cottages or buildings converted from non-domestic uses. Eschew the polyurethane and varnish: the doors will look perfectly well without any sort of coating.

There is one kind of door which I hope you will *never* use in an old building. That is the flush door. There are several types, ranging from expensive solid-core hardwood ones to the cheapest framed sort with a hardboard or thin plywood facing. All of them are completely unsuited to any old building, from a converted mill to a Georgian manor house – with only one exception. That is a baize-covered door to staff or kitchen quarters, or sometimes a room like a library. There, a flush door could be correct. However, such doors are usually panelled ones which have been flush-treated on one side to take the baize while the civilised side is seen in the hall or passage. Since there are no historical precedents for the use of flush doors, it is almost unbelievable that apparently knowledgeable architects regularly introduce them in old buildings.

Although French doors are often an anachronism, there is one honourable compromise, suitable for many buildings dating from the very late eighteenth century onwards. It is the pair of doors with marginal glazing in the upper two-thirds of their height. The tall and narrow glazed areas have fine glazing bars crossing over at the corners to form *sqaure* panes. Horizontal bars run across at nicely judged intervals. Below the lock rails are flush panels with flush beading around them. It is much the glazing pattern used in the traditional telephone box, give or take matters of proportion and panelling. This can also be the answer to glazing the upper part of a single exterior door, where light is an important factor.

Fire Regulations
Among the chief bugbears of all who deal with old houses are the stringent requirements connected with risks from fire. These normally arise from the Building Regulations but, in many instances, are connected with the Fire Precautions Act and similar legislation. For the most part, those regulations affecting doors will only trouble you if a building has some commercial or public use.

Unfortunately, a great many fine old houses become subject to the Fire Precautions Act when they are turned into offices, hotels or institutions. This frequently means the introduction of smoke lobbies and fire doors, not to mention self-closing devices for doors or the removal of the original doors themselves. Period doors may be cased in hideous new materials as a method of providing some additional resistance to fire. The damage being done to the character and appearance of thousands of old buildings in efforts to meet fire regulations is impossible to exaggerate. It is not merely the obvious vandalism of casing magnificent

hardwood doors, or substituting brash modern replacements. The threat is less obvious but more pervasive when you consider the simple ledged or panelled and painted doors in quite ordinary seventeenth-, eighteenth- and nineteenth-century houses. These doors and their frames and architraves matter so much that you should go to extraordinary lengths to save them.

If you cannot retain or upgrade the original door, then try to devise a method of replacing it with one of the correct type which does provide the necessary fire resistance. If you can do neither of these things and you would be spoiling an old building, then abandon ideas of using it for purposes which involve the destruction of interior detail.

The normal requirement for upgrading panelled doors to comply with the legislation is that they have framing members not less than 45mm thick, that the panels are sufficiently fire-resistant and the rebates or stop mouldings are at least 12mm thick. Also, the door must fit the frame properly. Many doors, especially those of the ledged type, or the earlier panelled ones of the William-and-Mary or Queen Anne periods, will be far too thin to pass muster.

Broadly speaking the way to upgrade a panelled door is to thicken the stiles, rails and muntins by gluing and screwing additional matching lengths of timber to one facc. The panels are thickened up with sheets of fire-resistant material and the mouldings replaced. If the stiles are already 45mm or more you may only have to deal with the panels. It sounds straightforward in theory. In practice it can be a nightmare. If the door has stuck mouldings you cannot just lift them off and re-fix them when the panels have been thickened. The door rebates will be integral to the frame so that they too cannot be lifted to accommodate the increased thickness of the door itself.

If you want to thicken the frame by applying matching strips of timber, you may find that the architrave, or part of it, is integral and may not be lifted and re-fitted. Mouldings around door panels, when replaced, will either stand proud of the framing or must be shaved down and spoiled so that they will go back again. If new mouldings are run, the sections may have to be modified to lie flush with the stiles and rails. If you have bolection mouldings, this difficulty may be overcome rather more easily.

Whatever measures you adopt, remember that the fire officer will only be concerned with upgrading one side of the door, and that will be the 'side of greatest risk'. This side is normally the one *within* a room rather than on the landing outside or in the hall or passage.

It would be unwise to specify the actual materials for covering door panels to upgrade them against fire. On the whole fire officers are looking for an added thickness of roughly 6–10mm of a suitable sheet material. At one time this would have been asbestos, but there are other products today which are less controversial.

There are also flame-retardant paints which may be used to upgrade woodwork, but consult your local fire officer about their suitability. If their use is allowed they may prove the answer to at least part of the problem. There are two kinds of these paints: one is 'intumescent' – it swells up when heated and hardens on the surface, so forming a barrier. The other paint is 'non-intumescent', and fuses itself into a heat-resisting layer, while also producing a fire-retarding gas. Intumescent strips may also be fitted to both door-framing members and door frames; when heated, these will swell and help to seal the gap around a door which is less than a perfect fit.

When applying intumescent paints it is very important that the manufacturer's instructions on preparation of the timber are followed carefully. The door may have to be completely scraped down to bare wood. The risk with such paints is that adhesion may not be good enough. Doors which have been oiled and stained, although stripped of any coatings, can sometimes leech substances which impair adhesion. Most fire brigades insist on such paints being applied by approved contractors who will certify that the work has been properly carried out. It should be remembered that flame retardant paints are designed to reduce flame spread, but are not regarded as a means of *fire resistance*. If officers are prepared to overlook the upgrading of old doors along a corridor, for example, they may require the introduction of a firescreen between this area and the staircase landing. Obviously it is much better for a listed building to introduce a solid door and architrave which matches the original ones. If it is quite essential that some areas of glazing be devised, the detailing should be faithful to the style of the building – panelling, glazing-bar profiles and mouldings must be accurate. The glass will usually have to be of special manufacture in two layers with a gell in between. Cornices may have to be returned across the top of the screen.

In some instances it may be possible to reduce the impact of fire measures on the building by doing a trade-off with the officers involving the discreet use of smoke detectors in ceilings. These, however, may in themselves prove repugnant. Sprinklers

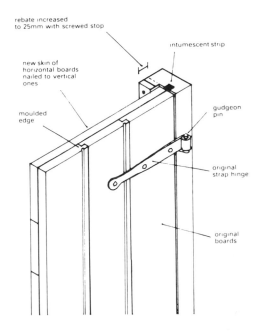

Fig 169 Old plank doors which must be upgraded to meet fire-regulation standards may be doubled in thickness by close boarding on the ledged side. Intumescent strips, which expand when heated, can be planted in the frame; and the depth of the rebate increased by screwing on matching hardwood battens.

as a fire-fighting measure can often do further harm to the building, and prevention is better than cure.

Many early buildings used as hotels, hostels, conference centres or country clubs have old ledged-and-battened oak doors. If they do not, and are of a period pre-dating the mid-seventeenth century, such doors have probably been removed by previous owners. Wherever possible they should be reinstated, especially those which must hang in arched or moulded stone or timber frames.

Ledged doors may be accepted by a fire officer if they are coated with fire-retardant paint. This treatment could be quite suitable for doors in a seventeenth-century building, for example. Natural hardwood doors, on the other hand, may have to be thickened up over the entire surface. This can be done by adopting the perfectly correct device of backing the vertical boards with horizontal ones from top to bottom (Fig 169).

Alternatively, you may decide to sandwich a fire-retardant material between layers of boards. Although not structurally required, it may be architecturally worthwhile to re-fix the horizontal ledges to the doubled-up door. If boards of such doors are only butt-jointed, fire-officers may want these gaps dealt with. One way is to plant cover strips, moulded in some authentic manner.

The new Building Regulations of 1985 include a section on 'Mandatory rules for escape in case of fire'. These requirements are the only way in which you may comply with paragraph B,1 (fig) of the Regulations. It is of interest that when the Building Regulations apply means-of-escape requirements to a building to be put to a designated use, the fire authority cannot normally insist on structural or other alterations being made as a condition of giving a certificate.

At present the rules must be applied to new buildings of three or more storeys, those of three or more storeys containing a flat, and ones containing an office or a shop. The fire brigade's officers will shortly give up the responsibility for issuing fire certificates in these cases and the duty will be taken on by the local authority's building control department. The fire officers will then have a purely consultative role. The mandatory rules in the Building Regulations will then be made to cover means of escape in all circumstances. From the old buildings point of view, dwellings affected are those which have been extended or materially altered, and which will have three or more storeys. This includes those of three or more storeys which have had a material change of use to a dwelling.

The rules say that: 'The means of escape provided need only, in the case of a dwellinghouse or a building containing a flat, afford escape for people from the third storey and above and, in the case of a building containing an office or a shop, afford escape for people from the office or shop.'

Loft conversions have special rules in Appendix B, including the need for a window or rooflight at least 850mm high and 500mm wide when open, positioned not more than 1.5m up the roof slope from the eaves, with a sill no more than 1.1m from the floor.

There are other rules which could prove very damaging to an old building. They include the need to fit fire-resisting doors, self-closing devices, wire glass in doors, fire-resisting enclosures for stairs and so on.

Most of the technical details for meeting the requirements of the mandatory rules are embodied in:

British Standard 5588 Part I 1990 – dealing with all residential buildings;

British Standard 5588 Part III 1983 – dealing with ofice buildings;

The Home Office Code – fire precautions in hotels and boarding houses;

Mandatory Rules For Means of Escape in Case of Fire – Building Regulations 1985;

Timber Research and Development Association (TRADA) publish a number of useful information sheets, including ones on upgrading doors;

Building Research Establishment is also

worth contacting on possible new answers to special problems.

Obviously, you will avoid being caught up with the mandatory rules if you possibly can. The danger lies in having to implement them to the detriment of the building, for instance, in the case of a warehouse conversion, change of use to an office or a shop or, most importantly, the conversion of the roof space of an old house into living accommodation.

The Importance of Doors

Although it is not difficult to persuade the owners of grand and imposing houses of the importance of their doors and other joinery, the occupants of more modest buildings are often less easily convinced. Even if the latter houses are not 'listed', their interior woodwork is every bit as crucial, in its own context. One hopes that planning officers will lend their weight to the idea that joinery in *unlisted* buildings matters and they have a duty to do so when applications for listed-building consent are involved. Either they can press the Building Control Department for more lenient requirements which will preserve the interior, or they can recommend that the proposed change of use which has made the measures necessary should be refused.

The first duty is to the building. After that has been established in the minds of the owner, the planners, the architect, the builder, the fire officer and the building inspector, everything humanly possible should be done to bring about a sensible, sensitive change of use. To sum up:

1 Never remove old period doors which can be saved.
2 Never put in new or replacement doors of the wrong pattern.
3 Never use a flush door.
4 Pay close attention to architectural details of doors and porches.
5 Make sure all door furniture is in the correct style and made of good materials.
6 Show integrity and determination when coping with fire regulations.

11

STAIRS AND WOODWORK

There are certain features in an old building which stand a reasonable chance of survival: the flues and fireplace *openings,* the principal roof timbers and the staircase. The reason must be that they are such an expensive nuisance to replace. Most house owners in former times had exactly the same iconoclastic itch to strip out the old and put in the new that pervades society today. The rich and the fashionable certainly managed to dispense with a formidable number of old staircases in the seventeenth and eighteenth centuries, but there is a wealth of delightful ones which they somehow overlooked. The Victorians, who were no slouches when it came to imposing their will upon elderly buildings, made some fairly severe inroads, but open wells and graceful stone cantilevers survived.

The last-named type of staircase is perhaps as good an example as any to back up the theory that plasterwork and windows, panelling and fire surrounds, were easy meat in comparison. The structural implications of digging out all those heavy interlocking stone treads would daunt the most sanguine Victorian paterfamilias. Although many excellent framed staircases were removed, anybody who regularly visits threatened buildings to report on their architectural contents and condition will agree that the staircase is frequently one of the few features which raise his flagging spirits as he wanders disconsolately from one ravaged room to the next.

The other place where he may learn something to his advantage is in the roof space, where ancient oak trusses survive beneath a garbled network of later structural timbers. Lacking the authority to chip away at the castellated oatmeal tiles

surrounding the fireplace, he never *does* learn that a vast stop-chamfered lintel lics behind.

If you have just bought an old house or cottage, do not despair if all you find there is an unappetising little flight of steps, with

All too few genuine cottage staircases survive the ministrations of the modernisers. Here is a typical example of minimal size tucked into a corner next to a gable-end hearth. It consists of three straight steps and five winders. The top one is of splendidly hazardous proportions. Stairs like these defy all building regulations so don't replace them unless you absolutely have to. If you do, you may become subject to the attentions of your local authority building control officers, who could quibble about substituting 'like with like'.

a quite decent handrail, sitting on top of a blank balustrade of eau-de-Nil hardboard. All may not be lost. Take a chisel and mallet. Prise the hardboard away and behold the unassuming but elegant square balusters, still triumphantly where they ought to be. Concealed by some 'improving' hand, a dignified turned newel post emerges like a diva, no longer young.

Staircase Development

Many very early houses and cottages had no need of a staircase; only fortified buildings, monastic buildings, and a few merchants' houses and manors usually had upstairs rooms. Until late medieval times (fifteenth century), the basic manorial plan included a great hall open to the roof timbers, with a solar and, perhaps, a minstrels' gallery. To reach the solar, you needed stairs and these took various forms. Early medieval buildings either had winding stone or brick newel stairs, which might be in a turret or rise within the thickness of the walls. Some great halls had internal timber stairs for access to a solar, but the usual scheme appears to have been a straight flight of exterior steps running up an outside wall. These might have a timber canopy or 'pentice'. Steps could also approach at right angles to the external wall supported on solid stone and rubble, or upon piers, or columns and slabs. Whether to reach a solar or a first-floor hall, these were the methods employed.

During the fourteenth and fifteenth centuries, stairs in domestic buildings continued to be stone or brick newel types, but in some instances they were constructed of timber, rising around a central pole or 'mast newel'. Some stairs took the form of straight flights of stone steps or steps made of heavy timber baulks. These were sometimes placed between parallel exterior and interior walls.

Until the second half of the seventeenth century when framed timber staircases became the fashion, stairs in the average small manor, farmhouse or town house were squeezed into any convenient corner available. A favourite position was the recess beside the hearth projection. Equally steps could wind around a central chimney, or you might discover a cramped little flight of stone winders forming half a circle, housed in a slightly thickened or projected wall. An external staircase turret of this kind is called a caracole.

Ladder staircases were always to be found, especially in small buildings, but few of these early ones survive; they have usually been replaced either with a framed staircase or some system of steps accommodated in a special wing built on to a rear wall of the house. Right through until the early nineteenth century, cottages for the very poor would have no more than an open hall-kitchen, with a kind of screened sleeping platform over part of the ground-floor room. This was reached by a ladder. Until very recently you could still find miners' cottages in the North-east with a ladder to the upper floor.

The Elizabethan development of the framed staircase began the open-well formation (Fig 171) which was to continue throughout the whole of the seventeenth century and in many houses during the eighteenth century. The sixteenth-century variety had closed strings, heavy turned balusters, short flights of oak steps and a fairly minimal well in the centre. Newel posts would have

Fig 170 Late seventeenth- and early eighteenth-century staircases began to have barley-sugar twist balusters and scrolled brackets decorating the open string.

ornamental finials – slightly flattened ball shapes, or patterns with faceting and mouldings. By the early seventeenth century, carved heraldic beasts were used to decorate newel posts if the house was a grand one. Balusters at this time were often cut in a rectilinear version of a column, or in lesser houses a number of flat boards were fretted out to convey the same idea in silhouette. These types are called splat balusters.

Seventeenth-century staircases became steadily more refined. Handrails which had once been either extremely plain, or in the form of a robust roll flanked by ogee or ovolo mouldings, now became broad, flat, and lightly moulded at the edges.

The average late seventeenth-century or early eighteenth-century staircase of the better sort remained a thing of solid, square, panelled newels, with balusters fixed to a closed string. The handrail was still jointed into the newels at every change of direction, but the upstand of the newel itself became less prominent. Newel caps were rather flat and echoed the moulding of the handrail. Balusters in the mid-seventeenth century could be massive turned vase-shaped affairs in the Italian Renaissance manner.

By the reign of William and Mary, grand staircases would often have graceful turned balusters shaped like vases, columns or barley-sugar twists. Handrails, at last, began to sweep over the tops of the newels, and beautifully executed swan-neck or knee-and-ramp sections were fitted to accommodate the changes of level between one flight and another.

Treads were now to be seen from the side in the cut-string formation; and often they were decorated by bracket-shaped mouldings on the string itself (Fig 170). Since the Stuart period very important houses had sometimes adopted an open well with a central bottom flight of steps, dividing into two further flights on either side. An alternative layout was a big full-width balustraded landing to the hall with matching flights descending from each end. Quarter-landings and further flights completed the descent. A variation favoured in many vernacular houses was a short

flight to a small landing. Thence the staircase would branch right and left, rising against the rear wall of the building (Fig 171).

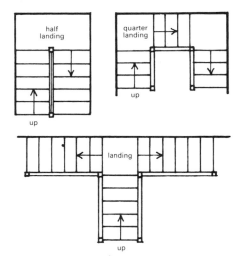

Fig 171 The dog-leg stair (*top left*) has never gone out of use since the eighteenth century. The open-well staircase with quarter landings (*top right*) started in the late sixteenth century and was common by the late seventeenth century. Seventeenth-century vernacular houses often had staircases which branched right and left at a small landing, approached by a first short flight of steps (*bottom drawing*).

By the middle of the eighteenth century most small houses had dog-leg stairs (Fig 171), although a great number of the more dignified ones still had open wells (Fig 171). Balusters became more and more elegant in their design, normally taking the shape of vases or columns, or combinations of the two. Cut-string staircases were the fashionable rule but closed strings never went right out, especially in rural buildings. Handrails almost always carried over the tops of the newels in more formal houses by the last third of the eighteenth century.

At this time a variety of stone cantilevered staircases of geometric plan were being installed. They had simple mahogany handrails and wrought or cast iron balustrades. Such staircases can be related to the neo-classical period of architecture which was so greatly influenced by the work of men like Robert Adam.

By the late eighteenth century and well into the nineteenth century, modest houses, and quite a few grand ones, had dog-legged stairs or open wells. They had plain mahogany handrails, thin square balusters and simple fretted brackets on the strings. The handrail finished on top of a plain column newel, rising from a curled-around bottom step. The rail was wreathed to reflect this final curve.

Throughout all the later phases of staircase development, cottages and farmhouses continued to have tiny flights of combined straight steps and winders, tucked in wherever a suitable space could be found. In the eighteenth and nineteenth centuries such stairs were often boxed in with a light stud framework which was occasionally panelled, but more often covered with thin boards or lath and plaster (Fig 172).

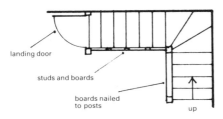

Fig 172 Many cottages in the late eighteenth and nineteenth centuries had stairs with winders to make maximum use of limited space. The stairs were frequently boxed in with boarded stud partitions.

Such staircase partitions were frequently closed with a ledged door at the top or bottom. Above the level of artisans' terraced cottages, later nineteenth-century houses frequently had slightly coarsened and elaborated forms of the eighteenth-century handrail and turned baluster. Throughout both the eighteenth and nineteenth centuries you would also have found many departures from the strictly classical staircases when owners decided to pursue some Gothic, late medieval, Elizabethan or Jacobean theme.

Repairing, Restoring and Replacing Staircases

The above brief account of staircase design shows the critical importance of using something in the correct style for the period. It may be a question of repairing or restoring an existing staircase, or even of putting in an entirely new one. Whatever you do should be of informed design and show great attention to detail. Do not, for example, fit a staircase with bold early eighteenth-century turned balusters in a little Regency rectory, just because you are offered some salvaged ones.

There are three common mistakes which people make when putting in a new staircase. The worst and most frequent error is to insert something of modern design, in harsh modern materials. The second is to put in a staircase which is a bowdlerised version of what might have been there. The third is to introduce a design which is reasonably accurate but from quite the wrong historical period.

As far as repair is concerned you will often find that it is the balustrading which has suffered. The very simplicity and structural importance of the steps themselves may ensure their survival, but handrails and balusters are another matter. The complications of fitting balusters which match those that have been broken, or getting a skilled joiner to shape a section of curving handrail, are enough to make some owners decide to box in the entire balustrade. (Fiends will even remove the balusters completely and shove in whatever stock items they can get their hands on.) All the same, the average old staircase balustrade is unlikely to have more than a few missing or broken balusters. Only if a building has been left derelict and open to intruders is much damage likely to have occurred.

Structurally there is not a great deal to say about repair work, since the basic methods of framing a traditional staircase have changed little in the past 300 years. The average competent joiner will find himself using familiar techniques.

However, there may be a few minor faults which the owner could cure. The most likely are treads which squeak or give underfoot, wobbly balustrades, loose newel finials and, perhaps, landing trimmers which have rotted in the walls, or have inadequate wall bearings.

On properly framed staircases you seldom see any sag across the width caused by dropped outer strings, probably because they occupy a space which is well ventilated and free from contact with sources of damp. Decay is, of course, the most persistent enemy of staircases, apart from major and flagrant overloading – the staircase scene by Rowlandson, with many drunk and portly persons tumbling down it in confusion, comes to mind. But that one was made of stone and was cantilevered.

Wall strings may suffer from dry or wet rot and so may any trimmers which are socketed into exterior walls. One reason, I believe, that stairs are usually so firm and well constructed is that this was one of the few tasks which no bodging builder could leave to his labourers. He had to employ a competent carpenter and joiner, who made up most of the staircase in his workshop. Consequently, with any luck, the treads fitted their housings, the joints between newels and strings were soundly constructed and the whole was rigidly fitted in place. Softwood carriage pieces, and brackets, provided extra support and balusters strutted firmly between handrail and string or tread.

Movement in stair treads often results from wedges having fallen out or become loose. Blocks glued in the angle formed between treads and risers may also have come away. New wedges may be cut and driven, and further blocks can be glued into position (Fig 173). Additional brackets may be nailed to the central carriage to support springy steps (remember to place the grain upright). If no carriage piece has been used it may be beneficial to put one in, if the staircase is in need of stiffening.

Methods of fitting balusters vary. Some are dovetailed into the edge of the tread; others are housed into the top of the tread, and others again are simply bevelled, tightly fitted and nailed to the string. Iron balustrades have a core rail slotted into the underside of the handrail and balusters are lead-jointed into sockets in the stone steps.

The whole secret of success with staircase repairs is an ability to make any carpentry or joinery work fit very firmly and snugly together. Play between different members must be avoided at all

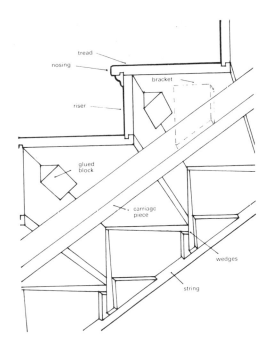

Fig 173 A faulty staircase may be stiffened by introducing a central carriage piece (shown off-set here for clarity). Springy treads can be supported by brackets nailed with the grain vertical to the carriage. New wedges can be tapped into the string housing joints, if necessary; and wooden blocks glued at the angle between treads and risers.

costs. Theoretically, you should be able to fit back a displaced piece of handrail, using special handrail bolts (Fig 174), but I should not advise amateurs to try it. You cannot afford to damage an old mahogany handrail, especially if it is wreathed in beautifully contrived curves.

Missing balusters may be turned to

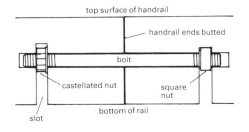

Fig 174 A replacement length of handrail may be tightened up against the existing rail with handrail bolts. The square nut at one end of the bolt is held firm by the surrounding timber, while the castellated nut at the other end is tightened with a screwdriver inserted in the slot on the underside of the rail. The new piece of rail is accurately located and jointed with wooden dowels.

match the remaining ones and the best wood for this is beech. Most areas have skilled men who specialise in wood-turning and none more so than the Chilterns. There, I recall it as a cottage industry carried out by men living in small isolated brick and brick-flint cottages deep in the woods.

Faulty softwood carriages with rotten ends should be replaced or repaired, and landing trimmer joists may have decayed parts cut away and new ones added. For this you can sometimes carry the sawn-off and treated end in a steel shoe, but, failing that, either cradle the end of the joist with mild-steel plates or bolt a new length, with plenty of overlap to the old one. If you follow one of the last two courses it would be well to use shear plates for steel repair or split ring connectors in the case of timber. The new joist should lap the old one by at least six times its depth.

Landing trimmer joists are, in effect, quite important 'beams' which carry the ends of numerous trimmed floor joists. Often trimmers are hidden by finishes, so that repairs may be kept out of sight. You will be unlucky if the problem of decayed trimmer joist ends is encountered when they bear on interior walls.

Open-well stairs may be provided with concealed props at their weaker points if they have panelled spandrels. Even if they have not, it may be possible to introduce a 100mm square softwood support below a quarter landing without doing too much visual damage. It depends on the visibility of the prop and the quality of the staircase.

Propping is bound to be an eyesore, if you are dealing with a curving geometric stair which will not take kindly to clumsy interruptions of that soaring and sweeping soffit. However, it might be worth devising a classical column to support part of an ordinary cantilevered stair, paying close attention to the orders of architecture used for nearby cornices, pilasters and door cases.

Fortunately, few people have to deal with this sort of dilemma. All the same there is an astonishing number of relatively small late eighteenth- and early nineteenth-century buildings with cantilevered stairs.

The removal of timber newel posts and the repair of the double-tenon joints between posts and strings is again no job for an amateur. But it can be done – with difficulty. Less demanding may be the task of re-gluing the joints and providing new dowels which make a tight fit. Balusters can be secured with angle irons screwed to the handrail at the top and to the string or tread at the bottom. The drawback is that these irons must be concealed. Perhaps the best way is to chisel out neat sinkings to a depth of rather more than double the thickness of the irons. Set aside the chips you have removed and glue them back to cover the angle iron when it has been screwed down.

The question of whether or not to use synthetic resin adhesives for repair work on internal joinery is often debated. These adhesives are extremely strong and those such as epoxide resins are almost impervious to water, acids, alkalis and solvents. On the whole it seems better not to use them, since they make it virtually impossible to part a timber joint should the need arise in the future. This danger is well known to those who repair antique furniture, and no responsible restorer would use an epoxide resin on the joints of a seventeenth-century court cupboard, for example. The traditional glue for timber joints has always been that derived from the bones, skins and hides of animals. It is very strong, but is not resistant to heat and moisture.

Joints made with animal glue or 'Scotch glue' as the best of it is called, may sometimes be steamed apart. Some old glued joints can be tapped apart with a mallet. Soaking in warm water is also a possibility for timber joints which one is not obliged to deal with *in situ*. Dowels may be removed by drilling out with a slightly undersize drill bit. The risk when applying steam and heat to a staircase newel or other feature of internal joinery is that you may damage a polished or grained finish. If it is essential for repair purposes to do some limited damage, you will have to try to match up the finish afterwards.

If you are using casein (milk-based) glue for timberwork, remember that it can cause dark-red or purple staining on oak.

Regulations concerning Staircases

Although the Building Regulations 1985 adopt a more conciliatory tone than those which they replace, nothing much has changed in the demands which they make on people with old buildings. The relevant requirement is K1: 'Stairways and ramps shall be such as to afford safe passage for the users of the building.' That sounds innocent enough until you look at the guidance in Approved Document K. It embodies *almost* every detail supplied in the former regulations of 1976.

The only possible let-out is the standard comment made in each approved document: '… there is no obligation to adopt any particular solution in the documents if you prefer to meet the requirement in some other way'. However, it goes on to say: 'If a contravention of a requirement is alleged then, if you have followed the guidance in the document, that will be evidence tending to show that you have complied with the Regulations. If you have not followed the guidance then that will be evidence tending to show that you have not complied. It will then be up to you to demonstrate by other means that you have satisfied the requirement.'

Presumably, the only way you could do so is by satisfying a court or an inspector that your own design and measurements are, while at odds with the guidance, still capable of 'affording safe passage'. Remember that sometimes your best bet may be to put in some form of ladder which is not counted as a staircase at all. It must be steeper than 55°.

If you can prove that a new staircase is 'replacing like with like' your work will be free of regulation. You could also claim that your stairs are not a 'material alteration' and do not 'adversely affect' the building under parts 'A' and 'B' of Schedule 1, which contains the actual requirements of the regulations (see Chapter 2). If your work requires that you observe the 'Mandatory rules for means of escape in case of fire', that contention will be impossible to support.

The hope must be that the law will be leniently applied and interpreted as far as old buildings are concerned, especially in the case of small houses and cottages,

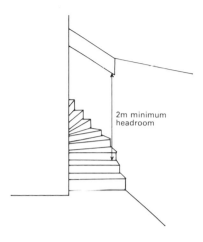

2m minimum headroom

Fig 175 To obtain 2m clear headroom above every part of a little winding staircase in an old building is often impossible; yet that may be exactly the right kind of stair in the most historically accurate or most convenient position.

where space is always restricted. If all else fails, you should seek a relaxation or dispensation of a regulation. Stairs in new extensions will clearly be regulated and so will those involving a 'material change of use' – barn conversions, for example.

Some of the main pieces of 'guidance' from approved document K are as follows.

A clear 2m headroom must be provided throughout the staircase. This is measured vertically above the pitch line and must be obtained over the whole width of the stairs – clearly impossible where low or sloping ceilings, old beams and other obstructions must be negotiated (Fig 175).

The pitch of the stairs must not be greater than 42°, which can mean that stairs project much too far into the available floor space of cramped cottages. This can also be a problem when converting small industrial buildings. In some situations a steep flight of steps may be absolutely demanded – a traditional stone flight partly built in the thickness of a wall, for example.

There is a regulation which requires a minimum width of 800mm between the wall and the edge of the handrail on the other side, or between handrails if there are two of them. If the stairs are wider than 1m – a common occurrence in Georgian buildings – then handrails must be provided on both sides. Imagine designing an eighteenth-century staircase with a clumsy and unnecessary handrail down the wall where the dado rail is meant to be.

Few regulations have given more headaches to people planning staircases for old cottages than the one which asks for 220mm minimum 'going' – the horizontal measurement from the nosing of one step to that of the one below. Also required is 220mm minimum 'rise' between the tread of one step and the next (Fig 176).

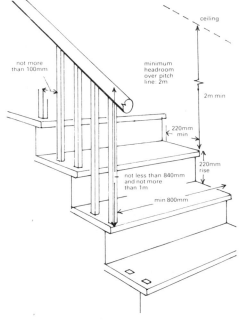

Fig 176 When the Building Regulations apply, they can present many problems for anyone designing historically appropriate staircases for old buildings. This drawing shows some of the more important requirements in private houses.

Handrails have to be at least 840mm and not more than 1m above the pitch line. Another restrictive regulation is the one which refuses any opening between balusters (or the steps of open-tread stairs) which will allow the passage of a 100mm diameter sphere. This could completely throw the design of a fine staircase with carefully researched balusters. If stairs have no risers you are obliged to close the soffit with boarding, or introduce little rails, rods or boards, at the backs of the treads, running between the strings.

All steps in a flight must have the same rise – another rule which can play havoc with your calculations when you are tight for space and trying to fit a staircase between existing floors. It is sometimes extremely convenient to have one step shallower than the others to complete a flight.

The rules for tapered stairs, such as might be found in a stone newel or timber mast newel staircase, require a minimum going of 50mm. On the pitch line (halfway across the notional width of the staircase) the tread must be no less than 220mm (Fig 177).

The sort of tiny winding staircase which leads up to one side of a projecting hearth could be an impossibility in terms of the Building Regulations. Yet it may be your wish to put the stairs back where they might have been.

It will be argued that by careful design you should be able to comply with the regulations, but a thoughtful study of old staircases and their measurements will show that the right solution cannot always be the one which best pleases inspectors.

The effects of the Fire Precautions Act on buildings with good staircases and landings can be severe. Smoke lobbies with self-closing doors may be required, and not infrequently fire officers insist on boxing-in stairs or landings. The best way around this problem is to devise a suitable bypass for the landing and staircase, giving independent access to another means of escape. It may involve sacrificing some space from bedrooms in a hotel, for

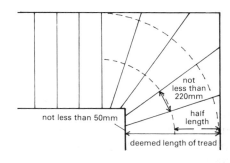

Fig 177 Staircases with winders may be necessary in old buildings as works of 'material alteration' or 'material change of use', or in new extensions. The Building Regulations' requirements for a flight which is 1m or less in width, are shown here.

example, but it is vastly better than ruining a beautiful stair well and hall.

Other Interior Joinery

The whole idea when repairing joinery is to make a firm, durable joint which retains at least as much flexibility as it had when first fitted together. This principle is especially apparent in panelling, since panels should be capable of movement within the framework of stiles, muntins and rails. Glued joints between the edges of panels and the rebates or grooves in the framing will inevitably lead to splitting of the wood when changes in temperature and moisture content bring about shrinkage.

The main forms of interior joinery which you will encounter, other than doorways, stairs and fireplaces, are panelling and shutters. Since panelling plays such a very important part in the repair and restoration of old houses, you need some idea of its historical development.

Panelling of the wrong kind is not acceptable; nor is failing to appreciate the importance of what is already there, though of course our ancestors were far less scrupulous, or perhaps inhibited, than we are today, and fitted whatever kind of wainscotting accorded with the fashion.

The idea of lining out rooms in framed timber seems to have caught on in the second half of the fifteenth century. Very few people have houses earlier than that date, so it is reasonable to take the fifteenth century as our starting point.

Possibly as a development of the post-and-plank form of screen or partition, it became customary to line or 'seal' walls with vertical boards, which might be crudely lapped, or rebated. Gradually, it became the practice to frame up true panelling, using horizontal rails and vertical stiles and muntins. Panels were frequently carved in the familiar linenfold pattern during the late fifteenth century. Mouldings on early panel framing were simple, and to avoid returning them at intersections, were either stopped, run out or allowed to butt straight up against unmoulded timber (Fig 178).

During the sixteenth century, panelling became more and more popular, and was

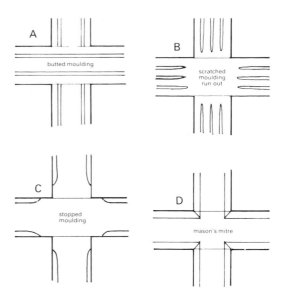

Fig 178 Craftsmen in former times had various methods of handling the difficult junction between moulded stiles and rails of panelling. The moulding of the stile might be butted against the rail (A). Scratch mouldings could be run out before reaching the rail (B). A chamfer might be stopped (C) or a *mason's mitre*, which is not a mitre at all, might be used (D).

to some extent standardised. Panels were fairly small and rather square, equating with the maximum width of board which could be obtained conveniently in any quantity. To begin with, mouldings were 'stuck' as the term has it: this means that they were run on the solid framing timber. Later it became common practice to 'plant' mouldings around panels, using the modern mitre technique where they met at the corners. Before that, intersecting stuck mouldings had to be joined with 'mason's mitres' (Fig 178), or one moulded timber might be scribed over another at intersections.

The standard framing joint for panelling was the mortice-and-tenon with pegs to secure it. The traditional framing method was to run the rails right across the wall and joint the vertical muntins into them. This is still the way it is done today.

The dates of basic techniques for moulding panels can overlap considerably, and apparently obsolete methods survived long after they had ceased to be fashionable. The simple scratch moulding, which overcomes the difficulty of treating intersections, was used long after

191

the mitre had become commonplace. The reason for this is that incised-groove decoration can be simply tailed off at critical junctions by lifting the moulding iron. It is rather similar in principle to the basic unmoulded stop with which you can allow a beam chamfer to run out at the ends.

The design of oak wainscotting became very plain after the Gothic influences of the late fifteenth century, and remained so until the late sixteenth and early seventeenth centuries. The latter change of decorative flavour coincided with the introduction of Renaissance details, many of them influenced by the mannerist pattern books published in northern Europe.

In grand houses, pilasters were commonly used both to divide up the wall area and to embellish window splays. Friezes and cornices were also employed. The pilaster details themselves tended to include fluting, often cabled at the bottom. Ionic and Corinthian capitals were in evidence. Friezes were elaborate and often more Gothic than classical, carved with floral and vine-leaf trails, or heraldic beasts. But Italian arabesques and symmetrical flat-ribbon designs also abounded.

Much use was made of corner blocks for panels, while diamond lozenges and squares of planted mouldings supplied further variety. Carved nail-head decoration was also used, again echoing the masonry rustication of Italian buildings. Arcaded panels were also fashionable, of a kind often seen in the doors of seventeenth-century oak court cupboards.

In the mid-seventeenth century the era of the planted bolection moulding (Fig 179A) began and it was to continue in fashion until just after 1700. These mouldings, of heavy and bold outline, were used in conjunction with tall, raised or fielded panels. There was now a definitely emphasised dado with horizontal panels divided from the main panelled wall surface with a chair rail. Friezes became steadily plainer and were frequently omitted.

By the reign of Queen Anne, panelling had adopted the simple form it was to retain for the next eighty years or so. The framing was usually moulded with a small

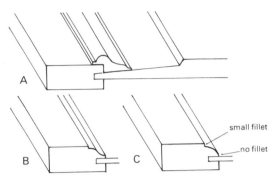

Fig 179 The bolection moulding (A) was frequently used for wall panelling and doors in the late seventeenth and early eighteenth centuries. *Stuck* ogee moulding (B) of various kinds was employed in many eras. The small fillet and stuck ovolo moulding, run out on the panel without an intermediate fillet (C), was very common in the early eighteenth century.

ovolo, often running out straight on to the unfielded panel, without a fillet (Fig 179C).

In the second quarter of the eighteenth century it was not uncommon to field the panels, but mouldings remained small and plain – an ovolo or perhaps a rather flat-looking ogee (Fig 179B). Eventually, the fashion for panelling gave way to rococo or neo-classical plasterwork on walls, and nothing remained except the useful visual punctuation of skirtings, chair rails and frieze rails.

Only in Victorian times was panelling reintroduced with any enthusiasm, and that was limited to houses which had been consciously designed in some former style. Panelling of a rather bland and meagre kind was reintroduced in the Edwardian era, frequently being terminated about two-thirds of the way up the wall and capped with a shelf. This was perhaps a throwback to the wainscotting of the Tudor great hall.

Partitions
Few partitions were needed in medieval houses, which were organised around the communal life of the great hall. Probably the earliest type of partition regularly fitted was the timber screen which took some of the edge off the draught from the main entrance door. This developed into a full-width screen forming a passage right across one end of the hall. Over it might

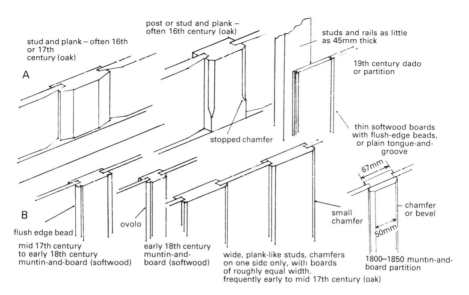

stud and plank – often 16th or 17th century (oak)

post or stud and plank – often 16th century (oak)

A

stopped chamfer

studs and rails as little as 45mm thick

19th century dado or partition

thin softwood boards with flush-edge beads, or plain tongue-and-groove

67mm

B

flush edge bead

ovolo

mid 17th century to early 18th century muntin-and-board (softwood)

early 18th century muntin-and-board (softwood)

wide, plank-like studs, chamfers on one side only, with boards of roughly equal width. frequently early to mid 17th century (oak)

small chamfer

chamfer or bevel

50mm

1800–1850 muntin-and-board partition

be a solar or minstrels' gallery. The screen partition was placed at the lower end of the hall, opposite the dais; and twin doors on the far side of the cross-passage led to service rooms, such as butteries or pantries.

If you have an early house, the original partitions will usually be of either post-and-plank, or muntin-and-plank, construction. A possible alternative is a heavy stud partition with an infill of wattle and daub or plaster, similar to the external walling of a timber-framed building.

After the sixteenth century, partitions could still be of muntin-and-plank formation, the width of the muntins becoming less massive in later examples. Eventually, in the later seventeenth century, such partitions might be made of imported pine, but oak was still preferred.

It is arguable at what point partitions first began to be plastered on riven-oak laths, but certainly this practice was well established in the seventeenth century. By the early eighteenth century it was quite common to use softwood lathing. If the laths of a partition are clearly seen to be sawn rather than split, they will probably be of late nineteenth-century or twentieth-century manufacture.

Whether a partition you introduce comes under the heading of joinery depends upon its construction and finish.

Fig 180 Post-and-plank or muntin-and-board partitions were used in all kinds of houses, especially up to the mid-seventeenth century. Drawing (A) shows a typical post-and-plank partition of the sort often used for screens passages, while (B) is an early eighteenth-century muntin-and-board affair which might be found in the attic of a Queen Anne house.

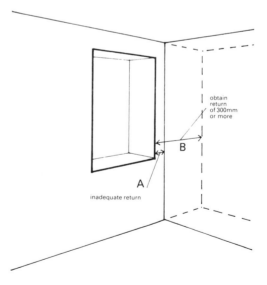

obtain return of 300mm or more

B

A

inadequate return

Fig 181 A major fault in positioning partition walls is to leave visually and practically inadequate lengths of return wall alongside openings (A). By placing the partition further to the right (B) the proportions of this room are likely to be improved and reasonable space is given for pulling back curtains or standing furniture.

193

The kind of timber screen, backing on to a cross-passage, which every old-buildings buff longs to discover. This one was found under layers of wallpaper by the DoE lister while inspecting the house. The screen is not of muntin-and-plank construction, but consists of scratch-moulded boards, chamfered, to allow each successive board to lap the next. The horizontal mark of what may have been a bench at the upper end of the hall is faintly visible. The boards clearly show that they were adzed to a finish. The screen could be early seventeenth century.

Panelling is joinery, while stud partitions for plastering are carpentry. The distinction is academic – what really matters is how partitions are placed.

Historically, partition walls were often butted up against window architraves, leaving virtually no return wall in the corner of the room. This precedent does not mean that it was a satisfactory method, either from the practical or the aesthetic point of view. If you put in a partition when adapting or converting a building, do your best to contrive at least 300mm of return between it and any opening in the adjoining wall. Cramped partitions not only look frightful, they also preclude any use of the wall alongside an opening for pulling back curtains, hanging pictures or placing furniture (Fig 181). This obviously vital factor is too often ignored by architects.

Another point to consider when designing a partition is how the quality of the finish you propose using relates to that of other walls nearby. A skimmed plasterboard affair on a stud framework may look horribly smooth and regular against old limewashed stone walls or some similarly mellow surface. Ways of getting over this difficulty when plastering are considered in Chapter 13. With joinery, it is worth noting that details like skirtings should be matched if possible, and that chair rails or dado boarding should be continued as elsewhere in the room.

Rooms lined out with matchboards reaching slightly above traditional dado-rail height may be found in many Victorian and Edwardian cottages. The boards can be vertical or horizontal, and a simple bull-nosed capping is fixed along the top. This system has the advantages of providing insulation, a visual break of texture, and a cover for pipes and cables or for damp-proofing measures. Generally

such boards are very thin with tongue-and-groove jointing. Sometimes they are butted plainly together without moulded edges; but often original ones have a thin flush bead along one edge of each board. The boards should be nailed to cross battens, themselves either masonry-nailed to the wall surface or plugged and nailed. Vertical boards are more often seen, since they make economic use of the timber, requiring no unsightly joints. Even on a stud partition they may require horizontal battens for fixing.

The total lining of rooms with match-boarding is frequently of Victorian date. The boarding would be attached to battens and was very often wallpapered over, although it was sometimes white-washed. Genuine, flimsy, matchboarding is difficult to obtain these days, but thin tongued-and-grooved boards may serve just as well, although they are wasteful of timber.

In cottages, boards often only covered one side of the partition studwork. In hall passages it was normal to leave the studs exposed, while the boarded finish was put to better effect in the parlour on the other side. Often the hall passage partition has a lath-and-plaster surface on one side and a matchboard finish on the other. Another place for a boarded partition is a larder under the stairs, or a dairy or pantry. These partitions are thoroughly in keeping with many old buildings and almost the only mistake you can make is to use V-grooved boarding when the house is too early in date to suggest it. Try to keep the V-groove type for late Victorian and Edwardian houses, when the boards are to be exposed. However, even in those, it will often be inappropriate.

What you should hardly ever do with exposed boarding, any more than with panelling, is to apply varnish or polyurethane finishes. Varnished boards may be all right for a 1900 timber bungalow in the Scottish Highlands or in Virginia Water, but should generally be avoided elsewhere. The rule with exposed softwood in 'rough' buildings is 'When in doubt, apply emulsion paint'.

Architraves and Skirtings

Architraves around window openings will often need minor repairs or re-fixing. Because they surround the window on the *inner* face of the wall, they are not always as rotten as you might have feared. However, builders have a way of ripping them out at the least provocation. They are, of course, important architectural features which should be retained or accurately replaced, at all costs.

Usually, they adopt the same mouldings as the interior door architraves, but by no means always. A favourite mid- to late eighteenth-century profile was a bold ovolo moulding between two fillets. But an ogee was also popular.

New openings for doors must be provided with suitable architrave mouldings, either matching those already to be seen around adjacent doors or designed in a style which suits the building. All you have to do is provide your joinery firm with a sample, a template or same-size drawing in section, and they will run off what you need. It is not necessarily a cheap exercise, but it does not have to be prohibitively expensive either. In old houses try to avoid using most of the stock architraves supplied by the builders' merchants. These are almost invariably bevelled or have a plain quarter-round edge, without fillets, and strike an unpleasing note.

Skirtings pose other problems. In most 'rough' cottages a plain square-edge board without any mouldings will often be suitable. Sometimes, you may think that a simple moulding along the top of the skirting, in the form of an ogee between narrow fillets, is preferable. When dealing with tall Victorian, or eighteenth-century, rooms in which other precedents have been set, you may need to design something more elaborate.

Tall rooms can require deep skirtings.

These are built up in tiers to obtain enough height for a visual balance with the other horizontal divisions of the wall. First a board is run along the floor, its top edge moulded to conceal and ornament the junction with the next one placed above it. In very big rooms, there may even be a third board, forming a three-tier, stepped-back skirting.

Skirtings are usually mitred at the corners, but in high-class work may be tongued together, loose-tongued or tongued-and-mitred. At internal angles, mouldings have to be scribed one over the other, so leave the work to somebody who is sufficiently neat and skilful to do it properly.

With some joinery mouldings it pays to have them run in hardwood so that they can be made especially crisp. But in most instances, softwood is quite good enough for any feature which is to be painted.

Never use modern types of skirting mouldings in old houses. They will ruin the effect. Also make sure that you make the skirting as high as it needs to be for the decorative role it has to play. Very shallow and meagre skirting boards along partitions in newly formed rooms, for example, can strike a false note which spoils everything else that you have done.

Early eighteenth- and late seventeenth-century skirting boards – usually in conjunction with panelling – are surprisingly low to the floor. They can be as little as 125–175mm overall, including any simple stuck moulding.

Repairing, Restoring and Replacing Woodwork

The treatment of classical panelling in old houses requires a certain degree of finesse, especially in the kind of finish to be used. The fashion for stripping fir and pine panelling back to the natural wood is one which pays scant attention to historical accuracy. Most softwood panelling in the late seventeenth and eighteenth centuries was painted. It may be argued that natural wood panelling looks well in an old house, and when it is of a fine hardwood that may often be so. In such instances the wood must be wax-polished; on no account treat it with varnish or clear matt polyurethane. It will be obvious if panelling has always been intended to retain a natural finish; when that is the case, it will usually be found to date from around the last quarter of the seventeenth century, or the first quarter of the eighteenth century.

Queen Anne woodwork was mostly painted in rather quiet, flat colours, a sludgy green being particularly favoured.

It is sometimes recommended that repairs to panelling are carried out on the bench after it has been removed from the walls. Theoretically this is a good idea, but in practice you should avoid disturbing panelling unless it is unavoidable – in order to treat dry rot on the back, for example. It is a golden rule in dealing with old houses that: *The more you disturb or remove, the more you will eventually have to destroy or abandon in the process.*

There are certain things to bear in mind when cleaning and repairing panelling which is to be left in its natural state. Never sand down old woodwork of this kind unless absolutely imperative; the old patina must be retained. There is no need to make panelling look of even colour, and free from marks, stains, dents, scratches and other signs of age.

Whether you replace missing mouldings is a matter for nice judgement. If their absence upsets the continuity of a classical design – a missing piece of architrave for example – then match the details and replace them, using well-seasoned wood of the same kind. The grain should run in the same way as that in the damaged timber, and any obvious figuring should be matched up as well as possible.

Experts argue about the 'honesty' of matching moulded work. Some take the view that new timber should be seen to be what it is; so they simplify mouldings or leave them out altogether. They also refrain from rubbing pigment into new wood to tone it with the original. Above all, they never distress the surface of the new wood by chipping, scoring, applying a hot iron, wetting with chemicals and so on. To some extent I think this view is wrong. The surface of panelling or other exposed interior woodwork provides a decorative finish, just like a paint or a paper. In the late seventeenth century, craftsmen went to great lengths to make humble softwoods look like finely figured hardwoods by painting on elaborate graining. None of that could be called 'honest'.

Since you may have to live with a large patch of bright new oak, staring out of an area of mellow and mature panelling, it could be worth running in a little very carefully matched and selected stain before wax-polishing. However, it has to be done with great discretion. Whether you decide to mix your own stain, using pigments and a medium such as linseed oil, depends on your abilities. I have found this a tricky business and it may be better to choose a good manufactured stain to blend in with the old timber. More often than not, it is better with new panelling simply to add some colouring matter to the wax finishing polish and keep rubbing it in until you get the tone you want.

The matching of mouldings and other details is nothing to do with 'honesty', which is an ethical consideration unconnected with successful repair work. All the same, avoid clumsy, brash and ill-informed faking. For example, do not distress new panels or their mouldings, but you could certainly have the finishing done with hand tools, so that an ogee or a chamfer does not look too mechanical alongside the original ones. If there are dentils missing from an Ionic cornice, cut some new ones and stick them back on. If one of the volutes on a pilaster capital is broken off, it will ruin the visual balance, so carve a replacement.

To my mind the idea of simplifying classical details to show that they are replacements is the worst sort of affectation, and any eighteenth-century craftsman or architect would laugh such notions to scorn. However, take the greatest care to preserve old or interesting detail or painted decorative design of the sort you may have the luck to find in late medieval buildings.

Nothing should be tarted up in a speculative manner. Where there is evidence of certain colours being used in a design, you have to think long and hard before undertaking major restoration. There is much to be said for gently cleaning such decorations, and otherwise leaving them in the state they have attained over the centuries. From an aesthetic point of view, many people would be appalled at seeing medieval paintwork looking as garish as it did when new.

If you have carved detail which obviously needs matching to continue a pattern, there is no harm in copying it, as long as the work is extremely well done. Equally, if three decorative stair finials have different carved decoration, I see no reason to prevent some skilled and imaginative craftsman from creating a fourth which echoes, without actually copying, the originals. He should sign and date his work in a discreet manner. However, any attempt to introduce a contemporary theme alongside others which are all of a particular period can look arch or even absurd.

After stripping wall coverings, which, incidentally, you should be at pains to preserve if they look at all interesting, you may find the walls are very plainly panelled indeed, with muntins, stiles and rails completely unmoulded and showing no signs of having been painted. This type of panelling, often seen in eighteenth-century houses, was probably designed as a framework for hangings, and is not entirely suitable as a wall finish in itself. You will often find that it has been covered right over with a canvas scrim, upon which layers of paper or other material have been stuck or nailed. Your best bet is to repair this framework, clean it down, and then re-cover it with a fabric or paper which accords with the period and flavour of the room.

When woodwork must be removed from the walls, take the greatest care. Never force anything which puts up more than mild resistance, or you will ruin what you are trying to save. You soon discover that the modern technique of fixing back dodgy timbers with a really big nail was popular in past eras as well. Work these out with infinite caution, tapping in a wedge behind the wood as you take up each new purchase. Do not assume that because one bit of panelling has been fixed with nails all the rest is the same.

Look out for hidden screws; and suspect a screw if one length of timber is more obdurate than another. Depending on the method of construction, you may be lifting panelling from a fairly substantial frame, or you may be pulling it free from the wall itself and taking the pegs or fixing blocks with it.

As with masonry, take great care to record and mark all the components to ensure proper reassembly. Flat-on photographs taken with a wide-angle lens are ideal for this purpose. They are marked up to give a key which corresponds with numbers on the woodwork (use self-adhesive labels for these).

Wood Decay and Insect Attack
Everybody dreads discovering dry rot, since it is hard to eradicate. Wet rot is a nuisance but does not involve the same elaborate precautions to prevent further spread. These, together with woodworm and death-watch beetle, are the four principal hazards to woodwork in old houses.

What you need to know is how to identify them, as far as this is possible without long experience. You should also understand roughly what is happening in cases of fungal decay and beetle infestation. Lastly, you need to know about the available remedies.

With the exception of woodworm (the common furniture beetle), these destroyers of timber all require damp conditions to get going. They are all treated with fungicidal or insecticidal liquids, and normally the decayed or infected timber is cut away before the rest is treated.

Dry Rot
Dry rot can be recognised by its smell, by the fruit body (sporophore) which grows on the diseased timber, by the mycelium, and by the moisture-dispersing fungal strands, which develop in it. It takes hold of wood which retains over 20 per cent moisture content and is especially encouraged by still, warm, badly ventilated conditions.

Softwoods are the normal candidates for dry rot. The wood goes brown and breaks up in a pattern of cubes, the cracks running with and across the grain. The

A very severe outbreak of dry rot (*Serpula lacrymans*) has split these timbers into deep cracks both along and across the grain. Timber affected by dry rot is often paler in colour than those attacked by wet rot. There is also a considerable loss of weight. (*Rentokil*)

Grey, violet and yellow fruiting bodies (sporophores) are characteristic of dry rot. These ones have started to climb the walls, forming typical bracket-like plates of fungus. They will generate new spores which will further spread the outbreak. (*Rentokil*)

When the floorboards were lifted the joists were found to be draped with sheets of dry rot fungus; and the cottonwool-like form of the mycelium can be seen on the vertical and horizontal boards below the area between the radiator and the door. This outbreak was caused by water leaking from the radiator, which raised the moisture content of the timber in the badly ventilated floor cavity. (*Rentokil*)

wood becomes soft and friable as the natural cellulose in its cell structure is destroyed. It also shows a marked loss of weight. The usual mycelium is often rather like cottonwool, when conditions are very damp. When dry, sheets of grey and violet mycelium spread themselves over all the building materials in the vicinity. Fruiting plates of yellow-brown fungus with violet and white edges appear on the timber, dispensing a red dust of spores which spread the infection. Sometimes the fungus puts out bracket-shaped protrusions.

The strands, which can form tubes up to 6–7mm in diameter, carry the moisture and infection for considerable distances, both on the surface and through the substances of softer materials, such as plaster and lime mortar. The fruit body, hyphae and an example of typically decayed timber are shown above.

Although dry rot is theoretically easy to identify, there will be times when you are not quite sure whether you are indeed dealing with *Merulius lacrymans* (now known as *Serpula lacrymans*) or the less worrying wet rot. The most likely cause of muddle is the decayed timbers themselves. If you are in any doubt, consult a specialist firm. As with most other kinds of identification it is as well to look for two, or preferably three, separate pieces of evidence. For example, you may guess the smell is of dry rot, rather than just that of damp and generally musty conditions. Then you notice the deep cross-pattern of cracking in the surface of the timber, and you note that it is not in an exposed position where it suffers continuous wetting. The dried larger tendrils leading away from the centre of infection are brittle. The spread of the decay appears to extend far beyond the limited area which would be affected by wet rot.

Wet Rot

Wet rot is a term which includes the cellar fungus, *Coniophora cerebella* (now known as

Coniophora puteana) and its close relations, the *Poriae*. It takes hold of wood which has at least 25 per cent moisture content and also develops a cubic pattern of cracking; but this is less marked than that of dry rot. The fissures are shallower and are more prone to follow the grain than to cut across it. The fungal strands are usually fine, black and rather cobwebby, unless they are very fresh, in which case they are yellowish in colour. More often than not there is no fungus to identify. The *Poria* type of wet rot, now known as *Fibroporia*

Wet rot or cellar fungus (*Coniophora puteana*), here shown appropriately in a wine cellar, develops a dark cobweb of fungal strands. It is normally localised in its effects and, once the source of damp has been thoroughly eliminated, further spread is unlikely. (*Rentokil*)

Wet rot also breaks down the timber in a cubical pattern and in larger timbers can look very like dry rot, although it is frequently much darker in colour – often almost as if charred by fire. The outer surface of the wood can retain an undamaged crust which is deceptive. Smaller timbers, such as sash-rails will often crack along the grain with only small cross cracks. (*Rentokil*)

vaillantii, is more easily confused with dry rot but differs from it in these important ways. It never has the violet or yellow colours in the mycelium, remaining genuinely white. The dry strands are not friable and are markedly less thick.

The colour of wood which has been badly affected by wet rot is normally dark brown to black. Sometimes, indeed, it looks almost as though it has been charred by fire. The decay of *Coniophora* remains in the actual area which has suffered the high moisture content. *Poria*, on the other hand, may spread a little beyond the really damp part of the timber, which is slightly lighter in colour. Unlike dry rot, it can survive at quite high temperatures, up to 36°C. Very often no sign of wet rot appears on the surface, and paintwork seems to be intact.

Since it is much the easier kind of decay to cure, let us consider wet rot first. The most important factor is to remove the source of moisture, whether it is penetrating damp or a leaking water pipe. Next you should allow the timber to dry out. You should then cut away any badly affected timber and either make good with a filler or insert new pieces of matching wood. Then treat with a fungicide.

The methods used for repairing decayed timbers after the rotten wood has been cut away will depend upon whether they serve a structural purpose; and the subject has been dealt with in Chapters 7 and 8.

Dry Rot Treatment

Dry rot requires much more extensive and stringent methods of eradication. Its effects can be both widespread and extremely damaging. Apart from cutting away infected timber you must also remove apparently good timber to a distance of 450–750mm on either side of the obviously rotten area.

All rotten wood must be removed immediately and burned. To prevent further outbreaks, apply a suitable fungicide to any nearby timbers which could conceivably be at risk. The strands of fungus may be traced and, where they find their way into the structure of the building by penetrating the walls, the latter should be thoroughly soaked in fungicide after plaster has been removed and the surface well brushed down. Sometimes, it may be necessary to irrigate the walls using a similar technique to that employed for heart grouting mentioned in Chapter 6.

Never forget the importance of cleaning up after cutting away damaged timbers. Wood chippings, insect frass, dust and dirt all help to promote damp and infection. Sweep and vacuum everywhere. Not only will you better see what you are doing if things are clean; but fungicides will not be wastefully absorbed in the dirt instead of penetrating the wood which really needs them. Some fungicides contain colouring matter which enables you to see where they have been applied, thus preventing double treatment.

There are times, however, when clear preparations must be used – exposed internal timbers being cases in point. New softwoods should be vacuum treated or brush-soaked with preservative. Vacuum treatment is done by placing a stack of wood in a pressure chamber from which the air is evacuated by pumping. The vacuum draws preservative into the chamber from a storage-tank. Preservative is then forced deep into the timber by means of pressure. Finally, the excess preservative is pumped back into the tank.

Preservatives

There are various types of wood preservative, including: those based on tar oil (for example, creosote); organic solvent preservatives, which usually contain pentachlorophenol; and water-borne varieties with a content of something like copper, chrome, zinc or mercury. However, the composition of a preservative is of academic interest to most people. The main factors governing your choice will be:

1 Is it fungicidal (ie kills the strands or stops them feeding)?
2 Is it water-repellent (ie does it stop moisture penetration)?
3 Is it toxic to insects, and if so which ones are most affected?
4 Will it leech out or is it reasonably stable?
5 Does it stain and can it be painted?
6 Will it smell, for how long, and how badly?

With the best-known products, the answer to most of these questions will be satisfactory. The solutions marketed by those firms which are household names are balanced to fulfil a number of functions at once. However, many do smell disagreeable, so if possible it is best to carry out treatment on a building when it is unoccupied. There are some aqueous fungicides which smell only a little after twenty-four hours.

The regulation of chemicals to be used against decay and beetle have become rather more stringent of late and some firms are selective about what they will use, others less so. Lindane and Tributyltin Oxide are said to be unusually toxic and might be ones to avoid. Permethrin and Organoboron have the reputation of being somewhat kinder. Do not forget that it is illegal to kill bats during the process of treating your roof timbers.

Despite the all-purpose nature of most products, it should be remembered that there are important differences. Some are intended principally as fungicides to kill hyphae in brick, plaster or stonework. Others lay emphasis on being water-repellent. Others are mainly insecticidal. The advantage of employing a specialist firm to carry out treatment is that they should (one hopes) correctly identify the problem, choose the right cure, and save you the unpleasant job of brushing, spraying or fumigating your building with noxious chemicals.

How much wood-decay treatment you do yourself depends upon the nature of the outbreak. Strands in walls may be eradicated by heating the surface with a blowlamp, but this can be a dangerous task for an amateur. Indeed, there are many situations in which a sensible professional would himself think it wiser to brush-soak or irrigate the wall with a solution. Specialist firms tend to charge high prices, but processes such as extensive dry-rot treatment (especially to structural timbers) may well be left to them. They may also provide a guarantee. The localised effects of wet rot, however, once properly identified, can be tackled by any competent amateur.

Be particularly careful in choosing a treatment for infected walls. Sometimes it may be necessary to isolate any remaining strands by sandwiching a coat of zinc oxychloride plaster between the rendering and the finish. Another method is to put the oxychloride plaster straight on to the brushed and dubbed-out wall surface. Paints containing this chemical may also be applied.

Woodworm and Death-watch Beetle

There can be few old buildings which have not been attacked to some degree or other by woodworm, otherwise known as the common furniture beetle (*Anobium punctatum*). It has a preference for the sapwood of softwoods which have been in place for a period of twenty years or more. However, it is unsafe to make many assumptions about where and when woodworm will be found. It can also attack the heartwood, and hardwoods.

The first signs of its presence will be 1.5mm diameter flight-holes on the timber and a scattering of fine chewed-up wood waste called frass, which is expelled on to nearby surfaces. Briefly, what happens is this. The adult beetle mates and lays eggs grouped in cracks in the woodwork or near the mouth of old flight holes. After about five weeks the eggs hatch into larvae, the little grubs which do most of the damage.

The grubs start tunnelling into the wood, keeping largely to the grain, but crossing over every so often into new annual growth rings. They tunnel for about three or four years, finally making their way to the surface in the spring. There they form pupal chambers just below the surface of the wood and adopt the form of a chrysalis. Six or eight weeks later they become beetles and bite their way out of the timber, leaving the characteristic holes. The beetles live about two or three weeks, during which time they mate and start the process all over again.

Unlike some other beetles, these do not require wood which has been softened up by fungal decay. Although woodworm

must be taken seriously it may not have greatly reduced the strength of the timber, provided that it is caught in time.

The remedies for most beetle infestations are straightforward, if messy and disruptive. You either brush on insecticidal solutions, letting the wood absorb plenty of the liquid, or you spray the timber and also inject the flight holes with insecticide. The smell of insecticide is a menace and the mess generated by treating insect attack in such spots as the joists of an occupied house makes people avert their eyes from tell-tale signs, just hoping the grubs will quietly go away. Sometimes they do, but never depend upon it.

An alternative to brushing, spraying or injecting fluids is the fumigation system of treatment. For this the building is sealed up and an insecticidal smoke or vapour is released. This penetrates, it is hoped, all those inaccessible corners and crevices. The method is ideal in tall buildings with open timbered roofs, for example churches, barns and medieval great halls. This is really work for specialists.

It often pays you to get a specialist firm to tackle beetle infestations. It has the right equipment, knows which insect it is trying to kill and, with luck, will do a more thorough and tidy job than you could. Obviously, if it is just a matter of brushing

Death-watch beetle likes to attack damp and decayed hardwoods. Look for it in the feet of principal rafters, the ends of tie-beams and floor beams, for example. Its flight holes measure about 3mm diameter. *(Rentokil)*

on insecticide when you are re-laying the roof of a cottage, you will do the job yourself (or have your builder do it).

When applying insecticides and wood-decay treatments, cover yourself with suitable protective clothing, especially if using any kind of spray. For the latter, wear goggles, a filter over your mouth and nose, a helmet and gloves. Professionals have special protective boilersuits and manage to cover up completely against the pervasive effects of sprayed chemicals.

The next type of insect you are likely to encounter is the death-watch beetle (*Xestobium rufovillosum*). Much of its activity is similar to that of woodworm, but it favours hardwoods which have suffered some fungal decay. Its grub stage, during which it tunnels the timber, generally lasts about four and a half years, but it can go on for as long as seven. The flight holes are roughly 3mm in diameter.

Death-watch beetle is particularly common in early buildings, since it is fond of oak, chestnut and elm. The former two timbers are frequently used for roofing. Oak was employed for beams, joists, plates and posts, while many floorboards were of elm.

Death-watch beetle expels bun-shaped pellets of frass, which feel gritty. The eggs are hatched on the rough surface of timber and the grubs then find their way into the wood through cracks or old flight holes. The three- or four-week chrysalis stage takes place in the summer and the beetle which emerges stays in the pupal chamber until the following April or May, when it bores its way out. The beetle cannot fly and this does much to ensure that infestations are fairly concentrated. The 'death-watch' tapping heard in the night is made by the adult beetle (whose life span is a mere two months) striking its head against the wood. It is believed to be a mating signal.

People working on buildings in Surrey and parts of Berkshire must comply with Building Regulation 7 (Approved Document para 1.9), connected with treatment of softwoods against the house longhorn beetle (*Hylotrupes bajulus*). This has big grubs 30mm long and leaves cylindrical pellets of frass. The flight holes are oval and measure roughly 10×5mm. It infests and reinfests dry timber.

The powder post beetle (*Lyctus*) is worth mentioning since it likes to eat the starch in the sapwood of trees such as oak, elm, ash, sweet chestnut and sycamore. It does not infest softwoods. The flight holes are similar to those of woodworm. You will be unlucky to encounter this insect.

The only other beetles one should mention are the wood-boring weevils (*Europhryum confine* and *Pentarthrun huttoni*). They almost always attack decayed and damp wood, leaving tiny, ragged flight holes measuring about 1mm in average diameter. They are largely found in London and the Home Counties, in both hardwoods and softwoods. They are dangerous in that they will continue to bore through timber which has suffered from a decay such as wet rot, *after* it has thoroughly dried out.

12

PLASTERING

The whole topic of plasterwork and renderings is fraught with reservations. While in one case it seems essential to take one course of action, in another it is vital to do something quite different. The technical merits of course A must never outweigh the aesthetic advantages of course B. This irritating state of affairs may be alleviated if you hold fast to the following maxim: *Employ the best practical method which will safeguard and enhance the character of the building.*

So we return to that most important of distinctions – whether the house is 'rough' or 'smooth'. An old cottage with undulating stone walls – sturdy and unassuming – may lend itself to soft and mellow renderings. An early Victorian villa could call for the use of a fine, hard finish in stucco, with boldly rusticated imitation quoins – pre-eminently a case for the 'smooth' approach. Those two examples are easy to assess. Other buildings may prove less so. The exterior might require a 'rough' treatment while the interior demands a 'smooth' one. There are endless variations. Try to analyse the flavour of the building, removing or adding in the mind's eye the textures, finishes and architectural details which it appears to demand.

I know, for instance, of a public house which is an object-lesson in how not to treat the interior of a 'smooth', nineteenth-century building with a stucco exterior. The stone walls in the bar have been stripped of their plaster, cleaned and nicely pointed. The effect is brutal and disturbing – totally out of keeping with the architectural character which one has been led to expect before one enters. There is no question that this hostelry shrieks out for 'smooth' plasterwork,

cornice mouldings, and all the other trappings of its period. Architects, owners and builders continue to strip the interior walls of old houses and expose the stonework when there is absolutely no historic precedent to support such action or visual advantage to be gained.

One is tempted to list the kinds of building which may be suited to the different types of external and internal finishes. Unfortunately there are too many exceptions to make it practical. Again, in the age-old committee member's cliché, 'We must take each case on its merits.'

Exterior Treatments
We have discussed the intricacies of pointing in Chapter 5 but we did not consider whether a wall should be plastered, as opposed to displaying the materials from which it is constructed. Many old houses were built in such a rough manner and from such poor stone or brickwork that they were clearly intended to have a plaster coating from the very beginning. Subsequently, owners with a love for the texture of old rubble walls or warm and elderly-looking brickwork have stripped the plaster and pointed what should really have remained hidden. There is no golden rule to follow, but there are signs to look for which may give a clue to the original builder's intentions.

If the stonework is exceptionally ill-coursed, with rough-looking timber lintels, you may suspect an original plaster façade. Should the lintels be hacked to form a key or bear the marks of laths, again you may guess in favour of plaster. Sometimes, stone or brickwork, while poor in general, stops at prominent quoins and dressings of superior workmanship or materials. There may be well-defined changes of colour between brick-work which was long exposed and that which had the protection of plaster. Awkward details or projections may show hack marks, suggesting a former plaster finish. In some cases, plaster may still adhere to the walling material and if it is original it will either contain lime, or it will be composed of hair or similarly rein-forced clay.

What you have to decide is whether the quality of the stonework or brickwork merits permanent exposure, both techni-cally and aesthetically. Many buildings will, of course, demand that old plaster is stripped off, revealing walls which have been subsequently rendered over as a matter of convenience, either to conceal crude repairs or to keep out driving rain.

Weather-protection is the main purpose of exterior plastering, and past eras have left us some grisly finishes to deal with. Perhaps the least pleasing render is pebbledash. This was applied liberally to many a pretty old building during the late Victorian, Edwardian and post-World War I periods. It is totally unsuitable for any kind of old house or cottage and is generally repugnant, even on buildings intended to have it in the first place. Pebbledash is effected by laying up a cement-gauged render and throwing on small stones with a trowel while the plaster is still soft enough to receive them. To frame and contain this fusty-looking finish, hard bands of cement mortar were frequently used to face openings such as doors and windows. These might be painted in some unfor-giving shade or left in their natural state, finally weathering to the colour of a ceme-tery wall.

A very close relation to pebbledash is the similarly applied but more rugged finish made with rough (often granite) chippings, small or large. This is much loved by modern architects who contend that it gives texture and natural colour to the walls of modern extensions. It is hideous and one can only advise that you avoid it.

Another variation to be seen on every hand is the Tyrolean finish, which is shot against the wall surface from a special machine. You turn a handle, rather as though using a mincer, and small, sharp pieces of gravel in a coloured cement-based slurry are sprayed on to the wall. Readers who have visited Austria will realise that it is indeed the homeland of Lukas von Hildebrandt and Jakob Prandtauer which appears to be the pro-genitor of this nasty concoction. Tyrolean finish, unlike pebbledash, is what is called a 'wet dash'. The adhesive plaster ingre-dient is applied with the aggregate.

This is also the case with roughcast – the only acceptable method for old buildings. At its best it can look delectable. Badly carried out, in some crude kind of 'spat-terdash' technique, it can look frightful. In Scotland it is called harling. Every year there are fewer good examples of old and beautiful roughcast for people to enjoy and copy. The necessary skills are dying out and tradesmen often replace rough-cast with a wood-float-finished plaster. The modern method employed for roughcast is to make up a mix containing about one part of cement, one of lime putty, three of sharp sand and three of well-graded stone aggregate. This is wet-dashed on to a well-damped and well-keyed backing, some-times consisting of a waterproofed plaster. The purpose of the latter is to prevent too rapid absorption of the moisture content in the roughcast mix.

In theory, all this sounds straightfor-ward enough, but in reality it is far less simple. Many factors must be got exactly

An example of the spatterdash technique, carried out with a very lumpy and cement-rich plaster. It does not achieve the desired mellow texture which a really good traditional roughcast displays.

right. Old roughcast gained its soft and attractive colouring from natural earth pigments contained in the muds which were used. Lime, unlike cement (which was not employed) made for a soft mix and did not adversely affect the final colour. The choice of stone used to reinforce and fill the mix was often fortuitously pleasing and is very difficult to reproduce with modern quarry-crushed materials.

The traditional method was to wet-dash the lime, sand and gravel mix on to a still-wet backing of lime plaster. If you use cement to gauge the wet-dash mix, it should either be toned with a suitable earth-coloured pigment, or should itself be a *white* cement. It is essential that the roughcast should not adopt a pale grey, cement-coloured appearance when it has dried. Experiments should be made on some handy piece of redundant wall.

The loveliest roughcast has the look of an old cob wall (of that kind which contains a fairly high proportion of small stones) as seen when the protective layers of limewash have flaked away.

The trouble with trying to reproduce old roughcast, plaster, and pointing for that matter, is often the necessity for using natural earth as a pigment – a practice discouraged in modern building technology. Earth, it is reasoned, not only contains injurious vegetable matter but also may leech the dreaded salts which cause so many other problems. However, salts would be continually washed off an exterior wall by the rain and, since roughcast of this kind is not meant to be painted, there is no risk of spoiling finishes.

Concentrated pigments which are earth-based can still be obtained from specialist suppliers and these should pose no inherent dangers, however slight those may be.

External Plastering
The type of plaster you use for the outside walls of a building will depend upon several factors – the formality of the façade, its period, the proposed final finish, the backing material, and the degree of exposure to driving rain.

As mentioned elsewhere, softness and flexibility are desirable qualities which can best be achieved by the minimum use of cement, the addition of lime and high proportions of a rather coarse aggregate in the mix. A typical mix for a first coat might be cement, lime and sand in the proportions 1:2:9. Where a slightly more durable mix is required, this might be adjusted to 1:1:6.

The advantages of using lime in the mix are that the rendering is made more workable during application, softer and more able to adjust itself to any movement in the background material of the wall; it is less likely to develop fine cracks on drying; and, lastly, it absorbs moisture evenly, on the sponge principle. This absorption property of a coarse and rather weak lime plaster has been referred to in Chapter 5 but it is worth mentioning the matter again.

Walls protected by soft, absorbent plasters tend to soak up moisture in a relatively even manner over the whole of their surfaces. The resulting penetration of the moisture is consequently shallow and capable of fairly quick drying by evaporation. Hard renderings, on the other hand, have many disadvantages. Most building workers greatly prefer to use very hard mixes for rendering, both inside and out. They like to avoid the bother of puddling hydrated lime, which is a messy nuisance, and to bash on with the job, using a rich, highly workable mix containing a plasticiser. They know that adhesion will be good and a rapid set will be ensured. When working with old buildings, you should insist on weak mixes containing lime, for the reasons given above. Cement-rich renderings have a way of parting company with the backing material. They also break up into fine cracks which drag in water by capillary action. They tend to convey a visual quality of smoothness and hardness which is often very unsympathetic, particularly to a rough building.

There is a certain amount of debate about the moral rectitude of employing hydrated lime in repair work on historic houses. It is not the traditional method, some say. Thus it is inferior to lump lime (quick-lime), especially when used for making limewash – a proposition which should not be discounted.

I do not myself see any advantage in the use of quick-lime for plaster mixes, although it may very occasionally be required to match early pointing or bedding mortar. The cost of labour, together with the difficulty of obtaining the material, seems vastly to outweigh any other factors. Quick-lime, if stored at all, must be kept in very dry and well-protected conditions or it can slake itself from the moisture in the atmosphere. The whole process of slaking the lumps and breaking them down into a usable putty, sieving and storing to fatten up, requires time and probably extra supervision. Protective clothing must be worn during the slaking process and precautions taken to keep unauthorised visitors out of the way.

Dry hydrated powdered lime generally appears to be a much better option. A big galvanised bin is filled with a few inches of water and the lime powder is added and stirred to a creamy consistency. More water is then added and further lime beaten in. The resulting putty is either kept moist in its container until it is needed or mixed with the coarse aggregate (usually sand) and left under damp sacks to be used when required. Whatever the circumstances, the lime putty should have at least twenty-four hours to 'fatten up' – become nicely workable. The longer it is kept, the better it will be.

Hydraulic limes, which have a high clay content, may now and again be required for special conservation projects, particularly where their ability to set and harden in moist conditions is an advantage. They are only available from one source in England and it is very unlikely that most readers will have any reason to use them.

Although plaster mixes have adhesive properties which help them to stick to the background material, whether it is brickwork, stone, concrete blocks or some form of lath, they must also depend upon a satisfactory mechanical key. It is self-evident that the rougher and more textured the background, the better will be the mechanical bond. Laths, for example, allow plaster to squeeze through and grip the sides and back of the metal or timber lathing material.

There is one other factor to consider and that is the degree of moisture suction exerted by the background. The background material for plastering must be sufficiently damp to prevent premature absorption of the moisture content in the mix, thus robbing it of the ingredient which induces it to set.

External renderings may be carried out in two-coat work, the first coat being scratched to form a key for the second. In most instances, the best finish for the second coat is one applied by working over the surface with a cross-grained wood float, just as the mix begins to set. This minimal degree of texture both looks convincing and can increase durability by discouraging fine cracks.

Plaster composed of cement, lime and sand, with a cross-grained wood-float finish. The plaster was in two coats, the second being textured with the float just before it began to set hard.

Cottages and 'rough' buildings should not have sharply formed arrises at window openings, doors and external corners. These look hard, monotonous and modern. You should therefore prevent your workmen laying up the plaster against timber rules. Get them to work the corners freehand, so that they have some degree of natural irregularity. The arrises should also be blunted a little as work progresses.

The general wall surface, if heavily

pitted with major hollows and indentations, should be 'dubbed out' with a weak mortar so that a reasonably consistent depth of key is provided. First-coat plaster will be laid up with a steel trowel, allowed to dry a little, keyed and then left, preferably for a few days, before the finishing coat is put on. The important thing to impress upon the craftsman is that the wall surface must not be made straight and flat, as if he were plastering a stucco villa or the interior of a 'smooth' building. He must let his trowel follow the natural undulations of the wall, so that without positively looking affected, the result will be a pleasing lack of conformity. It is usually a matter of great difficulty to persuade conscientious building workers to follow this simple course of action. They like things to look neat and tidy – a visual tribute to their innate ability for modernisation.

The reverse of the coin is that a plasterer may become excessively enterprising and, believing that he knows how to make something look old, will cover your building with one of the many repulsive fake textures which may be seen on every hand. These include the one which resembles birds' clawprints in the snow, the impetigo look, the brushed swirls of chocolate toppings, roundels, Snow White's bungalow encrustations and horizontal ripples. There is no end to the moderniser's ingenuity. Be vigilant and specify your requirements with great attention to detail.

The expression 'a stucco villa' is understood by most people to mean a preeminently 'smooth' building with neatly plastered walls. It is probably nineteenth century and might well display signs of Italian influence. However, in its strictest sense, there is a difference between stucco and ordinary plaster. Stucco in Italian Renaissance terms was a mix of lime and powdered marble, while plaster would normally consist of lime and coarse stuff (usually sand), reinforced with animal hair. Despite this distinction between stucco and plaster, it became fairly usual to describe almost anything as 'stucco', and the Italian and Swiss plasterers who carried out much of the finest eighteenth-

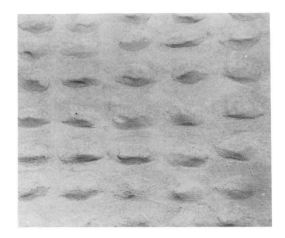

An extreme example of one of the numerous plaster finishes which people invent to give a wall texture and make it 'look old'. Such efforts always fail.

This wall has been rendered in a manner which attempts to emphasise the contour of the stonework beneath; but it fails because it has a slightly swirled texture of its own.

century decorative work were referred to as *stuccatori*. This obviously did not mean that they worked exclusively with true stucco as described above.

When dealing with exterior renderings for old buildings today, we should be careful to provide the kind of finish which accords with their date and the spirit of the architecture. Rustications in the form of raised and chamfered simulated stonework, banded rustication, and special effects like reticulation and vermiculation, should be reproduced if necessary.

Wall surfaces on such houses will be

kept straight and true, using screeds or rules as required. All the same it is unlikely that you will need to use hair reinforcement, although it may be useful when plastering timber-framed buildings. A lime and sand plaster, gauged with a small amount of cement, will develop quite enough strength and flexibility to provide a satisfactory undercoat on solid backgrounds. A mix in proportions 1:1:6 is a normal choice, although some people believe a plain cement-sand render (1:3) should be used in exposed positions. Personally, I think such renderings are rather too hard and, in some places, would even employ as soft a mix as 1:2:9.

In all cases, it is best to provide a wood-float finish to the last coat of plaster, since steel-trowelled surfaces are prone to craze and crack. It can be argued that cement rendering coats should incorporate a waterproof additive since this helps to prevent the essential moisture in the finishing coat being drawn out.

You should be a little wary of the temptation to use a lime and sand mix without any cement content to gauge it. It is certainly more flexible, more traditional and less likely to develop cracks. The disadvantage is that non-hydraulic limes take a very long time to harden and are dependent upon evaporation rather than a chemical reaction for their initial 'set'. Always remember that a rendering should be, if anything, rather weaker than the background to which it is attached; and subsequent coats of plaster should be the same or weaker than the mix used underneath.

When plastering walls, whether they are external or internal, it is important to plan the work so that obvious day-work joints are avoided in final coats. At the end of each work session, the plaster should have reached a suitable external corner, return wall, or other visual break. With rusticated plasterwork this can be made easy by finishing at a plinth or a new horizontal channel in the simulated stonework. If this factor is not taken into account, ugly joints which are almost impossible to disguise will be seen.

Rustication of plasterwork is not so

Fig 182 A tool used for pressing mortar into the joints of ashlar stonework when pointing.

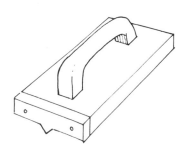

Fig 183 To V-joint plaster when carrying out rustications, a specially cut wooden template may be screwed to the end of a wood float, which is run along a guide batten. This is a very simple form of running mould.

difficult as you might suppose and well within the capabilities of any plasterer or mason/plasterer who has the required method and effect explained to him. Early nineteenth-century walls, especially those carried out on a stud framework and laths, may have shallow lines incised to simulate stone ashlar. This is done with a steel jointing tool (Fig 182) run along a straight edge to a depth of perhaps 3mm. Deeper forms of channelled work can be carried out using a suitably shaped wooden template, nailed to the leading edge of a float (Fig 183). Again, you run the float along a straight edge and so cut out the imitation joint from the soft plaster. Make sure that you do not cut *right through* the finishing coat or you will ruin its continuity and adhesion.

Buildings which have decorative external plaster features, like console brackets, moulded cornices, triangular or segmental window pediments, string courses and other details, should not be simplified in the cause of economy. It is no use thinking that the absence of a few mouldings will have no effect upon the overall impression. It will make a great difference, so go to considerable lengths to reproduce the correct mouldings or else to procure salvaged ones of similar date, scale and style.

In some parts of the country – East Anglia in particular – there is a tradition of decorative external plasterwork called pargetting. This takes the form of anything from a rather repetitive and primitive intaglio design to a sophisticated and inventive arrangement of swags, Baroque scrolls and figures in medium relief. One reason for its popularity in East Anglia was the fact that buildings were commonly timber-framed and were thus plastered to provide a weatherproof finish. The innate desire of craftsmen to express their own skills and individuality naturally led to various kinds of slightly naïve embellishment.

Like any other plaster during the late sixteenth, seventeenth and early eighteenth centuries, it was applied to a background of either wattles or laths. The mix consisted of lime and a coarse aggregate bound with animal hair. Incised (intaglio) motifs were often stamped with patterned blocks of wood, or were combed or scratched with improvised tools. Relief ornament was normally hand-worked *in situ* in much the same way that a sculptor models in clay. The final finish was a limewash, either in white or containing some kind of pigment.

The repair of plaster on timber-framed buildings involves two main operations. First comes the mending or replacement of the backing – usually riven laths or wattle panels. Then new plaster has to be introduced into these areas. It is generally best to do patching in the nearest possible approximation to the original material so as to ensure continuity, good absorption, evaporation of moisture and flexibility.

When it is necessary to replaster a timber-framed house entirely, after remedial work has been carried out on rotten studwork or large areas of decayed or worm-eaten laths, you may consider the use of expanding metal lath as new background. It has the advantage of durability, provided that it has been adequately treated against rust. Also, it can be

claimed that it is quicker to fix than traditional wooden laths. However, it may be less likely to accommodate itself to movement in the framing timbers.

When major replastering like this is done, an opportunity is provided to incorporate a vertical damp-proof membrane in the form of bitumen felt or some similar material.

A mix which many people recommend for both undercoat and top coat when a building is to be pargetted is 1:6:9 (cement:lime:sand). To this you add chopped hair, allowing about half a kilo to each load of a small concrete mixer, ie with a drum capacity of 90/60 litres or 3:2cu ft. Purists may demand hand-mixing using a shovel and mixing board.

Clay daub for repairing wattle-and-daub walls may be mixed by batching equal parts of lime, natural clay and cow dung, adding chopped straw to reinforce it. For ordinary rendering in cement:lime:sand the proportions could be 1:3:12. Long periods for drying and hardening will be required if the mix is not gauged with cement; for ornamental pargetting in relief, the absence of cement does allow plenty of time for a craftsman to model his designs.

While on the subject of ornament, it is worth mentioning the various eighteenth- and nineteenth-century methods of simulating carved stonework. These normally consisted of some kind of patent cement, mixed with suitable aggregates, cast in moulds and sometimes fired like a terracotta. Loudon, in his *Encyclopaedia of Cottage, Farm and Villa Architecture*, dated 1842, refers to 'the kiln-burnt artificial stone of Coade and Seeley, which is as durable as the hardest marble'. Much of the early nineteenth-century stucco was made with Parker's Roman Cement – a burnt calcareous clay, ground to a powder which could then be used in rather the same manner as a Portland cement. The latter came into fairly general use during the second half of the nineteenth century and, apart from ordinary building mortars and concretes, it was found to be ideal for casting the somewhat blowsy-looking ornaments which delighted our Victorian ancestors.

210

Colour and Limewash

There are three ways of colouring plastered walls. The first is to use aggregates and pigments to procure a self-colour for the actual plaster. The second is related to the ancient practice of *buon fresco,* used by the Renaissance painters in Italy; and long before that by the Romans and Cretans. The third method is to apply a colouring pigment to the dry surface of the wall in the form of a conventional paint.

It is exceedingly difficult to design a plaster mix which will dry to exactly the gentle and natural-looking colour which you are seeking. You are faced with the same problems which you encounter with pointing, except that you have a large flat area of wall to proclaim your ineptitude, rather than a system of recessed joints which may pick up the tones in the stonework, by default. The best advice is to do generous trial patches, using different mixes and writing careful notes of the ingredients. Your object is to prevent the grey of the Portland cement getting the upper hand over whatever colour may be in the chosen sand. This means finding a rich, warm-looking sand, preferably of gritty texture, well washed and thus free from salts.

Originally, in the eighteenth century, the idea was to make a plaster look as much like stone as possible. Grey Portland cement had not been invented and white lime was neutral in its colouring effect. Hydraulic limes or patent cements containing a good deal of clay could be slightly more assertive, depending upon the strength of the mix.

Today, you can either use a very small quantity of cement with lime and sand of good colour, or you can use the more expensive white cement, with or without lime. On the whole it is best to use lime, since it makes for a reasonably porous and flexible plaster, as mentioned before. Some of the best coarse sands are from granite, and thus have a tendency to dry to an unpleasing battleship-grey. These should not be scorned, however, since they are not hungry for cement and lime, and can be coloured with pigments available at builders' merchants.

Buff pigment powder is the most useful colour to employ, but it should be well mixed and sifted into the cement and lime in consistent quantities. If too much is used the resulting mix can turn out a disagreeable colour; if too little, it can fade. For reasons of durability alone, a sand of good colour is much to be preferred. Unless you are prepared to be painstaking in experimenting with the colour of the mix, it is not worth using a self-coloured plaster. The results can be too distressing.

Buon fresco or 'true fresco' in its accepted sense requires several coats of plaster, finishing with one called the *intonaco,* upon which the actual painting is carried out while it is still wet. The water-borne pigments combine with the lime in the mix and take part in the same process of carbonation, thus fixing themselves in the body of the plaster itself.

For the purposes of painting a wall in a single colour it is obviously not necessary to prepare a finely graded mix of lime and marble dust. Indeed this would be counter-productive when you are trying to obtain a rather gritty porous surface of the kind we have been discussing.

I have not myself experimented with applying colour in fresco, and enterprising readers will have to learn by their mistakes. I only put it forward as a suggestion for these reasons. The technique lends itself to a subtle type of colouring, using earth pigments, which will avoid the rather uniform and monotonous appearance of modern paints. When it weathers it will fade, one hopes, with a charming inconsistency, so affording just the right amount of colour while not looking unwholesomely decayed.

The traditional coating for most plastered walls and many natural stone surfaces, from medieval times onwards, was limewash. This is made by slaking lump lime (quick-lime) until a solution rather like skimmed milk has been obtained. It was then dashed on to the walls with long-handled brushes. Colour pigments were sometimes added while the limewash was still hot during slaking; and tallow, or some other binder, was normally included at this stage to improve adhesion and durability. It will be found that limewash without a binding agent will flake and brush off.

Limewash has the advantage of being suitably porous while looking natural and unassuming. On the other hand, even with tallow, it is far from permanent and needs fresh coats at regular intervals. When wet from the rain it often looks extremely patchy and unconvincing. Once you have used it, it may prove a poor undercoat for other forms of exterior paint and is tedious to remove or prepare. Despite these reservations, limewash is redolent of 'honesty' and unquestionably correct if you are trying to be historically accurate. Some people contend that a clear limewater applied so as to saturate decomposing and friable limestone masonry acts as an effective hardener.

Limewash should not be confused with ordinary whitewash, which is whiting (finely ground chalk) mixed in water, with a little size to bind it together. Although limewash was traditionally used both inside and out, whitewash is exclusively an interior paint. It is far from durable, tending to flake and dry to a powdery consistency. Because of its cheapness, it is constantly found in old cottages, particularly as a ceiling finish. Its condition almost always demands that it should be stabilised with a coat of size before any new decorations will adhere. As with limewash, pigments were sometimes added to whitewash to provide a coloured distemper.

Before dealing with interior plasterwork we should spare a thought for the process of disguising some of the more horrendous exterior finishes which may have been applied to old buildings. There are two main ways of doing this. The first is to use any unpleasant wall texture, such as pebbledash or modern-looking roughcast, as a key for re-plastering. A suitable render, almost amounting to a skim, can be laid up for purely cosmetic reasons. This can be wood-float finished and then painted with an emulsion, a cement paint or a textured and flexible masonry paint.

The other possibility is to camouflage the gritty surface with one of the above paints applied directly to it. Most people will probably go for the latter option since it costs much less and can be done by any amateur. As a compromise, you could always prepare a thick cement-sand slurry and paint it on with a brush, to act as a filler, using several coats if necessary. This, incidentally, is the best way to fill those lines of mock stonework with which Victorian and modern masons have embellished their pointing. Having expunged the incised lines you can then repaint the wall, as before.

Interior Finishes

Few old houses and cottages have been allowed to reach the present day without drastic changes to their interior wall and ceiling finishes. Bland hardwall plaster surfaces, with external corners like the crease in a guardsman's trousers, disfigure the parlours of 'rough' cottages; flat plasterboard partitions abut uneven whitewashed stone walls; ceilings are covered with hideous polystyrene tiles, unsuitable insulation boards, oddly textured and fidgety-looking paint or plaster finishes. There are so many ways of spoiling the interior surfaces of an old building that one can only hope to mention some of the more repellent varieties; for example, sheets of simulated timber boarding, asbestos panels, reeded hardboard, pegboard, chipboard, insulation boards, and faked structural timbers.

The first decision to make when thinking about repairing or restoring wall and ceiling surfaces is the usual choice between 'rough' or 'smooth' finishes. If the building has any pretensions to classical dignity then you will select 'smooth' forms of plastering. Decorative mouldings will be cleaned, repaired or replaced, and a fairly high degree of accuracy will be required when plumbing walls and ruling in corners.

If, on the other hand, you are dealing with a simple cottage, with irregularly placed window openings, low ceilings and big open fireplaces, then your plastering should be designed to look soft, free from rigidity and may be very slightly textured. The texture comes from finishing the top coat, of a cement, lime and sand mix, by working over the surface with a cross-grained wood float as the plaster begins to stiffen.

Cottage walls can be dealt with in four ways. If the building is fairly late – somewhere between, say, 1800 and 1900 – and has some of the qualities of a neat lodge or villa, it may call for equally neat, smooth and straight plastering. This, however, should be reasonably soft in texture and arrises should be blunted but not rounded. Dado boarding may be employed as mentioned in Chapter 11 and some rooms may require cornices. One method of avoiding the sharp straight edges which go with 'smooth' wall finishes is to plant a softwood return bead against which you strike off the finish (Fig 184). This is also called a bowtell moulding, a staff bead or a double quirk bead. In its crudest form you can nail a semicircular bead of about 25mm diameter to plugs in the masonry and simply plaster up to it. Correctly, however, the return bead should consist of a three-quarter circle of moulding with fillets or quirks on either side. The staff bead is ideal for creating that compromise between 'rough' and 'smooth' at window splays or external corners, since it is straight like the wall surface but can look suitably robust if the diameter is fairly generous.

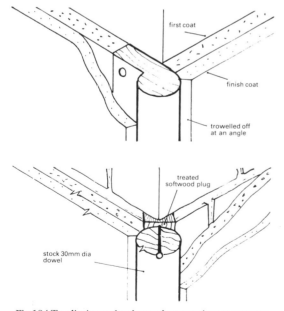

Fig 184 To eliminate the sharp plaster arrises at corners which so often turn repair and restoration into modernisation, angle beads may be used. The top coat of plaster is struck off at an angle so that it does not cover too much of the bead.

Small Victorian houses of the neat-and-tidy kind may sometimes have frieze rails or picture rails. Ceilings in parlours or dining-rooms could well include central plaster roses or medallions with Roman acanthus leaves and other ornament. All will depend upon whether the place has the flavour of a small house or that of a real workman's cottage.

Ceiling centres from the eighteenth century onwards might be of cast fibrous plaster, of papier-mâché, or modelled *in situ*. It requires very nice judgement to select something suitable of this kind. If in doubt, seek advice or stay your hand. A poorly designed ceiling centre or one of the wrong period, could prove a disaster.

The second group of wall and ceiling finishes includes soft cement:lime:sand mixes in proportions such as 1:1:6, 1:2:9 or sometimes weaker than that. The plaster follows the undulations of the background wall surface, whether it is brick, dubbed-out stonework, cob, wattle or lathing. The essential secret of the operation is to ensure that the undercoat is applied freehand and not floated off between screeds or timber rules. Corners should be allowed to wander just a little and should be blunted. This is a tricky business, since it is important that nothing about cottage plasterwork looks contrived. The best advice is to show the maximum restraint, always doing less rather than more in the way of texturing, blunting or freehand movement. This also applies to texturing with a wood float. I do not really recommend roughening with brushes and cloths, although they can occasionally prove effective if they are sandpapered afterwards to eliminate any actual marks.

Most interior walls in old 'rough' buildings, although of fairly uneven contour, look smooth as a result of many layers of limewash which have built up over the years. This means a steel trowel finish may often look more convincing. The important thing is to provide some visual movement and a sense of undulation to the surface as a whole. With cottage wall surfaces, the test is this. When people who know about old buildings look at the work they should not be immediately struck by the dexterity of your texturing. Preferably they should notice nothing at all, except that the walls look mellow and natural.

The third approach to cottage-wall treatment is what might be described as the whitewashed look. In this instance, you allow the shapes of the stones (or bricks) to be seen clearly through either whitewash, limewash or emulsion paint.

To achieve this, you first point the wall. This may be done in the normal manner, described in Chapter 5, but should be less 'hungry' than exterior pointing. The pointing should fill the joints flush with the arrises of the stonework. It should not, however, be buttered over the stones, unless you intend applying a number of coats of limewash, a skim or cement-sand slurry to the whole wall surface.

It is arguable that the limewashed or whitewashed look in which individual stones are not all seen, but rather implied, is the most truly traditional. When every stone is seen, depending on the kind of coursing used in construction, the effect is busier and more demanding to the eye. Different buildings or rooms dictate different methods.

Limewash and whitewash may need the addition of a very little colouring matter to give them life; and white emulsion, which tends to be whiter than white, can also benefit in this way. There are now some excellent off-white and cream emulsions available. In former times, internal surfaces might echo the colour of natural stone, might be colour-washed or lightly tinted. In nineteenth-century cottages it was not unusual for walls to be cream or buttermilk in colour; and occasionally rooms would be painted at dado level in something like a GWR brown. In recent times, this dado would be gloss-painted – an unwholesome practice which I suggest you avoid.

It may seem very unenterprising, but for modest cottages, there is no better colour than white. If you have beams or joists which can be left exposed in their natural state, then these may supply a nice visual contrast which reminds you of the structure. Do not, however, paint any such timbers in matt black, bitumen or similar dark finishes. If you find that somebody else has already done so coat the wood liberally in white emulsion paint, using something like an aluminium sealer over bitumen to prevent it leeching through. Better still, if the black paint will come off, strip it in the manner referred to in Chapter 11.

There is generally little merit in those very busy-looking ceilings which display a welter of softwood masquerading as oak, especially if all the structural timber is painted black. Beams may reasonably look dark, but there is no need to emphasise every joist. There is no rule, but at least *consider* painting the entire system of joists and board soffits white. This I particularly recommend for all cottages which have no cross-beams or spine beams, but only sawn joists laid on edge. The ceiling will have all the interest and sense of age required; the room will be lighter; and there will be no false references to 'olde oak'. The emulsion paint is purely decorative and in no way intended to preserve the timber, which may be two hundred years old or have been put in last week.

The fourth wall treatment for cottages is one which has little historic tradition to support it. It is the exposed-stone look. At no time, even in the early Middle Ages, was it the practice to leave stonework uncovered in its natural state. This is a romantic Victorian conceit, embodying rather unwholesome Gothic dreams of which the less said the better. Despite these strictures, it is not entirely inappropriate (if perhaps inaccurate) to have some exposed stonework in castles, converted mills, churches, engine houses, warehouses and places of that ilk. In nearly all cases, domestic buildings had interior plaster, however rough, or at the very least coat upon coat of limewash. As far back as the twelfth century it was normal to limewash the stones even in grand and important buildings, like churches or bishops' palaces, where considerable efforts had been made to employ expert masons and to provide well-detailed mouldings.

For many centuries internal plaster was in three coats. The first would be a rough render made from some kind of natural earth, or sand if available. The binding agent would be either the clay itself or an element of lime. This coat would contain some reinforcement from chopped straw, animal hair, small stones and anything else which struck the builder as useful for the purpose.

There might then be a second floating coat of much the same mix, but thinner. This could be plumbed and screeded to straighten the walls. Depending on the locality and date, it might contain lime and sand, the clay-and-coarse-stuff render being used only to dub out and provide a background. The top coat, or set as it is known in the trade, would be dictated by the refinement of the building. In grand 'smooth' houses it could be a very hard stucco or it might be lime and sand mixture (about 2 or 3:1 of lime putty and fine sand) or a $\frac{1}{2}$:1 plaster of Paris and lime putty mix. In a 'rough' building the finish could consist of just limewash.

It is a temptation to specify these traditional mixes for the different coats of plaster, but I shall resist it, since circumstances have always varied from one building to another and from one part of the country to another. In any case, you may wish to use gauged mixes containing either Portland cement or gypsum plaster. Whatever you decide to do, remember that each coat of plaster should generally be no stronger than the one upon which it is laid. Allow for the natural suction of the background material, which can rob the mix of its essential setting and curing moisture. Use fairly soft lime plasters in 'rough' buildings and in damp situations where no attempt is being made to supply a vertical damp-proof course. Such plaster can breathe and allow the moisture in both walls and atmosphere to evaporate.

Give each coat of plaster adequate drying time so that differential shrinkage does not take place between successive layers. Ensure that lime-plastered walls, including those gauged with small amounts of cement, are not decorated with oil paints or papers until they have thoroughly dried. Depending upon weather conditions and the season of the year, this process can take six months or more. During the drying period use emulsion paint, distemper or limewash as a decorative finish. Avoid limewash, however, if you do not mean to keep it permanently, since it is difficult to decorate over and a nuisance to prepare.

'Smooth' houses requiring hardwall plaster finishes may involve any one of a number of types of plaster from traditional three-coat work (using coarse stuff of lime, sand and hair) to gypsum and sand; or lime, gypsum and sand. Some finishes may require very hard traditional mixes such as true stucco or class D – Keene's cement. The latter is often used for vulnerable external corners.

For traditional lath partitions in early buildings, it is best to use three-coat lime plasterwork. It will look softer and more appropriate. There is no reason why formal partition walls with cornices and chair rails should not be formed of studwork with skimmed plasterboard, provided the house can accept it.

A truly difficult operation which outrages modern workmen, is the finishing of plasterboard, lath or blockwork partitions. These would normally be rendered and skimmed in the case of blockwork or, perhaps, given a traditional render, float and set treatment. Stud partitions would be clad in plasterboard and skimmed with gypsum plaster. Such subdivisions nearly always look hard, brash and new. One way of making plasterboard look as though it is old lath and plaster works like this. You skim the surface with gypsum finish, making sure that there is some slight variation in the thickness of the coating. Just before it has a chance to set, you brush it over very lightly with the tips of the bristles of a wide supple brush. When the plaster has fully dried out and hardened, you lightly sand it over by hand to remove the brush marks. The same technique can be used on blockwork as well.

The principal danger when leaving exposed stonework of any kind is that you will overdo it. In most cottages you should leave no exposed stonework, except perhaps a chimney breast or lintel. Even for chimney breasts it is much better to plaster and whitewash the stonework above the lintel, leaving only the jamb stones and lintel exposed.

Where they are of good enough quality in more important buildings like early

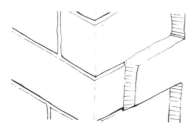

Fig 185 Plaster should never join exposed stone or brickwork so that it forms a steep escarpment like this.

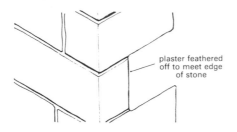

plaster feathered off to meet edge of stone

Fig 186 When plaster meets exposed bricks or stones it should be feathered off to a negligible thickness which does not draw attention to the junction between the two materials.

manor houses, it may sometimes be in keeping to leave the dressings around doors or windows exposed, and it is at this point that many people go wrong. Traditionally, such stones were usually given a coat of limewash which blended the dressings in with the general white of the walls, and experts can insist on this practice today. Personally, I have some doubts about its merits and think it may often be better to see the unadulterated stonework in its entirely natural state.

Skill and subtlety are required in effecting the transition from plaster to exposed stone. Nothing looks worse than old stonework surrounded and encroached upon by acres of white plaster which finishes in a steep escarpment, crudely outlining the stones, which are thus recessed below the general wall surface (Fig 185). You should so apply the plaster that it thins out to a feather edge where it meets the exposed stones. The junction should, if possible, be fractionally below the arris of whatever lintel or jamb it abuts (Fig 186). Pieces or areas of exposed stone, dotted around like stagnant pools in a marsh of plaster, may be counted as abominations.

Ornamental Plasterwork

'Rough' cottages seldom have ornamental plaster, other than the occasional cornice moulding or, for example, a rather rudimentary section, such as a plain cove. More refined buildings can contain anything from standard Victorian ceiling bands to superbly elegant musical trophies by famous eighteenth-century Swiss, Italian or English *stuccatori*. You may find nicely executed cornices carried out by competent rural craftsmen, following drawings from the pattern books of men such as Batty Langley; or crude but elaborate seventeenth-century ceiling centres loosely based on fashionable Baroque themes. Chubby cherubs in high relief will scramble clumsily in a playpen of bay-leaf garland. Ceiling centres of slender rococo arabesques, masks and swags, paterae, modillions, dolphins and dentils conspire to make restoration more difficult and the house itself vastly more entertaining.

Now, any fool can see that an elaborate and inventive piece of plasterwork may be worth saving; but beware the philistine architect who alleges that your modest little cornice mouldings are of such routine quality that they do not really matter. They matter very much indeed. There is, today, a positive army of professionals who are apparently impervious to the architectural features inside old buildings. For them, the absence of cornices in old houses which they have repaired or converted is no source of shame or regret, especially when they are dealing with what they consider to be rather ordinary little rooms of late eighteenth- or nineteenth-century vintage.

The replacement of plaster mouldings clearly presents problems. Firstly, materials may be difficult to obtain. It is not easy to buy hair for coarse stuff, for example. Plaster of Paris (class A) is usually not stocked by builders' merchants, nor is class D, Keene's cement. You might have trouble getting canvas scrim – normally, a rather wide-mesh hessian – for reinforcing cast mouldings. Wooden laths are readily available.

Your next difficulty will be to find a craftsman who is either experienced or adventurous enough to use traditional techniques. The majority of so-called plasterers are, in fact, bricklayers or masons whose duties include the task of plastering interior walls to modern specifications. They have no experience of making mouldings, whether cast or run *in situ*. However, it is my belief that many of them have the latent abilities to do the work if the architect or client explains what is wanted. You can always go to specialist firms for your plastering, but these are likely to be employed for every important conservation contract over a wide area. Their services are expensive and you may have to wait a long time before they can start on your job. The answer, as always, is self-help. Do it yourself or in partnership with a good tradesman, improvise and experiment until you achieve the results you want.

There are two main ways of making continuous mouldings such as cornices, architraves and ceiling ribs. You can run the moulding *in situ* – a common practice in the eighteenth century – or you can cast it in fibrous plaster by using a reverse mould. Some firms, of course, supply ready-made mouldings in fibrous plaster or modern synthetics, for example fibreglass; but with these, you will be very much constricted by the rather limited choice of styles. Much of what is available will be quite unsuitable for your purposes. Ready-made plasterwork often lacks crispness and is rather lifeless in design.

It is impossible to go into any great detail about casting and running techniques in a book like this one, but it is worth outlining the main principles involved.

To run a cornice *in situ*, for example, you prepare a background of wooden brackets and laths to fill out the intersection between the ceiling and the wall. The brackets, which are cut from thin softwood to adopt the general outline of the cornice, are nailed at the top to the sides of the ceiling or floor joists, and are sometimes notched over a batten, nailed along the wall, at the bottom. To these brackets you nail the laths with galvanised nails.

Next you prepare a horse or running mould, which merely consists of a short piece of board to which you attach a vertical timber batten called the stock (Fig 187). You make a cardboard reverse pattern of the cornice profile and trace this on to a piece of zinc or sheet iron, which is then cut out to form a template. At the same time, you also cut out a slightly oversized template of approximately the same profile, but roughly 5mm bigger all round. This is called a muffle.

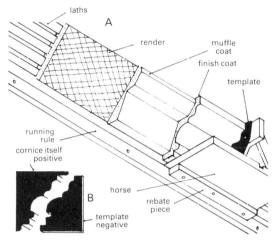

Fig 187 The principle of running a moulding *in situ* is shown here for a cornice in three-coat work (A). At (B) it will be seen that the template must be cut as a negative of the original moulding you are copying. Note that the outline of the muffle coat may be a much-simplified version of the finishing profile.

There are three main ways of obtaining the profile or section of a plaster moulding. The first, and easiest, is to use an adjustable metal template consisting of numerous fine teeth like those of a comb. You press the template against the moulding and it takes up the required shape. It is, however, rather inaccurate and will not adapt to fine fillets and tricky outlines. The teeth may also damage the surface against which they are applied.

The second method is to take a 'squeeze-mould' impression using a wad of a substance such as modelling clay (or jeweller's 'putty'), which is less easy than it sounds. The clay must be soft enough to adapt to the shape and stiff enough to retain it when removed. It must also lie reasonably flat on the paper while the outline is transferred with a pencil.

The third way is to make a relatively fine saw-cut through the entire moulding and

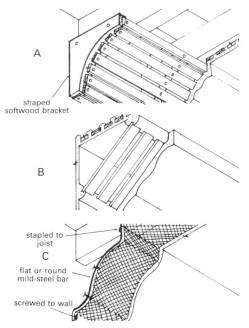

shaped
softwood bracket

A

B

stapled to
joist

C

flat or round
mild-steel bar

screwed to wall

Fig 188 Three methods of providing a background for cornice work. (A) the use of shaped wooden brackets nailed to the ceiling joists; (B) scotch brackets, made by bedding pieces of lath in plaster; (C) expanding metal lath wired to shaped mild-steel bars. Or use metal lath nailed to timber brackets, like those in (A).

slot in a piece of card upon which you can then pencil the profile. This is extremely accurate, but great care must be taken to avoid cutting through the background of the moulding or doing any other sort of damage. You *should* end up with a nice neat section on the card and an equally neat and easily filled cut in an inconspicuous length of the cornice.

The pricking-up or rendering coat of coarse stuff is first applied to the laths, being worked well between them to obtain a firm bond. In some cases you may provide a background of what are called 'scotch brackets' – laths bedded at either end into plaster screeds (Fig 188). Metal angles and expanding metal laths may also be used for this purpose.

Next you apply a second coat of lime plaster and, with the muffle template screwed to the stock, you run the horse along a timber guiding batten, thus shaping the muffle coat to the required outline. This coat is then scratched to provide a key for the final coat. The second coat is allowed to set and the

muffle template is unscrewed from the stock. A finishing coat of lime-gypsum plaster (perhaps 1:1) is then applied with a trowel and the horse is passed across the surface in one steady movement, so that the template begins to shape the plaster to its correct outline. This movement may be carried out several times, the plaster being filled and adjusted as necessary.

In some instances, particularly if the profile of the moulding you want is rather complicated, you can run it on the bench in much the same manner. If, however, you need more than just a short replacement length or two of damaged or missing cornice, you may decide that you should prepare a reverse mould for casting.

As you can see, mouldings run *in situ* require a negative or reverse template to cut out the shape of the final cornice, architrave or skirting. For the casting process, on the other hand, you must take a model or pattern or *positive* template. With this you can form a reverse mould, which is, of course, a negative. Then, you cast your positive, whatever that may be.

I shall briefly describe the traditional method of casting a simple late eighteenth-century coved cornice; but first, a word about working from models or patterns. Much plaster detail, especially if it was repetitive, was cast individually and applied to the entablature, fireplace surround or ceiling to play its part in the overall design. For this process a model or pattern is made, sometimes from clay or plaster, or carved in wood. In restoration work an original moulding may be used as a pattern.

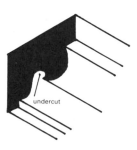

undercut

Fig 189 When moulding items like this Gothic beak moulding, which have a pronounced undercut, it is necessary to use a *piece mould* technique. The mould can be removed in separate pieces so that undercut features are not broken off.

Complex mouldings requiring ornamental embellishments, or those with undercut features, may need to be 'piece-moulded'. For this, the mould itself is made up of several pieces, each playing its part in the process, but capable of being removed separately without damage to the newly formed cast. An obvious example of an 'undercut' would be a Gothic beak moulding (Fig 189).

Obviously the subject of ornamental plastering is extremely complex, and the really skilled and specialist plasterer must be familiar with numerous techniques of modelling, running, casting and fixing. This does not mean that you cannot make a simple reverse mould to cast a replacement detail or some suitable lengths of cornice for rooms which clearly demand them. Anybody who can re-grind the valves of a motor car or make a dress from a pattern should, with a little care and patience, be able to cast some simple mouldings.

To make a mould for casting you cut out a positive template and run a reverse moulding on the bench, using a horse in much the same way as you would run a moulding *in situ*. The core for the mould is made up of any bits of plaster you have lying about, laid on a smooth flat bed of shellacked and greased plaster. A creamy mix of plaster is poured over the core material which is prevented from moving by nails buried in plaster.

Enough plaster is applied to the core to run the muffle across, bringing the mould to its basic shape. The muffle coat is then scratched for key, and covered with the final plaster coat, which is run with the positive template, the muffle having been removed. When the reverse mould has set and dried it is given three coats of shellac and greased.

Next, canvas scrim is cut to size and pieces of lath are made ready for use as reinforcement. An unretarded coat of plaster, called a 'firsting', is applied to the mould and the canvas reinforcement is pressed lightly into it. The 'seconds' mix, which contains some glue size used as a retarder, is then brushed over the canvas and 'edge laths' are bedded along the edges of the cast to act as screeds, defining and retaining the mix.

The overhang of scrim is turned back over the edge laths and consolidated with more plaster. Depending upon the size and shape of the cast, additional pieces of lath or suitable non-ferrous material may be inserted during casting to provide reinforcement. When the cast has set it is

Fig 190 Making a mould for casting. NB For clarity, the wet finishing coat which the template is shaping has been omitted from this drawing.

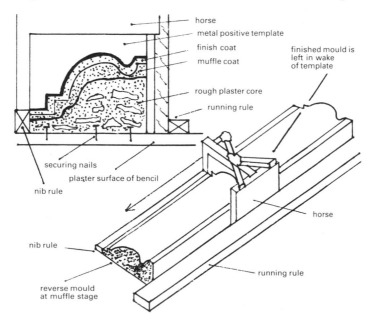

horse
metal positive template
finish coat
muffle coat
finished mould is left in wake of template
rough plaster core
running rule
securing nails
plaster surface of bencil
nib rule
horse
nib rule
running rule
reverse mould at muffle stage

removed from the reverse mould and fixed in its final position.

In former times, plastering firms used to own a variety of stock moulds from which they could cast the building owner or architect's choices to order. Some moulding, like the early to mid-nineteenth-century vine-leaf-trail ceiling band, was so popular that it could be obtained almost anywhere, as a ready-made item. Ornamental ceiling centres in the Adam manner were formed in papier-mâché in the eighteenth century and are still in production today. Modillions and console brackets for cornices, fireplaces, door and window entablatures were churned out by the thousand during the Victorian period. Some specialist companies still have their own moulds today; but should you want something of your own design or a patternbook moulding which is new to them, they will still have to make a mould, by this or some other method.

At the other end of the scale, the artists in stucco of the period which began in the late sixteenth century and finished towards the end of the eighteenth showed extraordinary powers of skill and invention. They combined the abilities of craftsman, designer and sculptor. The best of them could create almost anything in plaster from a wheatsheaf to a stylised rosette, a mouse to a mandolin.

In Elizabethan houses you sometimes find ribbed ceiling plaster carried out in what is frequently called 'finger-and-thumb work'. In other words it is pinched and modelled *in situ* by hand. The sculptural element in plasterwork should never be underestimated, whether it involves quite straightforward tasks such as carving small embellishments and cleaning up mouldings, or modelling masks, figures and heraldic devices.

Plasterwork Development

Historically, the principal trends in plasterwork may be summed up as follows.

Sixteenth Century

The most typical Elizabethan ceiling was one consisting of free-flowing or geometric and rather refined rib work, often with pendants and bosses at the main points of intersection. Friezes were decorated with arabesques and trails of fruit or flowers, cartouches and heraldic beasts. Elaborate chimneypieces with manneristic terms, herms, caryatides and Atlantes were employed in combination with Renaissance orders of architecture.

Seventeenth Century

Plasterwork early in this century had rather the same characteristics as were found in the sixteenth, but more emphasis was laid on elaborate flat bands for ceilings than fine rib mouldings, and the intervening spaces were often crowded with decorations. Strapwork became increasingly popular, as did ball, flute, pyramid and nail-head embellishments. Plenty of figures were in evidence and scenes from the bible, mythology and the chase were favoured. In vernacular buildings a liking was still shown for naïve and bold designs depicting flowers and foliage, especially over fireplaces. By mid-century, classical influences were gaining the upper hand and ceilings were outlined in panels with huge beam-like moulded divisions. A heavy oval garland of bay-leaves, fruit or flowers was popular. Within this were figures in very high relief of cherubs, gods and goddesses – often rather crudely executed.

Towards the end of the century, there was a liking for the *tête d'ange* framed by folded wings, heavy undercut swags of fruit and flowers, Baroque shields, cartouches and shells. Arabesques of elongated Roman-looking leaves were used in panels, but much relief design was extremely robust and coved cornices often contained shells and heavy acanthus motifs. For doors and niches, classical entablatures, often with pulvinated friezes and segmental pediments, might be seen everywhere. Broken pediments in the Baroque style were much in vogue; but most of the latter features were more usually executed in timber than in plaster.

Eighteenth Century

Queen Anne plaster ceilings tended to consist of rather simple ribwork, sometimes moulded to look like a rope. The layouts were usually based on a plain central circle or oval with echoing medallions at the corners. Cornices were frequently a simple cove or an ogee, but more important rooms could have quite accurately detailed Tuscan-looking groups of mouldings. Much early eighteenth-century cornice work was carried out in timber as part of a system of panelling.

Early to mid-eighteenth-century plaster remained classical with rococo motifs becoming increasingly popular. Dolphins, masks, key frets, hatched-rosette designs within asymmetrical scrolled panels, acanthus leaves galore and flat arabesques of geometric ribbon-work were all employed. By the 1760s a pleasing but rather stylised rococo was the norm – C and S scrolls, and lots of putti, trophies, and swirling ceiling centres.

The late eighteenth century, highly influenced by Robert Adam and the general European lurch towards neo-classicism, saw the introduction of symmetry, elegant little swags and trails of bell-flowers and husks, ribbons and medallions containing classical profiles or scenes from mythology. Oval and circular paterae decorated every other fireplace surround, and all the normal repertoire of acanthus, egg and dart, dentils and modillions might be found. From the 1740s, a number of Gothic enthusiasts were at work, using pointed arches, fan-vaulted rib decoration, trefoils and quatrefoils, cusps and bosses. This all became fashionable with Horace Walpole's Strawberry Hill – an act of whimsy which the more earnest old-buildings experts today would consider 'dishonest', if asked to comment upon a similar modern Gothic adventure.

Nineteenth Century and Edwardian
At the beginning of the century, plasterwork, like most other forms of design, tended to be either Regency Gothic or traditional Roman, or to contain elements of the Greek Revival. Key frets were popular, as were palm-leaf ceiling centres, guilloches, hard little rosettes, spiky acanthus leaves, vine-leaf and grape or floral perimeter ceiling bands. Except for the centre feature, and the borders, ceilings tended to be plain in everything but the most grand domestic buildings.

Walls were austere, with only a tall skirting and a frieze rail to divide them up. The cornice mouldings in small houses were often enriched but were fairly minimal, the main decorative effect being provided by the ceiling bands. When carried out in plaster (as with the much more usual wood), door architraves, fireplace surrounds and the like frequently displayed roundels at the corners and reeding down the sides. The reeding was either set flat or curved like an engaged column. Incised-line decoration, terminating in Greek key patterns was much used on plain pilaster panels.

Victorian plasterwork, when the architect meant business, could be wildly Gothic or monumentally Greek, public-baths-Roman or of the most accurate and restrained classical derivation. In some cases interiors might become French in either the early 'château style', or Second Empire 'Louis Quinze' and 'Louis Seize'.

Finally, everything quietened down to a slightly anaemic neo-Georgian, with the Arts and Crafts movement and Art Nouveau thrown in for luck. Plasterwork did not exactly flourish in the late nineteenth and early twentieth centuries. Nevertheless there were enough splendid exceptions to provide us with an amusing legacy which is well worth saving.

A typical late Victorian cornice might consist of a bland build-up of rather fussy looking ogee, ovolo and cavetto mouldings, divided by fillets. A favourite Edwardian ceiling had a wide perimeter band of variegated parallel mouldings, one of which tended to be rolled and undercut in the Gothic manner. Expensive houses could boast elaborate, often accurate, essays in various period styles.

Repairing Plasterwork
Probably the worst fault you will encounter with plaster is collapse resulting from separation from the backing. In old buildings this usually results from decay of the laths, rusted fixing nails, or sagging or broken joists. Ceilings may be bulged, cracked, soggy or stained. Cornices sometimes pull away from the wall or ceiling as a result of differential settlement.

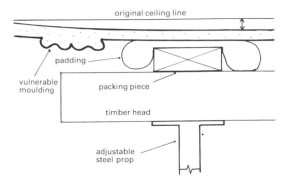

Fig 191 Care must be taken when propping moulded ceilings to pad the plaster so it is not damaged by the timbering. More vulnerable and elaborate mouldings should not be directly supported.

For the most part, faults which can be put down to bad workmanship and materials are infrequent in 'smooth' houses. Structural deficiencies and neglect are the chief enemies. These should be remedied first; the plasterwork can be dealt with afterwards. However, it is vital that plasterwork should be given adequate temporary support during structural repairs – the better the plaster detail, the more careful you have to be to ensure that ceilings, for example, are propped from below with a system of adjustable steel props, head trees and scaffolding boards. Upon this platform straw-filled sacks, polystyrene chippings or similar soft bedding material should cushion the ornamental mouldings, and with the aid of timber packing pieces and pads you should ensure that the *unmoulded* plaster takes the weight. Delicate relief plaster should have no pressure of any kind applied to it (Fig 191).

Where the ceiling plaster has come away from the joists it requires a good deal

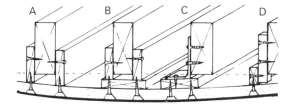

Fig 192 Sagging ceiling plaster may be screwed to boards which are also screw-fixed to the joists (A). Packing boards may be introduced where depth of sag increases (B). Boards can be screwed to angle brackets (C) or battens attached to short wooden hangers (D).

of thought to decide whether you will attempt to jack it back into contact with the structural timbers, or accept some sagging. In the former case you may be able to re-secure the plaster to the joists using countersunk brass screws and nonferrous discs. In the latter, it may be possible to introduce packing pieces of softwood into the widest gaps between the laths and joists or to run a framework of secondary ceiling timbers – perhaps of 50 × 50mm or 75 × 50mm stuff – using galvanised brackets, plates and hangers to screw them to the sound joists (Fig 192).

These remedies will not be sufficient if there is a major breakdown in the laths themselves as a result of decay and woodworm attack. This problem must be met by digging out as much of the rotten lathwork as possible, vacuuming out the dust and dirt from between the joists and treating with a preservative *if this can be done without the solution making stains on the face of the plaster.*

The next step is to provide a sound, reinforced backing, keyed and bonded to

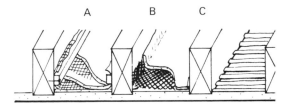

Fig 193 To support ceiling plaster which has sagged because the laths have rotted and broken away, new plaster may be cast in layers between the joists, using canvas scrim held by battens as a reinforcement (A). Expanding metal lath can also be used in the same way (B). The key formed by the original plaster by squeezing between laths can be seen at C.

the old plaster. The standard practice is to run new plaster over the rough, corrugated surface of the old plaster. The ridges of plaster which have squeezed between the original laths make an excellent key, but the whole surface should first be painted with polyvinyl acetate solution to kill the suction (Fig 193).

What you are really doing is casting a screed between the joists, into which you will press a canvas scrim in two layers. The scrim is then fixed with battens to the joists or looped over them. Wire-netting

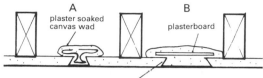

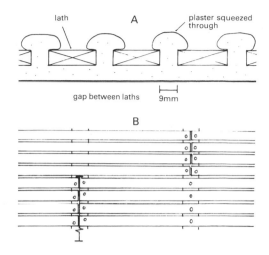

Fig 194 Ceiling plaster which is still firmly attached to the joists, but has sagged due to inadequate key, may be fixed back by pushing canvas wads, covered with plaster of Paris, through dovetailed holes (A). For bigger areas, back up the hole with plasterboard or metal lath secured with a layer of grout (B).

Fig 195 Laths for ceilings and walls should be about 9mm apart (A). In *butt-and-break* lathwork (B), the breaks at which the laths are butt-jointed over joists or studs should not continue for more than 900mm before moving the butt joint to the next fixing ground. Modern sawn laths measure about 1.5m × 24mm × 6mm.

or expanding metal lath can also be used as reinforcement and secured to support the loading. Another method of tying the new backing cast is to press in wads of plaster-saturated scrim and nail these to the joists. The standard mix for casting this kind of repair is plaster of Paris, lime putty and glue size, although other mixes can be used.

When lathwork is sound and still secured to the joists, but the plaster has sagged through failure of the key, you can repair it with plaster-of-Paris-soaked canvas wads, pushed through numerous dove-tailed holes in the ceiling. The edges of the holes, which are about 35mm in diameter, are painted with PVA against suction, and the wads are prodded through gaps in the laths before they stiffen (Fig 194).

Actual holes of any size where plaster has come away in chunks should be trimmed to a neat, splayed edge, again treated with PVA to stop suction. A new backing must be provided in the form of expanding metal lath, or any other material which gives adequate key (Fig 194B). This can be secured from above or below, according to circumstances, being screwed to sound timber or bonded with a pour of new plaster. The patch is then made good with coats of plaster as required. Normally, lime plaster is not recommended for patching, since it shrinks when it sets, as do cement-gauged mixes.

During all ceiling repair work, keep hammering to a minimum and use screws to avoid further cracking or disturbance.

When inspecting ceiling plaster remember that it may look innocent of faults even when it is in fact ready to drop on your head. Lack of proper key because of laths

being placed very close together cannot be seen except from above. The correct spacing of laths, for both ceilings and partitions, is about 9mm. Lath ends are butt-jointed over studs or joists, with the line of joints broken every 900mm. This is called 'butt-and-break lathing' (Fig 195).

Another possible fault to look out for in ceilings is cracking of the plaster key between laths as a result of excessive shrinkage of an over-rich rendering coat. This tends to be found in later houses, since lime, sand and hair coarse stuff do not usually err in that direction.

Repairs to walls may involve the widening, cleaning up and filling of cracks with either a ready-made filling powder, plaster of Paris, or Keene's cement. Damp well the edges of the existing plaster and the background, if possible undercutting the edges of the crack to dovetail key for the fill. The trouble with all cement-gauged and lime plasters is that they shrink on setting. This produces fine cracks at intersections between walls and ceilings and also in the plaster generally, should insufficient drying time be given between coats, or if the strength of coats varies too much. In old houses, however, most cracking of this kind should have taken place shortly after they were built. Cracks seen today will sometimes have been caused by background movement, thermal effects, or wetting and drying due to rising or penetrating damp. Cracks often have to be widened to permit filling.

The treatment of larger patches will depend on the type and condition of the background key and the qualities of the original plaster. Obviously, hard mixes should not be used to patch soft plaster.

The adhesion of plaster will be affected by the amount of suction in the background, which is one reason for using a rendering coat as well as a float and set. It helps to even out suction as well as filling up the hollows. Weak rendering coats sometimes fail in adhesion or because too strong a finishing coat is placed over them.

Many faults in the plaster of old houses are caused by the use of incorrect repair or replastering mixes in much more recent times. An obvious example can be the formation of efflorescence on damp walls – a white powdery covering of salts deposited when the moisture containing them evaporates. Many old walls contain salts but they do not show themselves until a new source of dampness leeches them to the surface. Repairs carried out with mortar or plaster containing salts in the sand component of the mix will lead to efflorescence. Damp walls which have been stripped of old plaster, or revealed by removing internal linings, can produce a white bloom of salts when air and warmth evaporate the surface moisture. They should be brushed off and the basic source of damp eliminated. When the wall is replastered, magnesium salts can theoretically be kept back by using a cement: lime:sand mix. The lime component reacts with the salts to stabilise them. A gypsum plaster mix, gauged with lime, also has this effect. However, this method can equally result in the salts forming behind the plaster and pushing it off the wall.

Although exterior wall surfaces may be repeatedly washed to remove salts, this is more likely to make matters worse inside most buildings. Really, the best answer is to stop the source of dampness. Remove badly contaminated plaster. Allow the walls to dry thoroughly and brush away the salt crystals until no more appear. You may then replaster with a soft mix, such as 1:2:9 of cement:lime:sand.

If the conditions are such that efflorescence recurs, even when the source of damp has been eliminated, it is best to provide a dry lining, either by plastering over a bitumen lath or using battens and panelling, plasterboard, expanding metal lath or any other material which breaks the contact between the salt-bearing walls and the finishes. Normal measures to avoid decay in timber should be taken.

Occasionally one can treat the existing wall surface, after the damp has been dried out, by painting on a porous alkali-resisting primer. This can then be decorated with an emulsion paint.

The use of an adhesive barrier to salts, whether it is a bitumen solution, a waterproofed cement-sand plaster, or special laminated lining papers and impervious foils, can be risky since, again, the salts may crystallise behind the barrier material and push it off the wall.

Ceiling plaster is sometimes affected by a condition known as pattern staining, caused by dust in the atmosphere of the room adhering to the cooler parts of the ceiling plaster, which correspond to those which are least well insulated. Thus the plaster under joists will be least affected. It will be rather more so under the laths, and most of all between them (Fig 196).

Obviously, the way to treat this is to clean the surface and redecorate, having insulated the back of the plasterwork with a glass-fibre quilt, vermiculite or expanded polystyrene.

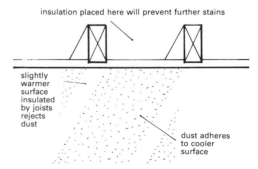

insulation placed here will prevent further stains

slightly warmer surface insulated by joists rejects dust

dust adheres to cooler surface

Fig 196 Pattern staining occurs when dirt adheres more readily to cooler areas of a ceiling, represented by the spaces between the joists. Insulation between the joists can prevent a repetition of staining after redecorating.

Cleaning and Repairing

If you need to clean plaster mouldings and enriched work before redecorating, be extremely careful to avoid inflicting damage which could prove worse than the existing build-up of paint. Enriched cornice mouldings of egg-and-dart details (on ovolos) or acanthus leaves (on ogees and concaves) will need scraping and picking out to restore their sharpness. Valuable plasterwork should definitely be left to experts; but workaday stuff you may be able to tackle yourself. The older and the better the quality of the plasterwork, the greater should be your caution in setting about cleaning operations. Some people, whether amateur or professional, have the kind of controlled dexterity to clean and repair plaster. Others do not; be honest in deciding to which category you belong.

It cannot be said too often that every scrap of original decorative plaster should be saved, whether it is a 4in length of architrave moulding or the shattered fragments of an ornamental trophy. As with china repairs, nothing is too unimportant to save, place in cardboard boxes and label for future use.

The two traditional types of paint you will encounter are oils, consisting of linseed oil, lead pigment and turpentine; and distempers, made of whiting (chalk) bound with glue size water (or casein), a byproduct of milk curds which needs lime to make it into a soluble solution.

Every job will be different, but the likelihood is that the more recent coatings of oil or emulsion paint will be the most difficult to remove. Cleaning plasterwork needs three things – caution, skill and patience. Avoid using metal instruments as a general rule since they can do so much harm so quickly. Wooden or plastic spatulas are advised by some experienced repairers.

Modern paint-strippers will often be too harsh for decent plasterwork, and water is then the recommended medium. Depending on conditions, it may be sponged on, or applied with a fine spray of the sort used for garden pesticides. The water should normally be cold, and brushing can be carried out with an ordinary toothbrush. One of the most success-ful methods of lifting obdurate layers of paint is the application of a poultice made from damp newspapers. These should be wetted, scrunched up into something akin to papier-mâché, and then held against the plaster with boards and props. Firm contact for twelve to fourteen hours should sufficiently loosen the paint layers. A manufactured system on the poultice principle is available. In most instances, abrasives should not be used, although there may be some advantage in rubbing with a domestic pan-scouring cloth.

There will be times when you will need to replace or repair small plaster details – a rosette from a frieze, the tail of a dolphin. Your rule is to repair if you possibly can and only replace as a last resort. Replacement may be done in several ways. You may take a squeeze mould in modelling clay of an existing detail and cast one or more copies, using plaster of Paris. Where you have a pattern which can be taken to the bench, you can make reverse moulds from beeswax and resin, gelatine, latex materials, plaster, clay or glass-fibre.

For small details, which are not undercut, you will probably use what is called a fence mould. The model or pattern is placed on a smooth moulding ground, shellacked and brushed with oil. Around it is placed a fence of clay or of softwood, smooth, well greased and at least 12mm higher than the highest part of the model. The moulding material is then poured around and over the model until flush with the fence. When set, you have a reverse mould from which you can cast a number of copies from any suitable material (Fig 197).

Another means of replacing small details is to key the area to which the ornaments will be attached. You then build up a rough simulation in a medium such as Polyfilla and, when this reaches a suitable degree of set, you carve the final detail with scalpels and other tools. Larger ornaments may need armatures consisting of galvanised wire, wood or any other kind of reinforcement which will not rust and can be made to adopt the required shape. An armature also acts as a means of fixing to the background.

Although many early plaster ceilings in

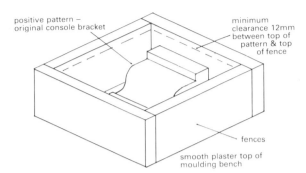

positive pattern –
original console bracket

minimum
clearance 12mm
between top of
pattern & top
of fence

fences

smooth plaster top of
moulding bench

Fig 197 Fence moulds on this principle are used to make reverse moulds from original models or positive patterns. The ground should be very smooth and wiped with moulding oil. The wooden fence should be held in place by weights or nailed battens; it should clear the pattern by at least 12mm.

Jacobean and Tudor houses were painted and gilded, it requires the uttermost discrimination to try to repeat this process today. The only occasion when you might need to use colours for plaster detail is in works of restoration where existing colours can be followed. Even then you must be advised whether the colours you see are indeed original or speculative Victorian embellishments. Obtaining a good match for early earth and mineral pigments is a job for experts and it is almost certain that modern paints will be unsuccessful. Always remember too that modern paints are extremely difficult to remove once they have been applied. For this reason a water distemper is often better than modern emulsion – especially for earlier ornamental mouldings.

Evidence shows that eighteenth-century plaster was often painted with gentle colour washes; but the relief detail was normally white. Again, the choice and use of colour on good plasterwork is a matter upon which you should be advised. Often the enrichment of mouldings was gilded and to repeat this process today is extremely expensive. There are two kinds of gilding: mourdant or oil gilding; and water gilding. In oil gilding, gold leaf is applied to the required areas, which have first been painted with a special gold size. The leaf is pressed on gently like a transfer and sticks to the size.

Water gilding is a much more refined process and is the one most used in past eras. The surface to be gilded was painted with several layers of gesso – a mixture of chalk and glue. This was brought to a very fine surface and then painted with a red earth pigment called 'bole'. Over this was laid a thin film of water, probably containing a very small amount of rabbit-skin size. Then the ultra-thin sheets of gold leaf were picked up from a gilding cushion, which the gilder held like a palette, his thumb through a leather thumbpiece. To lift the fragile leaf he would rub a gilder's tip (camelhair held between pieces of card) against his hair and immediately touch it against the leaf, to which it would adhere. Then the leaf could be transferred to the water-sized surface.

Basically, this is still the method used today. Early gilding was carried out with thicker sheets of leaf than was possible from the eighteenth century onwards; and there is evidence to suggest that a tip was not used with such heavier leaf. The great advantage of water gilding is that it can be burnished, for example with a piece of agate. Never use gold paint instead of gold leaf. A cheaper form of leaf called Dutch metal, made from brass and zinc, might be used for iron railings etc, but will lose its colour unless varnished.

I have described the basic process because you might need to gild chipped areas or a replacement moulding.

Summary of Principles for Plasterwork

1 Save or retain every bit of old original plasterwork you can.
2 Always return plaster mouldings on new partition walls.
3 Do not be afraid to put a cornice of a correct pattern in any room which visually or historically requires it.
4 Show extreme caution when repairing or restoring early plasterwork and above all never over-restore.
5 Pay attention to traditional plaster mixes.
6 Never straighten up the walls of rough buildings.
7 Make top coats of plaster weaker than their backgrounds.
8 Weak flexible mixes are usually the ones to choose.

13

FLUES AND FIREPLACES

It is said, and it does seem to be true, that there are *architects* who have actually stripped out the original fireplaces from their own period office buildings. Everywhere, there is a tendency to pay lip-service to the idea of conservation, while simultaneously committing acts of vandalism upon old buildings. Fireplaces are one more item in the list of victims of those who indulge their itch for cost-effective modernisation.

Almost everybody appears to hanker after the nostalgic joys of an open fire, but all too many enthusiasts for the crackle of logs upon the hearth are insensitive to the quality of the fireplace in which they are burned. In a cruder context the soldier of yesteryear would insist that he didn't look at the mantelpiece when poking the fire; unfortunately there are many worthy citizens who take this injunction literally when it comes to their treatment of old houses.

Development of Fireplaces

Early buildings, with the exception of a few very grand or important ones, did not have fireplaces as we understand them today. The usual system was to burn logs on an open hearth in the centre of the room, whether that was the hall of a medieval manor or the single living-room of a peasant's hovel. In the former, the smoke found its way out via a louvred timber turret straddling the ridge of the roof, or through louvres high up in the end gables.

A bed of ashes was allowed to build up on the beaten-earth, brick or stone floor and the logs were burned as in a bonfire. Sometimes the fire would be slightly raised on a platform of bricks a few inches high. Wrought andirons or dogs were used to support the timber while it burned, so inducing a better circulation of air. For wall hearths, big andirons, used in pairs, usually had metal brackets attached to their stawkes, upon which horizontal spit rods could rest, while meat roasted in front of the fire.

Basically, some kind of 'down-hearth', as it is called, was the principal means of cooking in every kind of dwelling until the introduction of the kitchen range in the late eighteenth century. Water was boiled in kettles and pans hung on pot-hooks, or rested on trivets. The pot-hooks themselves were suspended either from a triangular-shaped chimney crane socketed into the back or side of the hearth, or from a stout iron bar set across the bottom of the flue. Only for baking was another device employed. An oven, consisting of a stone vaulted chamber, was usually placed in the thickness of the wall at the back of the hearth or in the recess at the side of it. Hot embers from the fire were placed on the floor of the oven and bundles of twigs and sticks on top of these. This fire was permitted to burn right through and the ashes were then raked out. The prepared dough was put in the oven and the iron door closed while baking was done by residual heat.

Towards the end of the seventeenth century, cloam (earthenware) oven linings were manufactured and sold for building into the walls of the hearth. They continued to be manufactured right up to the twentieth century. Cloam ovens were used in exactly the same way as stone vaulted ones, and the heavy removable earthenware door was often sealed in place during baking with a daub of clay.

The very earliest fireplaces to be set against walls, with their own smoke holes or flues, can be seen in a few twelfth-century buildings, but the wall hearth, with a corbelled hood and conventional chimney, only came into its own in the fourteenth century. The fireplace with a big chamfered or moulded lintel, flush with the wall, was in use everywhere by the end of the fifteenth century. Most surviving examples of early medieval wall hearths are to be seen in monasteries and fortified buildings.

During the Tudor and Elizabethan era, which ended around 1600, fireplaces of quality tended to have lintels formed in a shallow four-centred arch – typical of the time. These lintels often consisted of two stones, butted together at the centre. They might be moulded with a substantial roll moulding (fifteenth and early sixteenth century) or a cavetto. The spandrels of the arch would be ornamented and the jamb stones moulded to continue those of the arch or lintel stones. In simple houses or cottages, the most common fireplace lintel was a massive oak baulk, often crudely chamfered and stopped with some device such as a quirk-and-tongue.

Sixteenth-century fireplace lintels were frequently cambered, partly for decorative reasons, but also because the timber used might already lend itself to that shape. It is also possible that the camber was considered beneficial from a structural point of view. Thousands of such lintels can be seen in modest houses. More often than not, they have been hacked with a chisel at some time to provide a key for plastering. Many sixteenth- and early seventeenth-century stone lintels were deep and rather plain with a chamfered bottom edge. The jamb stones of the fireplace opening were also chamfered and stopped at the bottom with a tongue, a pyramid, a quirk, or a fillet and faceted ornament.

Elizabethan and Jacobean fireplaces in grand buildings might adopt rather manneristic Renaissance designs, with a somewhat naïve mixture of classical mouldings and medieval Gothic decoration and sculpture. Everywhere, there were caryatids, Atlantes, herms and terms.

Around the middle of the seventeenth century, fireplace surrounds echoed the features of contemporary panelling with bold bolection mouldings. At this point, the structure of the fireplace became hidden and was therefore no longer ornamented. If a lintel or jamb is moulded or chamfered you can be sure it was meant to be seen; and is earlier than any system of architrave, frieze, cornice and shelf which has been planted over it. Usually, but not always, a bolection-moulded seventeenth-century fireplace did not have a chimney shelf. These were often added later. There was normally a panel above the fire in which a painting could be fitted, or an artist could carry out a topographical painting directly upon the wood.

In the early eighteenth century, after the reign of Queen Anne, the Palladian fashion held sway – classical surrounds in the form of 'continued chimneypieces', making an assembly of the opening, its surmounting panel and scrolled or triangular pediment. The broken pediments loved by Baroque architects also remained in use. Large console brackets or volutes might be employed at either the sides or the front of the opening. As the century went on, rococo themes were used to decorate the fireplace frieze, and decorative slips of marble, about 3–6in in width, were fitted inside the timber architrave of the surround, partly for decoration and partly to separate the inflammable joinery from the fire. The rococo, with its C and S scrolls, was followed by the filigree precision of neo-classical decoration – paterae, swags of husks and ribbons, in the Adam manner. Next came the universal early nineteenth-century theme of reeded architraves with roundels at the corners. Often the fireplace might consist of three absolutely plain pieces of marble with corner blocks and roundels, capped by a simple shelf. Frequently, the marble slabs had incised decoration terminating in a key pattern or anthemion, very much like the pilaster panels favoured for stucco façades.

Throughout the eighteenth century, the logs or coals were burned in a free-standing basket with an incorporated iron fireback. These baskets were often beautifully contrived, with finials of brass, steel bars and pierced steel aprons. There were numerous broadly classical designs. The earliest fire basket appears to be a sixteenth-century invention, laid upon the billet bars of the fire dogs. It was a crude affair of wrought iron made up by the blacksmith.

At the end of the eighteenth century, the main innovations were the hob grate and the kitchen range, which were both designed on much the same lines. The fire basket, raised high above the hearth stone, sat between cast-iron panels and had flat ledges upon which one could rest kettles and pans. The hob grate developed in three phases (Fig 198). It started in the late eighteenth century and finished around 1850. Also, in the last third of the eighteenth century, kitchen fireplaces were often adapted from down-hearths to brick hobs with simple improvised iron baskets (Fig 199).

During the nineteenth century, kitchen ranges began to be used in most houses

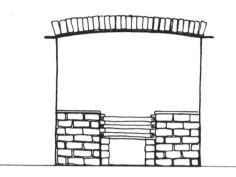

Fig 199 In the late eighteenth century, big down-hearths in kitchens were often adapted to coal-burning hob grates by building up the sides and placing a wrought-iron fire basket between.

above the level of a cottage, becoming increasingly sophisticated as time went on. Eventually, by the 1880s, almost all the artisan dwellings in towns had cast-iron ranges, patterns varying in different parts of the country. North-country ranges tended to retain the open fire basket with an iron bar over it for hanging pot-hooks. The fire box of the other main type of range had a cast-iron plate over it, and smoke funnelled out at the back into the flue.

Unless you have one of the few entirely unspoiled buildings, it is almost certain that you will need to tear out the vile legacy of oatmeal-tiled fireplace surrounds, concoctions of crazy-paving or pseudo-looking Tudoresque products from modern manufacturers. Squat and remorseless arches of vivid brickwork will glare at you from panelled walls, interesting late Victorian grates, splay-lined with tiles depicting Highland glens, will vie shamelessly with the elegant simplicity of Georgian surrounds. It is useless to repair and restore the other features of an old room while leaving an unsuitable grate or fireplace. The examples just mentioned should obviously be swept away. If the fireplace has any age, it will pay you to remove it gently and sell it to somebody with a house of the right period. The tiles alone can be worth money and specialist dealers and antique shops have a brisk trade in them. If you have the immense good fortune to discover original blue-and-white eighteenth- or seventeenth-century tiles – or

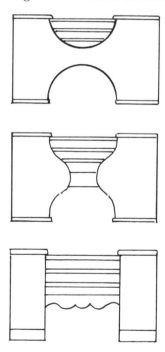

Fig 198 Cast-iron and steel hob grates were first manufactured towards the end of the eighteenth century. First, they had a single front plate with opposing hoop openings *(top)*. Two ogee-shaped plates with steel bars for the fire basket gave a more flexible design in the last years of the eighteenth century and the early nineteenth century *(middle)*. By adjusting the length of the bars, the grate could be used in fireplaces of various widths. The third phase had rectangular plates with linking bars and an often elaborate apron *(bottom)*.

People often talk about 'duck's nest' grates, when they are really referring to the early nineteenth-century hob variety. Here is a genuine 'duck's nest' exactly where it ought to be – in a Buckinghamshire pub. Such grates did not have to be fitted and always sat close to the floor of the hearth. This one has been slightly raised on a course of bricks and steadied with bricks to either side of the hobs. The fireplace itself is interesting with an inglenook on one side and a fireplace window on the other. The lintel has been chopped short and supported on a somewhat inadequate iron stanchion. The brickwork has been taken back to complete the operation of providing extra space and light.

any other kind of early tiles for that matter – in a fireplace, guard them with your life.

Despite the natural impulse to take a sledge-hammer to whatever is ugly and inappropriate, use extreme caution. Much which fits this description may be old in itself and, *while not original to the room, demands to be left where it is.* To what extent should you enforce period accuracy as far as fireplaces and grates are concerned? There is no absolute answer, but here are the ways to approach the problem.

First ask yourself whether a later fireplace is in tune with the general tone of the surrounding features. If it is not, decide whether it has any intrinsic age or merit.

Only when you are satisfied that it plays no historic or aesthetic role of importance should you think of replacing it.

If your fireplace survives from an earlier phase of the building's development, it is most unlikely that you will remove it. Indeed, in most instances it would be criminal folly to do so. You might, on the other hand, be obliged to return the rest of the room to something approaching the style of the fireplace. However, more often you will be advised to settle for a mixture of styles and periods, in the same way that you would accept an oak court cupboard in the same room as a set of Regency sabre-legged chairs. All the same, these questions require great discrimination. Much depends upon the degree of classical unity and formality of the room. The main point is that you should replace any missing fireplace surround with something well researched which accords with the predominating architectural style of the room.

Where grates are concerned, you may be rather less purist. You will notice that the majority of old houses of quality have had numerous fixed grates in the form of hobs, installed within their classical fireplace surrounds. This trend continued from the late eighteenth century until about 1850. After that, a variety of register grates with hoop or horseshoe openings were used until the late nineteenth century. Although classical fireplaces which pre-date the hob-grate era should. strictly speaking. be supplied with free-standing basket grates, it must be admitted that the majority of cast-iron hob grates, especially the more graceful and early designs, look very well as an alternative. Conversely, the early nineteenth-century 'sarcophagus' grate, with its heavy paw-footed end standards, is a free-standing type which would not look particularly out of place in a William-and-Mary fireplace.

There is one kind of very early *fitted grate* which appears in some late seventeenth-century houses, consisting of a tall, narrow opening, very like a modern one, with a generous surround of plain or faience tiles. It has a little coal-burning, low-set fire basket and a small brass smoke hood, a bit like those so often seen in Edwardian houses. These are a rarity and should never be confused with the twentieth-century ones, which doubtless derive from them.

There was a thriving business in reproduction cast-iron hob grates during the period before World War I and it is sometimes very difficult to tell what is circa 1800 from what is nearer 1900.

Choosing Grates and Fireplaces

The type of grate or fireplace you choose is of vital importance. Even the most avid old-house fanatic will shy away from the prospect of burning an open fire without a flue pipe in the middle of his or her medieval hall. He or she would have to live with smoke permanently in the eyes, in imminent peril of a major conflagration, and would drive the building inspector to a frenzy.

The main choices are as follows. First, there is some kind of a down-hearth, which means you burn the fire upon the hearth stone without a basket and employ fire dogs. This system is very suitable for big open fireplaces in cottages, early

manor houses and farms. The inglenook fireplace with its recessed seats is a down-hearth, as a rule. Sometimes, however, a basket grate made by the local blacksmith from stout wrought-iron sections may be placed in these. If you are lucky, you might find a duck's-nest grate, which is like a low and portable version of the hob grate. Used for cooking, it was favoured in pubs and farm kitchens (page 230).

Then there is the free-standing basket grate, which can be an elegant polished steel affair of mid-eighteenth-century date, or something which you have contrived for yourself, using a basket or steel tray, rested on dogs. The dogs can be of total simplicity or quite elaborate, according to your whim and the formality of the fireplace. This type of grate will burn coal or wood or a mixture of the two. It can be placed well back in the opening to avoid smoke getting in the room, and may be raised on a stone or brick plinth to lower the smoke line. In formal and classical fireplaces it often looks wrong to raise the grate on a plinth; but for a big open fireplace under a wide stone or timber lintel it may do very well. The basket grate is very adaptable from both the practical and aesthetic point of view and will burn more slowly than a hob grate, unless the latter has an adjustable throat to the flue.

The hob grate is a fine choice for any more formal fireplace of between about 1780 and 1850. They are fun to buy because there is such a good selection of old patterns. Some can be identified by the cast maker's marks at the foot of the

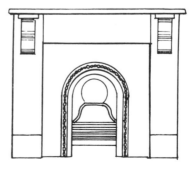

Fig 200 The utilitarian hoop-shaped cast-iron register grate was very popular from about 1860 to 1890. This one has a removable shutter at the back to facilitate sweeping. The basket was set fairly low. The fireplace surround here is typical of the period.

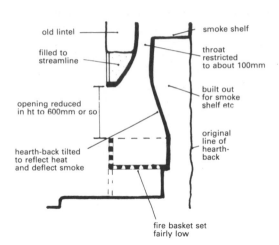

old lintel

filled to streamline

opening reduced in ht to 600mm or so

hearth-back tilted to reflect heat and deflect smoke

smoke shelf

throat restricted to about 100mm

built out for smoke shelf etc

original line of hearth-back

fire basket set fairly low

Fig 201 This section through the back, lintel and hearth of an old fireplace shows how they were often adapted to follow Count Rumford's rules.

hob plates – normally on the outer side of the plate. Later hob grates, from 1842, may have coded registration marks which include a date.

For an artisan's house parlour, early Victorian farmhouse, rectory or similar building with a more or less formal fireplace surround, you could opt for a cast-iron register grate. This has a register plate which fits the fireplace, its opening in a hoop to accommodate the grate and fireclay cheeks (Fig 200). There are hundreds of patterns, from the most ebullient Great Exhibition rococo designs to severe, not to say dull, versions with coarse and rather stylised mouldings on the archivolt. To all intents and purposes these register grates are modern ones, with all the features we expect. The inside of the grate is splayed, the back may be angled inwards, and the throat of the flue is normally restricted to the 4in (100mm) or so of width which was first advocated by Count Rumford in his famous essay 'Of Chimney Fireplaces', dated 1796 (Fig 201). Count Rumford might be called the father of the modern fireplace, except that he always sounds too much of a scallywag to be described with such gravitas. He started out in America as a shop assistant and schoolmaster, and married a rich widow fourteen years his senior, whom he abandoned when he left hurriedly for England. He served as an officer in the American army, and became a knight and a Fellow of the Royal Society in England. He experimented with gunpowder, cannon-founding, agriculture, landscape gardening and much else besides. He gained his title of Count of the Holy Roman Empire while serving the Elector of Bavaria in various capacities, including that of War Department chief. He drained marshes, reorganised the army, popularised potatoes, started a military academy, introduced poor laws and laid out the English garden in Munich.

It must be admitted that Rumford's ideas about fireplaces have proved extremely efficient. They ensure the best use of the heat while promoting a good flow of combustion air, and encourage the smoke to draw up the flue rather than out into the room. Often, however, they have nothing much to do with the kind of fireplace you hope to have in an old building. Even the hob grate does not permit more than a nod in Count Rumford's direction, although some of these did incorporate an adjustable flue-shutter operated by a key and ratchet system. You can identify such grates by the round keyhole pierced in an ornamental boss in the steel plate over the opening.

With any luck at all, a Victorian or Edwardian house or cottage will have at least one original fireplace for you to take as a pattern for any others which you may have to replace. The register grate with a hoop opening is not difficult to obtain – they are being hurled into skips as rubbish by an army of modernisers. In the second-hand market, the Edwardian grate, with its tiled splays, is paradoxically more popular than the early nineteenth-century hob. This must be because the tiles themselves are amusing and appeal to the youthful nostalgia of a generation which docs not associate them with terrifying food-halls and fern-infested restaurants. The tiles are often interchangeable. Broken ones may be replaced, if necessary, by using an entirely new set, of perhaps ten in number, salvaged or bought from a dealer. These grates are unsuitable for burning wood; the logs have to be chopped into silly little pieces which tend to burn too quickly unless you make coal the basis of the oper-

Fig 202 The kind of straightforward design that may be made up by a blacksmith or a firm of steel fabricators. It is a simplified version of the mid-eighteenth-century basket or dog grate.

ation. All the same, a log or two for good cheer can never be a bad thing.

When designing or procuring a free-standing basket grate for an eighteenth-century fireplace, your objectives should be either accuracy or simplicity (Fig 202). You can buy excellent reproductions of Georgian steel basket grates and your only care should be to ensure that you do not pay antique prices for them. Quite a few already have some age about them and were made around 1900. It is extremely difficult to be sure of the date of such grates, but it may help to inspect the iron back plate to see if it suggests an earlier period. The grating upon which the fire is burned should be a bit crude, not like one of those dished and evenly regimented contraptions which you find in the bottom of any present-day fireplace. However, it must be remembered that the latter type of grating has often been used to replace the old one in a genuine basket grate. Any sign of patent marks will tell you at once it is not original.

Probably the best alternative choice for a free-standing basket is to have one made to your own designs by a good blacksmith who is capable of forging iron in the fire,

Fig 203 For a big open fireplace in a cottage or seventeenth-century house, something of this sort can be made quite easily from stock mild-steel sections.

as opposed to the general metalworker who welds up standard steel sections. Avoid all fussy detail and make the height and proportions echo those of the old grates of this type. The design shown in Fig 203 is the sort of thing I mean. It may be stoveblacked if wrought iron is used, or could be constructed of burnished steel. Or you might combine the two materials.

If you can find a small foundry which will undertake 'one-off' castings, there are all kinds of useful things you can have made from your own wooden patterns. A type of grate which is ideal for seventeenth- and sixteenth-century open fires consists of a sway-backed cast-iron grating of bars resting on simple dogs (Fig 204). If you cannot find a foundry, get somebody to weld a flat steel grating of suitable standard sections. Do not worry about hand-forged work for gratings if they will hardly be seen.

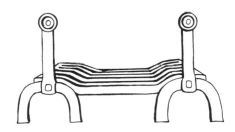

Fig 204 Another useful grate for an open hearth consists of a sway-backed cast-iron grating rested on simple wrought-iron dogs.

The important benefit of fire dogs carrying a grating is that you can raise the height of the fire, thus lowering the smoke line to a manageable level. In a big inglenook type of fireplace it may be necessary to increase the height of the hearth by building a brick or stone platform upon which the dogs and grating may be stood. The smoke line can be lowered to meet the fire from the top by introducing a steel plate, stiffened with standard flat-section mild-steel bars, riveted around the edges of the plate. This may be polished or stoveblacked, according to how civilised the fireplace is meant to be.

This general system of raising the hearth level and lowering the smoke line below the lintel is applicable to most big, log-burning fireplaces, even if they are not large enough to contain inglenooks. A typical cottage or farmhouse fireplace may either have been used as a down-hearth, or have contained a kitchen range. A fireplace designed for a planted surround, shelf and marble slips, will have a relatively small, unadorned and low-set lintel. In brick areas, a crude arch was constructed, usually supported by an iron chimney bar. This form of opening may also be seen as early as the late seventeenth century, especially for down-hearth kitchen fires.

There are two pitfalls to avoid when creating or reinstating 'rough' fireplaces, big or small. They must not vie in character with the rest of the room. It is anathema to see a pointed stone chimney breast and jambs in a room with otherwise 'smooth' features such as panelled shutters and doors, cornice mouldings and

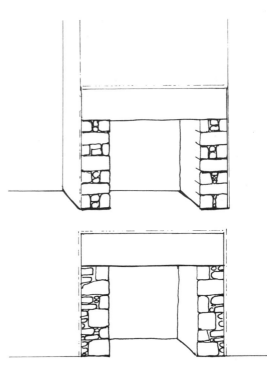

Fig 206 A fireplace lintel and the rubble stonework of the jambs may be pointed and exposed in many 'rough' buildings; but the rest of the wall and its recesses plastered (*above*). A similar fireplace without recesses can have the plaster feathered-off to meet the exposed stonework (*below*).

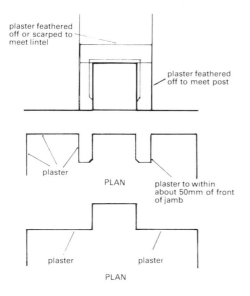

Fig 205 Stone fireplace jambs and lintels should be exposed where the quality and period suggest they are worth seeing. But the adjoining walls are best left plastered (*top drawing*). Where the jambs of the opening are flanked by recesses, the jambs can be exposed stone at the front, but plastered on the outside to within about 50mm (*middle drawing*). When there are no recesses at the sides and the stonework of the jambs is of poor quality, it may be best to plaster right up to the opening (*bottom drawing*). The lintel, however, may be left exposed as a feature. In many eighteenth- and nineteenth-century houses. an appropriate surround with architectural details must be fitted.

chair rails. However much you like the stone or brickwork, it must be covered up with plaster, and a planted fire surround should be fitted which is appropriate to the period.

Even cottage fireplaces should not display acres of exposed stone. It is quite sufficient to feature the lintel and the jamb stones, provided that these stand clear of the general wall surface, having recesses on either side. It is exceedingly difficult to expose the jamb stones of fireplace openings if they lie in the same plane as the walls (Fig 205), since it is impossible to effect the change from plaster to stone without it looking contrived and ugly. You *can* feather off the plaster to meet the stone but this should not be done in the way shown in Fig 206. A solid pier of exposed stone should be left intact if the coursing permits. Usually it does not.

Refer to Chapter 12 on plastering when considering the treatment of exposed stone in relation to fireplaces; and always apply the principles of good pointing for hearths, as for any other part of a building.

Free-standing fire baskets for early hearths (pre-1650) should never incorporate frills of Gothic embellishment, such as fleurs-de lis or spiked palisades. There *are* some original fire baskets in this style, but attempts to reproduce them invariably suggest the Tudor roadhouse rather than the Elizabethan manor at the end of the lime avenue. A plain and robust box of iron bars is what you require.

The only other alternative to the down-hearth, the basket grate, the fitted grate with or without hobs and the register is a stove. Here, the selection is tremendously varied and there are some splendid designs available, many with impeccable origins.

Stoves may be purchased in patterns which vary from a sober, horizontal Norwegian enamelled box on legs, to a riotous, brass-embellished bombé affair, which looks as though it has come out of a Victorian brothel. There are stoves of the sort which you associate with a Second Empire studio full of besmocked impressionist painters; and elegant burnished steel creations quite well suited to an eighteenth-century house.

There are also stoves of such grotesque and hideous design that beholding them in the showroom you 'make an excuse and leave'. There is a stove which reminds one of a medieval jousting dwarf with his visor down; and there are numerous highly efficient and bland-looking modern designs which will do little to enhance an old fireplace.

Stoves provide excellent heating, are clean compared with open fires, safe, can be used to keep food warm, to cook, to air clothes, and for supplying hot water. They will stay in much longer than an open fire and are generally more economical of fuel.

The disadvantage of any slow-combustion stove is that it can produce very high levels of condensation in the flue, and, when used in conjunction with cement-bedded brick chimneys, may induce the production of salts, and expansion and contraction of the joints.

Faults and Technicalities

Probably the commonest fault of open fires is sending smoke into the room instead of up the chimney. Occasionally a fire just refuses to burn with any vigour. Often flues leak smoke into lofts and upstairs rooms. Some flues are unusually difficult to brush out because of their convoluted shape, while others are so cavernous that water pours down them in wet weather. Finally, there are fireplaces with flues and chimneys which are in some way hazardous because they spill sparks and burning cinders into the room or are in too close proximity to combustible building materials.

Smoking Fireplaces

All fires need a certain amount of combustion air to burn at all. If there is too much they may roar away in a thoroughly uneconomical manner. Seal a room too efficiently from draughts and the fire may refuse to burn at all. In many instances, lack of air is not the problem. The average room in an old house lets in enough air, under doors and through cracks, to keep a fire burning. If you try to encapsulate yourself in a futuristic environment, swathed in fitted carpet, draught excluders and double glazing, you are more likely to need a special supply of air to the fireplace. It is also worth remembering that you can actu-

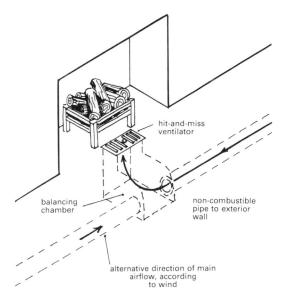

Fig 207 Combustion air may be conducted to the front of an open hearth through pipes buried in the floor. The hit-and-miss ventilator must be far enough back to avoid danger from burning fuel.

ally be killed by fumes if you hermetically
seal a room with a stove or fire.

A supply of combustion air may be
obtained by laying a 100mm or preferably
150mm pipe under the floor between the
foot of the hearth and an exterior wall. To
avoid suction, you can vent the hearth
with two pipes – from different sides of the
house. The pipes meet in a little brick or
concrete box, called a balancing chamber,
from which the air is ducted to the base of
the hearth. For control purposes, a
butterfly valve may be fitted to adjust the
air flow or shut it off altogether (Fig 207).

The pipes, which can be of any material,
provided that they are non-combustible,
should be protected by grilles or gratings
to prevent the entry of mice, birds and
dead leaves.

The efficiency of a fire and its chimney
is governed by several, often conflicting,
factors. An important one is the ratio
between the area of the flue and that of the
fireplace opening. A big opening requires
a big flue for the obvious reason that a
small one cannot cope with the volume of
air flow which is trying to find its way
upwards, taking the fire smoke with it.

Smoke and combustion air gather at
the mouth of the flue and trickle back into
the room before they can escape. The
answer to this problem may be to restrict
the size of the fireplace opening so that
less air need be admitted to the flue.
Alternatively, you might increase the size
of the flue itself. The latter remedy is
frequently out of the question since it will
mean rebuilding the chimney.

In cases where there is insufficient air to
draw smoke up the flue and additional
ventilation to the base of the fire is impracticable, the best answer is to fit an
enclosed fire, in the form of either a stove
or a freestanding grate with cheeks and a
pipe passing through a register plate into
the chimney. By so doing you will greatly
decrease the amount of air required and
will channel the smoke straight from the
fire to the flue without it having a chance
to escape.

This free-standing enclosed fire or
stove may well prove to be the answer to an
obdurately smoky open hearth which has
too small a flue to serve it. Sometimes, the

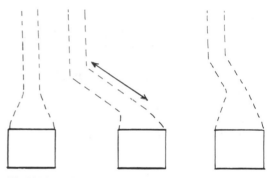

Fig 208 Straight flues (*left*) cause least problems. Where
there is an offset, it should not be long or at a shallow
angle (*middle*); it should be short and steep, preferably
30° max from vertical (*right*).

inside of the flue is very roughly
constructed, presenting obstructions to a
smooth and free passage of the smoke, or
it is built with offsets which hinder efficiency (Fig 208). Generally, offsets should
be high, steep and short – that is, they
should start reasonably high above the
hearth, should be as steeply angled as
possible, and their length should be
limited to the minimum which will enable
the flue to be carried to a point where it
can join the main chimney.

The flue arrangements in Georgian
town houses are notoriously complicated,
as anybody looking at the regiments of
chimney stacks in cities such as London
and Bath can see.

These massive stacks, each with its ranks
of chimney-pots, are the outward and visually obvious culmination of a tortuous
system of flues, many of which have to be
offset to find their way around those from
fireplaces in other rooms. The use of
brickwork enabled Georgian builders to
form very small flue openings and to
engage in alarming bends and kinks
which seldom improved the flow of
smoke; although such measures were
sometimes thought to prevent down-
draughts.

On the whole you can be fairly certain
that sharp changes in the direction of the
flue will do far more to impede smoke
getting out than to prevent gusts of wind
blowing back; and this will lead to another
cause of smoky fireplaces. But, first, let us
consider the question of whether or not to
line an old flue.

Lining Flues

The main fault which really demands re-lining is that of smoke leaking into upstairs rooms through the walls of the chimney breast or the stack. Officially the chimney stack is that part of the chimney which is above the roof covering. However, from a structural standpoint it must normally start from the loft space.

If experiments do not reveal repairable chinks and faults in the stonework or brickwork through which the smoke is escaping, lining can be carried out in the following ways.

You can open up the flue at about 1m intervals, starting at the bottom, and fit conventional clay or concrete flue liners. These can be rectangular or circular but in most cases you will find circular ones easier to cram into the often very limited space.

The outside skin of brickwork or stone adjoining a flue is usually very thin perhaps as little as half a brick thick (about 100mm). So opening-up is not always such a problem as you might imagine. However, care must be taken to rebuild these temporary openings with attention to the structural bond of the materials and to their ultimate appearance.

Some big open-hearth flues permit you to climb a considerable way up the inside to carry out flue-lining operations. It is not impossible to clean off the stonework and build up a lining by first dubbing out with pats of mortar and then applying a steel-trowelled rendering. This is difficult and limited by the extent of your reach and the width of your shoulders. Even if you manage to carry out the operation to quite a reasonable height, you will almost certainly have to complete the final length of flue with lining pipes of some kind.

Old flues can be lined with clay or high alumina cement and kiln-burnt aggregate pipes. Your aim, with these and any other kinds of liner, is to achieve the biggest diameter you can fit up the chimney. For an open fire you cannot expect to employ anything less than 200mm diameter, and 225mm is much to be preferred. Stoves, on the other hand, may operate well with 150mm diameter flue pipes.

To link a fire hood or stove with the main flue you might employ a length of cast-iron pipe, a mild steel one at least 3mm thick, stainless-steel pipe or one made of vitreous enamelled steel. These could be carried up to a gather some feet above the hearth or fitted through a hole in a metal register plate. The Building Regulations regarding fires, stoves and other appliances are fairly complicated and do not lend themselves to abridgement. If your work comes within their orbit it is essential to study them carefully. At other times do what is safe and practical.

Clay flue liners may have socket joints like drain-pipes, but are normally provided with a not too generous rebate. The liners are placed with the socket upwards and are pointed with a 1:4 cement and sand mortar. This mix is also recommended for modern parging, but old flues were parged internally with a cement:lime: sand mix in the proportions 1:3:10. The latter is roughly what you should use for rendering the insides of old flues and it is much less likely to crack through thermal expansion and contraction than a hard mix. However, it is less capable of preventing flue gases from condensing and leeching through to the walls of adjoining rooms.

If a slow-burning stove is to be used in connection with a flue, you would do better to employ a cement-sand mix for the above reason. Better still, line the flue with suitable pipes. Some companies market specially made stainless-steel flue pipes with an insulated core. The pipes are socketed together and each joint is secured with a bolted metal collar. These are usually too expensive for lining an entire chimney unless there are special circumstances; their normal purpose is to link a stove with the main flue. All the same, a great advantage is that they provide a warm flue – a prime requisite for any efficient system, but especially one serving a slow-burning appliance.

Once a flue has thoroughly warmed up, not only will it draw better, but it will also reduce condensation. A newly lit stove with a cold flue pipe can produce incredible quantities of condensation when the fuel is damp oak, for example. This means that provision may be needed for catching the condensation liquor and removing it (Fig 209).

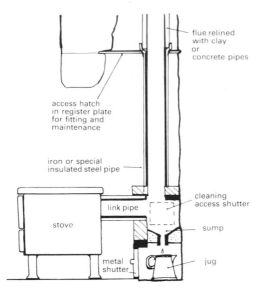

Fig 209 When fitting a wood-burning or similar stove in a big open hearth, this kind of system may be adopted. Condensation from the flue lining is caught in a jug below a specially constructed brick or blockwork smoke-box and sump. The link pipe should not be more than 150mm long (Building Regulations).

Although it is tempting to line a flue with long sections of pipe, they are extremely difficult to fit: some obstruction, slight bend or narrowing of the ōld flue can cheat you of success. What you thought might be achieved with three lengths of pipe eventually involves opening the flue and inserting clay liners here, cast iron there and parging the bit at the top.

One of the best methods of lining a difficult flue is to employ the services of a specialist flue-lining firm. An excellent system involves an ingenious form of shuttering which keeps the required part of the flue open while a lightweight concrete made from vermiculite gauged with cement is poured or pumped around it. The shuttering consists of a strong and flexible pipe which is filled with compressed air. When the lining concrete has set, the air inside is released and the pipe collapses, allowing it to be withdrawn easily from the re-lined flue.

Lightweight aggregate concretes are excellent for lining flues since they put relatively little additional loading on the chimney structure and provide very good insulation. However, they are hungry for cement, and mixing has to be very thorough or you may find that you have poured a mix of what is effectively just wet vermiculite. This finally dries out and escapes where it will.

Once the flue is re-lined, the danger will be that its reduced area is inadequate for the amount of air and smoke being drawn out of a big open fireplace. It is sometimes said that the size of the fireplace opening should not exceed six times that of the flue. This rule cannot often be implemented, since well-made open hearths are served by massive tapered flues, which start very wide and are 'gathered' in as they ascend. A calculation based on the six-times formula will seldom make much sense. Even a quite modest opening of say 1.2×0.7m would require a flue lining measuring about 375mm^2 or a circular pipe of 410mm diameter. Both these are much bigger than the standard sizes for liners. However, such a fireplace may work perfectly well with 225mm diameter liners.

A large open hearth could easily measure 1.5–2m in width and be 1.3–1.6m high.

If you are going to use a big open hearth in the manner originally intended, it is best to leave the flue alone, if you can. The men who built these hearths frequently had the dimensions about right. Many must have smoked abominably, but I know of plenty which seldom leak any appreciable amount of smoke into the room, and that only if the wind is blowing ferociously from the wrong quarter.

Many down-hearths were used for cooking and the fire was never allowed to go out. At bedtime, the red cinders were raked into a compact heap against the cast-iron fireback. They were then covered with a domed metal curfew which deprived them of air so that the fire could be revived in the morning. The practice of keeping the fire in permanently had the added advantage of preventing the flue from ever getting really cold.

After considering various adjustments to the ratio of flue size, opening size and the amount of combustion air, think of preventing the escape of smoke into the room by lowering the smoke line above the hearth. In some cottages this was achieved by dangerously pinning a valance of cloth to the chimney shelf or the lintel. In others, an iron plate was fixed across the opening.

The hearth can be raised on a plinth of bricks or stone and even further by using a tray grating or basket resting on dogs. The conventional effective height of a modern fireplace opening (derived from Count Rumford's principles) is about 550–600mm. By raising the hearth and lowering the smoke line with a steel or fire-resistant-glass shutter, you can operate what is really a big open fireplace, while keeping within this limit.

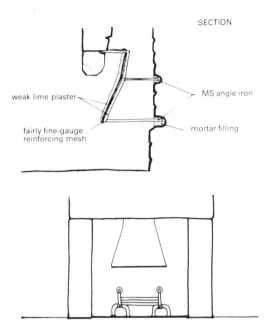

Fig 211 Reinforced plaster fire hood. Another method of providing a smoke hood, based on an early form of chimney construction. Make up a gathered steel framework covered with fine-gauge mesh as reinforcement for a weak lime plaster applied on both sides.

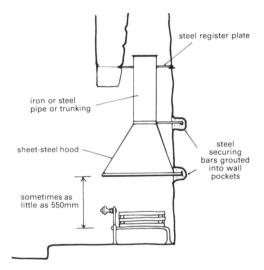

Fig 210 In principle, a fire hood to prevent smoke escaping into the room is fitted in this manner.

Many people with big open fires blank off the flue with a steel register plate. They then suspend a smoke hood low over the fire and connect this to the main flue with a short length of iron or steel pipe (Fig 210). When made of burnished or stove-blacked steel in some simple shape, these can be both effective and visually unobtrusive. However, there are a good number of such hoods, often in beaten copper, which look extremely suburban.

One kind of hood worth considering is made from a soft plaster, reinforced with a material such as steel mesh or galvanised wire mesh (Fig 211). It has a strong affinity with the primitive smoke hoods traditionally used in cottages and early buildings,

not to mention the wattle-and-clay hoods and flues seen in Breton farmhouses of quite recent vintage.

In some cases, it may help to introduce an adjustable throat; but to do this you will have to cut down the size of the flue opening by thickening the fireback, using corbelled brickwork, reinforced concrete or specially designed steelwork made up for you in the local steelyard or blacksmith's shop.

Remember that big open flues of the kind which serve a traditional down-hearth are much easier to rod out than ones which incorporate features such as smoke shelves, offsets and throats, or have been fitted with register plates. If the fireplace and flue are not run smoothly together there will be awkward corners where soot can build up. This is especially so of flues blocked off with register plates, so if you use one, be sure to include some kind of access-hatch in your design. If it is possible to set the hatch vertically in the exterior wall, flush with the top of the plate, it will be far easier to clean out. Clawing about from below through an access hatch set in the plate itself is a great nuisance.

When deciding by how much you must lower the smoke line below the fireplace lintel, cut a big piece of stiff cardboard, a length of board or something of that kind, and fix it across the opening. You then light a fire, preferably with rather damp wood on a windless day. You observe how and where the smoke gathers inside the fireplace, and keep lowering the cardboard until the smoke ceases to trickle out into the room. If you are burning your fire on the hearthstone itself, you can try raising it temporarily by burning it on a base of concrete blocks or a piece of steel laid across some bricks. If the fire is in a basket grate or on a grating, you can stick some bricks or blocks under the dogs. It is a matter of trial and error.

Never forget the possibility that smoke is gathering in the void behind the lintel as a result of a general lack of streamlining in awkward corners and under ledges. Such difficult areas can be filled with concrete, fire cement, fire bricks, steel plates and other improvisations to help the flow of the smoke (Fig 212). You will normally find that old timber fireplace lintels were at least bevelled off at the back to lessen the chance of catching smoke

eddies. Incidentally, old timber fire lintels are nearly always charred.

All these remedies are based on the notion that your fire has a continuous tendency to smoke into the room; but there are fires which do so because of downdraught in the flue. Down-draughts are of two kinds. The most irregular one is caused by winds from a certain quarter literally blowing down the chimney. Normally this happens when a chimney is close to the foot of a hill, a tall building or trees. When the wind blows over the top of the trees, for example, it sends the smoke

Fig 213 A house situated in the lee of a hillside may have a direct down-draught blowing straight into the top of the chimney. A patent cowl, or a covering slab on stone or brick supports, may be the only answer.

straight back down the flue. The best answer is to fit a patent cowl which induces the right aerodynamic effect or to place a big stone or slate slab set on brick or stone piers to protect the flue opening (Fig 213).

The other cause of down-draught is suction. This occurs when the top of the chimney is in a high-pressure zone, and the fireplace is in a related area of low pressure. The classic situation is a house with its back to the prevailing wind, and a laterally placed chimney stack terminating well below the ridge of the roof on the windward side (Fig 214). If, as often happens, there are few or no windows on this back wall to help create a balance of pressure between the two sides, the suction from the windows and doors at the front can be severe.

If the room can be provided with a supply of air from the windward side, that may do the trick. Failing this, the answer may be to increase the height of the

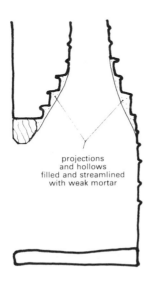

projections
and hollows
filled and streamlined
with weak mortar

Fig 212 Smoke is often obstructed by ledges and irregularities in the gather above an open hearth. These should be cleaned up if possible, and streamlined with weak mortar.

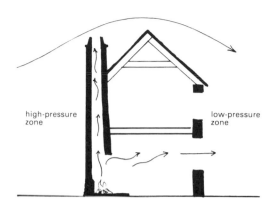

Fig 214 When the top of a chimney is in a high-pressure zone, below the ridge line, and on the windward side of a building, the low pressure on the lee side may draw the smoke back down the flue towards doors and windows.

chimney stack so that it is a metre or so above the roof-line. However, that could prove impractical, since very high chimneys can lack structural stability. You will have noticed that they quite often have tie-bars running into the roof to help brace them. A much-heightened chimney could also be totally out of character with the building. Lateral fireplaces, for instance, are very frequently seventeenth-century or earlier, and are supplied with chimneys which are distinctive architectural features of the house.

In the last event, you may find that you can cure the problem of intermittent down-draught by fitting a suitable cowl designed to counteract such difficulties. Be warned that some of the metal ones can sing in the wind like demented furies and may do little for the appearance of the building. An obvious clue that smokiness is being caused by suction is provided by smoke drifting towards windows or doors giving on to a low-pressure area outside.

From the historical point of view, chimneys which pre-date the eighteenth century are unlikely to have had chimney-pots. The top might be drawn in with specially cut stones, or a pair of slates might be butted together to form a tent over the flue outlet. Very early chimneys of grand buildings were occasionally provided with elaborately designed chimney tops simulating turrets, spires or other architectural features. The subject of clay

chimney-pots is a specialist study in itself and the variety of designs made in the nineteenth century is astonishing.

As with every other part of an old building, attention must be paid to detail when rebuilding, repairing or pointing the chimneys. If they are of brick, they will normally be provided with some over-sailing courses at the top, and these should never be eliminated in favour of modern concrete cappings. Generally, flaunching should be inconspicuous, which means that you do not heap a great mound of mortar around the base of the pot, unless there is a good architectural precedent for doing so.

Some argue that you often do not need a pot at all, and it is true that fires frequently burn well without them. Certainly do not put a pot on an early chimney if it can be avoided. A pot may help keep rain from pouring down the flue but it can equally impede the smooth flow of smoke by gathering it under the slates, or whatever else has been used to set the round pot on the rectangular hole (Fig 215). If you do need to use a chimney-pot, make sure it is in the correct style. Do not, for example, use black chimney-pots on cottages or houses which should have the red clay variety.

The weathering of the abutment between the roof tiles or slates and the

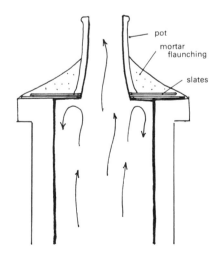

Fig 215 The ledges formed to seat a chimney-pot can catch and turn smoke back in the flue. This does not happen when flue-liners are employed. Sometimes removal of the chimney-pot may help.

241

chimney stack can be a tiresome business. Before the eighteenth century, lead soakers and cover flashing were not used. Instead, slates, tiles or stones were built into the stack and these projected over the roof, providing some protection. The commonest form of junction was one consisting of a mortar fillet sometimes itself weathered with tiles or slates bedded in it. Many houses, of course, originally had thatched roofs which were snugged under projecting stonework or brick from the stack, and the leadwork was added later when the roof was replaced with a different material.

Building Regulations

If you are making a new fireplace or doing work on an old one which requires approval, there are certain points to bear in mind. The hearth itself must be at least 125mm thick and must extend on each side of the fireplace opening by at least 150mm. The hearth must also project at least 500mm in front of the jambs, and must, of course, be made of non-combustible material. This usually means concrete; but in old buildings you will also, no doubt, incorporate stone, brick or slate as a finish. Although the Building Regulations 1985 do not require it in old original hearths, it still makes good sense to see that wooden floor finishes, or any other material which might burn, do not encroach nearer than 300mm from the front of the fire or any stove with doors which can be left open (Fig 216).

If the fire is not to be burned in a wall recess – if you are to have a hearth served by a flue and hood in the middle of the room, for instance – that hearth must measure at least 840 × 840mm (Fig 217).

A fireplace itself must have a back and jambs at least 200mm thick in a material, such as blockwork or brickwork, which will not burn. If the back wall is of cavity construction, the first leaf must be at least 100mm thick. Where the back of the fireplace is part of an external wall, the back may not be less than 100mm thick.

Chimney tops (excluding the chimney-pot) should discharge at least 600mm above the ridge if 600mm or less horizon-

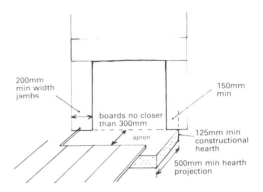

Fig 216 Most of the more important requirements of the Building Regulations 1985 for fireplaces are shown here; plus the sensible one from the 1976 Regulations regarding how close floorboards and other combustible materials may get to the front of the fire opening.

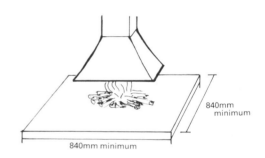

Fig 217 When a fire is not in a fireplace recess, the Building Regulations require that it should have a *constructional hearth* measuring not less than 840 × 840mm. This rule applies to a centrally placed open fire or one that is placed against a wall with a smoke hood and an insulated metal chimney, for example.

tally from it. If it is more than 600mm from it horizontally, it must discharge at least 1m above its highest point of contact with the roof slope. In the case of a flat roof of less than 10° it should discharge at least 1m above its highest point. A chimney may not terminate less than 1m above the top of any openable part of a window or skylight within 2.3m of it. This would apply particularly to a new chimney near a dormer, for example.

The regulations referred to here include some of the more important ones which may apply to work on old houses, but by no means all of them. It is essential that you read either the regulations themselves or one of the standard explanations such as that by Powell-Smith and Bill-

242

ington. When you are merely repairing or improving the efficiency of an old fireplace, the Building Regulations should not be applicable, unless, perhaps, you are obtaining a grant. Then it will be a matter of not doing anything to the fireplace and flue which makes them contravene the regulations more than they did before.

Firebacks and Fenders

Up to the advent of the free-standing basket grate, most open fires in more prosperous houses were burned against a cast-iron fireback. These became particularly popular in the sixteenth century, when they were usually designed with more width than height. In the seventeenth century their shape took on upright proportions, suited to classical fireplace openings.

There are plenty of cast-iron firebacks on the market, many of them made from original patterns or original examples. The copies reproduce all the signs of age and are almost impossible for anybody but a specialist to identify. Since firebacks were just leant against the back of the hearth, you should be suspicious if you see one with threaded sinkings for bolted connections. However, an iron fireback is an attractive, correct and practical way of protecting the stonework or brickwork from the heat.

Take good care of any original hearth cranes, ratchet pot-hooks, spits or dogs which you have the luck to have inherited, and try to ensure that they are so fitted and positioned that they can do the work for which they were intended.

Fenders were an eighteenth-century invention and original ones of pierced steel or brass normally fitted straight across the fireplace opening. Generally, it is the Victorian or early nineteenth-century fenders which protrude in front of the fireplace. All the same, one cannot imagine failing to use a pretty brass fender which looked the part just for that reason. Presumably the protruding fender resulted from the arrival of the fitted grate, which ran almost flush with the jambs of the opening. Club fenders with padded tops upon which you can perch are both decorative and useful.

If you are obliged to leave an open fire for any length of time, always place a sparkguard in front of it; if you are burning a sparking wood, such as pine, it would be reckless to leave the fire unshielded. It is quite sensible to provide anchorage points on either side of a new open hearth so that a sparkguard can be latched on. There are various special guards available for open hearths in old houses, some of them traditional, others clever or inoffensive. One type is in the form of a sort of chainmail curtain which can be pulled across the opening. There are also some awful sparkguards in pseudo-looking wrought iron.

14

EXTENSIONS AND CONVERSIONS

Most of this book concerns matters of fact and the history and technology of repairing old buildings can prove to be common ground among specialists. It is when one comes to what might rather grandiosely be described as the philosophy of the subject that friends tend to fall out.

At no point in dealing with an old building is the 'philosophy' more contentious than when you are converting a barn into a dwelling, or adding an extension to an old cottage or farmhouse. There seem to be two ways of thinking, among people who care. The first of these might be called 'modernist'. A trend became evident after World War II which embodied a modernistic approach to the conversion and extension of old buildings. Architectural history began to be neglected in training institutions, and new materials techniques and Bauhaus-inspired forms of design became fashionable. In fairly recent years this fashion has been partially arrested, if not actually reversed.

Proponents of the modernist view of how to convert and extend buildings are often earnest, idealistic, sociologically minded and also surprisingly pragmatic. They are usually interested in the materials and archaeology of old houses, but less in such factors as style and atmosphere.

The other persuasion tends to appeal to hedonists and dreamers, who are also possessed of a streak of purism. So closely do they identify with the past that they are repelled by anything which might shatter the illusion. They are Watteauesque in spirit, but by no means lacking in realism when it comes to the practicalities of achieving their goals.

Like all generalisations, these group-ings may appear both arbitrary and inaccurate, and most people will want to imagine themselves adopting some reasonable middle course. But be warned that the tendency to embrace moderation seldom works with old houses. The truly moderate person usually ends up with a bland and half-hearted building which does no more than tantalise you with speculations about how fine it would have been had the owner shown some nerve and sense of purpose. You need an element of ruthlessness in your determination to safeguard or possibly reinstate the character of a building.

The people who support the idea that extensions to old buildings must be identifiably of their own time have usually been the victims of a prolonged and worthy propaganda campaign. They have been taught to hate 'pastiche' and the 'reproduction' of styles from past eras. Above all they have been inculcated with a revulsion for the precise and harmless word 'restoration'.

If you question a modern architect about what he or she means by 'Georgian pastiche', it usually turns out that the words are automatically assumed to have a pejorative sense. They imply what has been done is insensitive, untutored and lacking in everything which makes a genuine Georgian building agreeable. The detailing is ill-informed, the scale outlandish and the finishes repugnant. However, there are numerous architects who claim to deplore 'pastiche', yet extol the quality of Victorian or Palladian architecture. Now there are many excellent eighteenth-century 'Palladian' buildings which are undoubtedly pastiches of the country villas which Palladio himself built in the Venetia. Likewise, one hundred years later, William Morris, the arch exponent of the honesty creed which decries restoration and pastiche, in 1859 employed Philip Webb to build him a house which seems to contradict much of what he was later to proclaim. The Red House is a mass of eclectic features. It is, in a word, a pastiche – and a rather interesting one at that.

What comes out of all this is a deal of humbug, coupled with the general misappropriation of perfectly respectable words with definite enough meanings. Fortunately, the majority of architects who revile the word 'restoration' do not adhere too closely to their own philosophy, or most of the buildings which they purport to be merely 'repairing' would look very strange indeed.

The trouble with the word 'restoration' is that it has come to imply 'over-restoration' – a practice for which Victorian church architects had a genius. William Morris, aghast at what they were doing, decided it was time to put on the brakes: hence the famous manifesto which all of us sign when we join the Society for the Protection of Ancient Buildings – an organisation which is rightly respected for its influence and expertise. Nevertheless, it is dangerous to take Morris's words too literally and out of the context of the day. In the 1870s the concrete brutalities which the twentieth century has had to endure were undreamed of; and everybody was drawing on the architecture of past ages with frantic zeal. They freely employed the idiom of former periods in the most unselfconscious manner, sometimes with great effect and equally often with pompous crudity.

The trouble with so many modern extensions and conversions is that they are wilfully designed to be just a little at odds with the original building. This demonstrates, presumably, that nobody has been caught copying, and the designer has a mind of his own. Rather than sublimate these egotistical passions, he builds a feeble travesty, or a brutal rejection, of the old house. It neither blends gently into its setting, nor strikes a telling and attractive note of its own.

'Yes', you will say, 'but what about the charming Queen Anne wing added to the medieval manor; or the pretty Regency addition to the late seventeenth-century parsonage? Will none of these do?' Mellowed by time and shielded by custom, these ill-assorted companions often give us great pleasure. Maybe one is not being consistent. It is certainly a great dilemma. If our ancestors could get away with extensions which were patently in the style of *their* age, why should we act otherwise? The answer can only be tentative, and is perhaps a sentimental one. The process of ageing, in some strange way, seems to soften and unify apparently disparate pieces of architecture, especially if each has its own kind of quality. I believe that the effect is not entirely a material one; but whether we are subconsciously making allowances for the house, or the building is conveying something to us, I have no idea.

Whatever the answer may be, I see no reason to add anachronistic extensions to old buildings today, just because other people have got away with it in the past. They may have been either lucky or clever. Think of all the patently vile additions to old buildings created in the nineteenth century: no feat of justification can dull their clamour, only the forgiving foliage of honeysuckle and rambling rose.

People who declare that it is 'dishonest', not to say philistine, to carry out new work in some earlier style will tell you in the next breath that they are admirers of Sir Edwin Lutyens, who has rightly been seen once more for the creative genius which he undoubtedly was. Yet no matter what our admiration for Lutyens, we can hardly accuse him of ignoring the grammar of former architectural styles.

The truth of the matter is that you can, with perfect propriety, add to an old building using the details, materials, scale and character of the original. Not only *can you*, but in a great many instances, you *should* do so. What you must *not* do, as a general rule, is get the *details* wrong, while conforming more or less in your handling of the easier elements like roof coverings and walls. Try to be accurate in your choice of windows, doors and interior features as well. This does not mean that you merely copy what is there already. You may be inventive and imaginative within the limits dictated by the original building.

The modernist method is often nothing but an easy option. It allows the architect to congratulate himself on his mellow stonework and well-chosen roofing tiles, while avoiding the complex study involved in selecting glazing bars, door frames and cornices. His approach is pragmatic and, dare one say, cheaper for his client. When an old cottage, for example, is given a nicely coursed and pointed new extension with blank contemporary windows and a fully glazed hardwood door, we are assured that this treatment is entirely 'honest'. As already pointed out, the word 'honesty' denotes an ethical value, and it is hard to see why it should be dragged into the argument. We may believe that the architect is an 'honest' man who can be trusted not to pocket the church collection, but is he a sensitive one, with a feeling for old buildings? The qualities we require him to have when designing an addition to an old house are knowledge of architectural history, taste, consistency, invention and competence – we are not asking him to prove himself a worthwhile member of society.

If there were hard and fast rules for art and architecture, every man could be his own Veronese or Vanbrugh overnight. Since most people are constrained by their own lack of knowledge and originality, there are times when it is best to tread a safer path. When adding to old houses I firmly believe that path should lead in the direction of natural materials, well-researched detail and attention to scale.

On this point, the conventional post-war architect who loves old buildings and his or her more hedonistic counterpart may agree. Nothing looks worse than an extension to an old building which dwarfs the original structure. Nor does it look well to add wings which assimilate so well with the earlier house in both size and roof-line that its identity is destroyed. Extensions ought generally to be smaller

than the original, lower, and preferably joined on to the least sensitive part of the building. Nor must extensions be added to the detriment of original features. You would not, for example, cut away cornices, or pleasant old windows and doors, to graft on new extensions. You might, however, seek to pull down an exceptionally ugly Victorian extension and replace it with something in the spirit of the older part of the house.

It is impossible to lay down exact guidelines, since too draconian an attitude may preclude an addition which would be perfectly sensible and suitable for one building, while desecrating another. In my view you should aim to reflect the style of the part of the building to which you are

If proof were needed that fakes can be fun, here it is in Farnham. This splendid brick façade, with its shaped gable, Jacobean square panels, niches, arcading and giant order of Ionic pilasters might be a merchant's house in Holland or the cross-wing of an English mansion. It is, in fact, a more than somewhat imaginative 1930s reconstruction of a late seventeenth-century building which stood on the site. The brickwork is laid in English bond, with modelled, moulded or sculpted decorative detailing.

actually adding. It is an unadventurous piece of advice, I know, but the imagination boggles at what many architects might devise for you on the understanding that they were merely to exercise their own talents.

They might respect the scale, and employ the correct brickwork, and have a great deal of fun echoing the ratios of window and wall spaces. When it came to the glazing, however, like many professionals, they could totally lose their grip of the situation. For the modernist, a sash window is a sash window. It does not make a lot of difference whether it has twelve panes, four or just two. I refer you again to Chapter 9.

'Truthful' to themselves and to posterity, there are people who will provide repugnant wall textures, modern eaves details, patio doors and various other enormities, which kill the original building stone-dead. Walking up from the orchard on a glittering May morning, you will see the new extension through the lichen-covered branches – a blank-eyed, moronic intruder from a brash and graceless world, which you had hoped to escape.

I promised to give both sides of the question, so here is the architect's 'good taste' alternative to working in the manner of the old building to which you are adding. The rules are as follows.

You preserve the scale of the original – make it 'human'. You handle the stonework or brickwork in a sensitive and informed way, provided that the structural design allows you to do so. If it does not, then by *force majeure*, you express the structure.

An example would be the use of stretcher bond for the cavity brick walls of an extension to an old house built in Flemish bond. If you have cavity walls it would be both costly and 'dishonest' to employ snap-headers to give a Flemish bond appearance.

An architect who is only halfway to salvation is likely to build the walls of such an extension in Flemish bond with salvaged bricks, and then put in blank, centre-pivoted windows, or two-pane double-hung sashes. He will pay attention to the quality of the pointing, and the roof pitch will relate to that of the original house. The doors will be modern, but of expensive quality, usually full of glass since clients are assumed to like plenty of light and air. Many do, but they should be persuaded of the advantages of warmth, shade, privacy and atmosphere.

Old Wings and Extensions

Let us now take another path through the same minefield, and think about an approach to the wings and additions built by other people in past centuries. Two main categories will be encountered. One is the original rear wing which pre-dates the main building. This may be more or less intact, but generally it has been greatly altered, either at the same time as the newer house or later. The second type of extension is that which has been wholly added to an older house.

In general, I do not like to see the demolition of parts of the building's historical development. If they can be repaired, then as a rule I think they should be. They provide useful extra space in a world which is constantly restricting your elbow room.

Earlier wings should hardly ever be demolished. Newer ones should, perhaps, be treated on their merits. No one would resist an application to demolish a violent red-brick Edwardian servants' wing added to a graceful Georgian rectory; nor could I conceive an amenity society which would gainsay an owner who wanted to remove an overblown stained-glass and iron conservatory from a medieval timber-framed manor house. These, however, are extreme cases and relatively easy to determine. Many others will prove the subject of passionate argument between opposed factions. All you can do is assess the building, apply some aesthetic judgement and refer to history. After that it is a matter of trying to convince the other side that your view is the best one. They are questions about which anybody with knowledge and conscience is likely to agonise.

Cosmetic Changes

Apart from the decisions to demolish or retain a clamorous addition to an old building, there is one more option which you should consider. Since it is important to avoid the temptation of sophistry, I do not flinch from describing this process as 'cosmetic'; that is exactly what it is. It is an exercise in softening the blow which some previous improver has already delivered.

The basic idea is that you retain this structurally sound, but ugly and unsuitable, extension by revamping its detail and disguising its more repellent features. Both the words 'cosmetic' and 'revamping' have been chosen with care, and I am inviting you to crunch this apparently nasty medicine before swallowing it.

Most of the processes have been described in previous chapters. Roof coverings may be substituted which better accord with the original buildings. New windows and doors can be introduced in place of those which offend. Verge and eaves details can be adjusted, exterior walls plastered, weatherboarded, tiled or slate-hung. Chimney stacks can be rebuilt, pots and ornamental ridge tiles removed, gutters replaced and downpipes re-sited.

How you do all this is of vital importance. Almost everything which is mass-produced and advertised on the basis that it will give your house a new look (or a new old look) is likely to be disastrous. I am talking about the informed use of natural materials, architecturally correct joinery and convincing textures.

For example, let us suppose that you have an early nineteenth-century stucco villa with an extension wing built in the 1880s of hard and relentless polychrome brickwork. The roof is composed of thin violet-coloured Welsh slates, with pseudo-medieval clay ridge tiles. The downstairs windows are in the shape of Battersea bays with feeble, four-pane double-hung sashes. The door to the garden is a well-made, panelled affair with four panels surrounded by rather bulbous mouldings.

Your well-mannered little Regency house should not be required to accept this monstrous neighbour, and you might

It is hard to imagine that anyone could so abuse an old house, even a late nineteenth-century or Edwardian wing. Part of this huge open-plan conversion incorporates one of the downstairs rooms of the eighteenth-century part of the building, although that area does not appear in the pictures. The walls have been fitted with fake timber studs, hacked to look old and bizarre panels of brick nogging have been placed between. The timbers are varnished and the floor is screeded over, presumably as a basis for acres of fitted carpet. First floor joists are exposed to view.

very well carry out some cosmetic alterations as follows:

1 Leave the roof slates alone. They do not please, but may be little different in size and tone from those on the earlier building. Remove the ridge tiles and replace them with something which matches those of the villa. If you can find old tiles, so much the better.
2 Plaster the brickwork to match that of the Regency house, paying attention to the mix and finish.
3 Should it prove worthwhile and technically feasible, provide a closed soffit at the eaves, and even fit eaves brackets to match those of the main house. I am assuming that the late Victorian roof has protruding rafter ends without a fascia board.

The other end of the room shown below sports a corner fireplace of startling ineptitude, relating to no known style or period. The ceiling soffits between the joists have a viciously textured plaster finish, with downlighters inset. The exterior wall of the late nineteenth- or early twentieth-century wing has been partially demolished to form an enormous bay window with a lean-to roof. The casements have sham lead glazing. The present owners are bracing themselves for a major assault on these infelicities.

4 Replace all the windows with sashes with refined glazing bars like those of the villa. The number of panes will be dictated by the width of the window openings, but you can prevent them being offensively tall by raising and replacing the sills. Avoid being forced to use pairs of panes.
5 Remove the bays altogether and replace them with sixteen-pane or twenty-pane sashes, again with the right glazing bars. You may need to make up the width of the opening with fixed marginal lights, consisting of single upright rows of panes. If you decide to retain the bays, rebuild their roofs, if necessary, and replace the windows. This is the easier option, but the proportions may be unconvincing and evade disguise.

When you have finished, you will have an unobtrusive addition, which will harmonise with the Regency house, even if some elements – the roof pitch perhaps – make old-buildings experts look thoughtful. Such measures are not a complete answer, but done with discretion, they can mean that you are no longer required to co-exist with a brute.

As a final gesture, if you are really keen, you might replace the door, planting flatter and more suitable mouldings around the panels. Trellis can be nailed to the walls, and with patience you can contrive a riot of wisteria, which further conceals your 'cosmetic' malpractices.

This wonderfully sensitive transformation has a sting in the tail. What do you do about the interior? Is it to be Victorian, as before, or do you get to work on that as well? If it is listed and the planning officers have given consent for the outside changes, they can scarcely refuse a stylistic conversion of the rooms within. All the same, the procedure begs many questions. Was this Victorian wing so dreary that it would be heartless to insist on making anybody endure it? Or does one take the view that its commonplace vulgarities have their own sort of horrid charm? Maybe it is so bland and lacking in detail that a bold and well-informed essay on a Regency theme will prove a considerable asset. It all depends.

Remember that, as a general rule, the interior of a building should reflect the style of the exterior, otherwise you may be creating nothing more than a stage set. I think that this cosmetic work should be carried through with conviction or not undertaken at all. If the interior is too good, in its own way, to lose, it follows that the exterior should be left alone, except for minor elements of superficial disguise in the form of paint and Russian vine.

New Extensions
If you are now won over to the notion that a new extension may be carried out by the restrained use of some former style, you should pay some attention to the irritating giveaways which can ruin the effect. The worst of these concerns the thickness of walls.

Should you build the new wing in proper stone – and I do not mean sawn 100mm cavity-wall facework – you will end up with a satisfying amount of substance. This will allow you to splay the insides of the openings and you will not have the sensation of living in a doll's house.

However, should your architect advise you that thick stone walls will be beyond your means, you may be obliged to opt for a conventional modern construction, consisting of two thin skins of blockwork or brickwork, enclosing an insulated cavity. The result will be that your lovely new sixteen-pane sashes will be set in flimsy-looking walls which from inside instantly betray their recent origin.

Even if you are not worried about giving the game away – and some would say that 'honesty' dictates a candid admission – you will find that thin-looking modern walls can be very tiresome to live with. They permit no softening dispersal of the light by traditional splaying; they prohibit the fitting of window seats, or sills big enough to accommodate a jug of wild-flowers; shutters cannot be employed, and so on. Refer again to Chapter 9 on windows.

The best plan in these circumstances is to form two skins of concrete blockwork, laying the inner skin flat, with the length at right angles to the outer ones, which are bedded upright in the normal way. This will slightly increase the cost, but greatly improve the sense of substance. Ensure that windows are not set too high in relation to floor surfaces, unless authenticity dictates something like a highbacked window seat and a long horizontal sliding sash or a range of early mullions, for example.

Many old buildings have inherently thin walls, which more or less equate with the modern cavity construction in brick or blockwork. Early timber-framed buildings are unlikely to have walls of more than 250mm thick, even when they have very substantial main posts and the infill panels will be much thinner. The walls of stud-framed eighteenth- and nineteenth-century houses will be perhaps 150mm thick. Vernacular buildings in brickwork are often one brick thick (roughly 215mm). Walls as thin as these are just capable of accommodating a sash window. Clearly, it would be missing an opportunity if you failed to make the walls of a new extension thick enough for decent splays, even if the original brickwork was wretchedly flimsy.

Pay great attention to floor finishes and favour flagstones, wood, tiles or bricks.

Use panelling, vertical dado boarding, soft and forgiving plaster, cornices, chair rails and all the other interior features and fittings mentioned elsewhere in this book. What you do not want is a bland little modern extension; if you are going to be modern, be so with flair and conviction, providing future generations with something which they can really enjoy – not a squalid and lifeless addition into which they step from the old house with a sigh of resignation.

Barns and Other Conversions

Changing patterns of agriculture and industry have made many fine old buildings redundant in recent years. Some of the best of them should not be converted to new uses at all, others require the most sensitive handling. Even the least interesting are worthy of the kind of treatment which retains the sense of their former purpose. Local authorities usually have policies which set out the criteria for dealing with 'change of use' applications and, if buildings are listed, further constraints will be imposed. This is not the place to rehearse the arguments for and against the conversion of barns and mills, warehouses and chapels. The issues are complex, the pressures formidable and there is no golden rule about *whether* the thing should be done. *How* it should be done is another matter.

The first principle to observe when converting any such building is this. The appearance, both inside and out, should be changed as little as possible. Window and door openings should be left intact, roofs preserved in their original form and materials, walls repaired rather than rebuilt whenever it is remotely practical. Features which tell the tale of how the building was used, like exterior steps, waterwheel machinery, horse engine beams, blacksmiths' fireplaces, ventilation louvres, canopies for hoists and so on must be retained. The whole idea is that a reasonably well-informed observer should be able to read the building in its intended role, despite the fact that it has acquired a new lease of life in a different one.

Clearly there are some working buildings of such quality that it would be an act of vandalism to attempt a domestic conversion – repair and sensible use of the available space is the only decent option. A beautiful sixteenth-century timber-framed tithe barn, for instance, might defy conversion, which is not to say that you could not make it into a theatre with minimum disruption. What you could not do, with honour, is turn it into a satisfactory dwelling, unless the sleeping arrangements, kitchen area, bathroom and other facilities were managed with a great deal of subtlety, to the exclusion of such things as intrusive partitions and new upper floors.

A second principle in conversion work is to employ the existing structure and the spaces it affords to the maximum advantage. The temptation to make new extensions here and there must be resisted. Most local authorities are keen about this principle; but are much less Draconian when it comes to the handling of the interior. The majority of working buildings like barns, warehouses and mills typically have a wealth of open space and very little subdivision. This is the quality which any decent architect or owner will be zealous to preserve.

Principle three concerns materials and textures. Where there is cob, let there be cob. If floors are constructed of naked boards or flagstones, so they should remain. The great risk in converting barns to dwellings is that architects of deeply suburban persuasion will turn them into smooth-walled, plasterboard-partitioned, carpeted, overheated and over-windowed abominations. What they are doing, the monsters, is to provide some comfort-loving philistine with a plush modern house in a place where he would never normally be permitted to build one. In some cases, the strategy is to accommodate several families of similar ilk by conversion to flats.

If working buildings are to be converted, for whatever purpose, the interior texture of old brickwork or stone, the simplicity of whitewashed walls and timber-boarded partitions, exposed roof trusses, ladder staircases, little narrow ventilation-slit windows and broad plank doors has to be retained. Otherwise go and build a bungalow. If you want to live or work in a converted barn then you have to be prepared for more austere surroundings than you would find in a smooth-stucco Regency rectory, let alone an urban flat. Texture and detailing is of the greatest importance; so is subdivision – the fourth principle.

Your aim should be to subdivide as little as you can. You must on no account break up the main open spaces of hay lofts, threshing floors, byres, workshops, and chapels by introducing floors where there are none, or partitions for conventional rooms. It is essential that the whole ethos of dwelling-house subdivision is jettisoned. Your converted barn does not have to include an entrance hall, a separate kitchen, a dining-room, three bedrooms with their own bathrooms *en suite*, walk-in fitted cupboards, landings and formal staircases, French windows to a patio from a 'lounge' with a fake Adam fireplace. If you think I am fussing unnecessarily, you should see some of the converted barns I have visited and many of the application drawings it has been my unhappy duty to assess.

The best bet with a barn is to have your living area upstairs, if there is an upstairs. If there is not, then it should occupy a truly generous amount of the ground floor and be open to the rafters. It is possible to introduce some kind of gallery to link, for example, floored areas in the bays at either end of the barn. However, such plans should respect the original design of the building. Not infrequently end bays of barns were floored for storage but the open part should usually monopolise most of the space, as it would in a chapel for that matter. If, for example, you have a barn with a first floor right through and cow stalls under, you will probably want the living area on the upper level and minimum subdivision below.

The original window and door openings will dictate your main entrance, which may be either on the ground floor, or, if the barn has exterior steps to the upper storey, there could be two or more entrances. Windows are likely to be small, not very numerous and in all the least convenient positions. You are going to

have to live with *less light* in some parts of the building than your mother-in-law thinks you deserve. Forget all this need for light and air which people say they require. They don't, but merely think they do. All you want is *enough* light to see by and enough ventilation to permit you to breathe in the normal way. The ordinary cracks and faults in an old building, together with a few windows which open, will give more than enough air. Light is required in varying amounts and should be distributed in different ways, according to the character of the building and the purpose of the area concerned.

Where you need hardly any light is in bathrooms, lavatories, storage rooms, laundries, dairies and halls. In many instances they are better used and enjoyed with the narrow castle-turret type of window which a barn may afford. I have seen people put substantial windows in modern house bathrooms, which then demand obscured glass and present no wall space for a mirror to shave at. Halls can be relatively dark without being in the least depressing. Imagine the door of a cows' house, low and fairly wide – perhaps four feet. You have partitioned bits of the ground floor for a couple of bedrooms, a bathroom and a laundry. This means you require some sort of entrance lobby containing the stairs, so that you do not have to go through other rooms for access.

Each case will be different, but the overall method and choices may be somewhat alike. Stairs in a barn can often be in the form of an open-tread ladder of the simplest kind. They will not have upright balustrading but horizontal timbers below the handrail which will be just a plain pole or rounded-off rectangular section. The newel post will probably rise to support the joist, trimming the opening in the floor above. The lobby or hall could be partitioned off with wide horizontal boards, nailed to studs, with whatever thermal or fire insulation you care to sandwich behind them. They will be painted, preferably as though whitewashed and they will echo the kind of crude divisions used in farm buildings of all kinds. Another good partition for work buildings is formed from whitewashed brickwork, but do not, on any account, leave the timber boards, bricks or stones exposed. They absolutely must be painted, whether in limewash (very authentic but tiresome), emulsion or eggshell. Gloss might be all right historically for doors and windows, it depends on the building.

Although from a structural point of view brick partitions can be only half a brick thick in stretcher bond, I suggest you increase the thickness at openings to give a sense of substance. Everything in a working building should look rather sturdy and capable of sustaining heavy loads. Some partitions may be carried out in stone walling appropriate to the locality and this, like all other stone walls in such buildings should be pointed flush and painted over. Exposed stone, although sometimes authentic in work buildings, especially agricultural ones, was limewashed over more often than not. A big open barn space rising away into the roof trusses is horribly gloomy without paint and you must remember that with fairly minimal windows you depend to a great degree on reflected light.

A third type of partition is made by finishing a stud framework with fairly wide vertical matchboarding. These boards may have a flush bead run on one edge. It is best to avoid lath and plaster. Skirting boards will usually look too domestic, and architraves also suggest a much more sophisticated kind of living. Door frames must be substantial, with mortice and tenon joints at the head secured by pegs or dowels. Doors will almost always be made from wide, flush-beaded boards held together by generous ledges which are nailed not screwed. Hinges will be of the strap variety hung upon steel pins, which may be either spiked into the frame or screwed to it by means of a back plate. As to patio doors for working buildings, they are a total anathema, especially those terrible neo-Georgian ones with pairs of panes. If you absolutely must have some kind of glazed door, it should be of the most stark and sturdy design, and the hope will be that dark paint may help to make it less conspicuous.

It is extremely difficult to produce suitable windows for barns and working buildings, but there are a number of patterns and finishes which should be avoided like the plague. All those with glazing bars are suspect, unless as in some warehouses and factories, they were part of the original design. Ordinary vernacular farm buildings tended to have solid frames with central mullions and shutters. They might be subdivided by vertical timber bars and these were frequently set into the sill and head like the diamond on a playing card. The frame might have a horizontal transom about two-thirds of the way up and this is a good device for reducing the height of the casements you may be obliged to put in the frame. Tall, narrow casements in barns look terrible. Try inserting diamond-set bars or flat timber muntins (about 50mm wide) in the transom light and leave the casements below as plain glass. The transom light may be glazed behind the bars, but it is better to rebate them for individual panes. A cheap method of glazing workshop windows, which is worth considering, consists of a solid peg-jointed frame with vertical mullions, set one pane's width apart. These are glazed with overlapping panes of glass fixed in much the same manner as courses of slates. The overlap, however, is minimal.

Most working buildings should have doors and windows painted, preferably using some dark, receding colour – black, dark grey, dark green, brown, navy blue or deepest crimson, come to mind. White should scarcely ever be used for converted agricultural buildings, although it might be correct for some mills, factories and warehouses. The purpose of the receding colour for joinery is to play down the fact that you have been obliged to glaze apertures which would normally have been open to wind and weather.

Narrow ventilation slits or tiny windows in brick or stonework are very typical of working buildings and they provide a perfect answer to the problem of allowing maximum light with minimum disruption to an elevation. Where walls are thick, the interior of these slits may be widely splayed, like the windows of a Norman church. This greatly increases the barn-like appearance. Glazing in these openings should be plain glass, set straight into the masonry without timber framing. If they must open, then use a very narrow rebated steel frame with a steel casement of the kind employed in late medieval and sixteenth-century houses.

Exterior doors should be of solid plank-and-edge design, or stable-door variations on the same theme. Big doors in barns can have narrow wicket doors set in them, allowing easy access. Transom lights can be introduced over some wide barn doors to give light, but areas of glazing down the sides of solid doors have a most uneasy appearance. The notion of filling tall, broad working-building openings with plain glass is tempting but seldom gives the intended impression of being open to the outside air – glass always shows because of the reflections. Consider framing in a solid pattern of studs, rails and transoms with generous areas of glass, but retaining the heavy boarded doors as well, so that they can be closed over the glazed framework.

Some barn doors, and windows, may be glazed by having wide muntins with glass of equal width in between. This provides a ventilation slat appearance. You can then fix a horizontal sliding shutter with matching vertical muntins on the inside or out. This, when pulled across, covers the glass panes if the shutter muntins are in exact register with them.

In no circumstances should you fit stained or varnished doors and windows, no matter whether they are hardwood or softwood. They draw attention to themselves in the worst possible way. Occasionally, as in the case of early timber-framed buildings, you may have natural hardwood joinery, but this is never oiled or treated in any manner which shows and should only be a matching adjunct to the framing of the building itself. Everything about a converted agricultural building should look unassuming and authentic, so that it does not strain credulity to imagine cows swaying in for milking or pitchfork loads of loose hay being tossed up through an open door.

Attention should be paid to eaves' details and, as a general rule, barns will have clipped eaves and verges without fascia or barge boards. Gutters will be cast iron or some forgiving substitute, on brackets spiked into the masonry or screwed to the feet of rafters. The roof materials themselves should be chosen to match what was, or would have been there originally. That means stone-slates, slates, tiles, pantiles or, sometimes, corrugated iron. Ridge and hip details should be carried out in the ways described in other sections of this book.

Exterior balustrades should be of simple iron sections, with a minimum number of stanchions. In many cases, close-set vertical iron balusters will look fussy and contrived. If you can possibly convince the building inspectors, put a couple of thin round or flat sections below the handrail in the manner of park railings. Sometimes, solid timber balustrading of the sort used for interior stairs, mentioned above, will be appropriate. Steps may be of stone slabs or of slate or timber, according to the local precedents.

The yards or other flat areas around working buildings should not be made to look domestic. Try to get away from the idea of lawns and garden flower beds. All you need is a few tubs or granite troughs, a bench here and there – nothing that spells the message of a nicely ordered garden. Under foot, flagstones, bricks, close-set pebbles, cobblestones and other traditional local materials make all the difference. What you cannot contemplate is a lot of tarmac, concrete or garden centre paving stones. Walls should, if they are relatively high, be pointed correctly and then blinded with sieved topsoil just before the mortar 'goes off'. This will encourage organic growth and greatly soften the appearance. Low walls may not require bedding mortar or pointing and will normally follow the traditional laying technique used in the neighbourhood. Large areas of hard-standing for vehicles, drives and turning places may be laid in hardcore, followed by well-rolled-in toppings and graded blinding. Properly drained, this system will stand a great deal of wear and looks vastly more appropriate than acres of horrid chippings or gravel, which suggest the terrace of a grand

Georgian manor house

Remember that few agricultural buildings had chimneys so, more often than not, stoves are the answer since they only require rather stingy flue pipes sticking up through the roof. If you must have a chimney it should be placed in the least conspicuous position and be authentic in its design and materials.

When it comes to exterior lighting for working buildings, simplicity is the name of the game. Eschew all those terrible little fake coach lamps, Victorian street lamps and other so-called 'architectural fittings'. Farmyard bulkhead lights cleverly positioned are the best bet in most cases. If you are keen on *son et lumière* and your building is an inordinate source of pride, you might conceal a few floodlights here and there, but these do not give you light to see by so much as set you and the building upstage for a performance.

Bathrooms, Heating and Ventilation
Central heating is a specialised subject which lies outside the scope of this book, so I shall not go into details of header tanks and pumps, heat-exchangers, direct and indirect systems. But be warned that everything connected with central heating is fraught with dangers for anybody who wants to retain the character of an old building.

As in every other field of endeavour, there are regular changes in fashion governing advice on central heating. I earnestly implore you to resist the one that dictates placing all radiators under windows: they look awful and make it impossible to sit on a window seat without singeing the backs of your legs. Radiators, almost without exception, are brash and ugly intruders, which take up valuable wallspace better used for furniture; and mercilessly shrink and crack the veneers of everything from a satinwood bonheur-du-jour to a Queen Anne tallboy.

The fact remains that almost everybody wants some kind of permanent heating to at least take the chill from the air and combat dampness. I am at a loss to advise you. The only central-heating systems which are visually satisfactory are either those which involve warm air ducted to grilles in floors, walls or skirtings, or radiators hidden behind vented panels of some kind.

For those who dread the kind of cold flagstoned, brick or tiled floors which I am always advocating, it is worth considering the expense of an underfloor heating system. There is a particularly good German one which consists of 13mm alkathene pipes (laid about 225mm apart on a 50mm polystyrene membrane) through which warm water is pumped. They are covered with 100mm of mortar screed and the flagstones are placed over this as a finish. The floor feels faintly warm if you walk upon it barefoot and it provides splendid background heating for the room. Clients of mine swear by it and wish they had used it more freely.

Your choice of boiler is entirely up to you, provided that it is not placed in a position which spoils an old room or persistently encourages conditions which may damage the structure. My personal preference is for a solid-fuel appliance in the form of something like a stove or a kitchen range. After that, I think I might select gas, if it was available. Oil-fired appliances, while clean and efficient, require special storage tanks, are expensive to run and are a prey to any hiccups in delivery arrangements.

Check up on the possibility of installing a multi-fuel central heating, cooking and hot-water system, which enables you to switch if necessary from one type of fuel to another. Central heating can be a godsend, but I cannot help loving hardy idealists who tell me that they dislike it and are refusing to put it in their old cottages or houses!

At the risk of offending the Electricity Board, I should mention storage heaters. I have lived with one or two of these for many years, but find them costly and capricious. My bills seem to contain a disproportionate percentage for storage-heating, and the radiators themselves, which can only be recharged overnight, are marvellously hot in the morning when I do not need them, and tepid in the evening when heat is wanted. Manufacturers keep updating the design of electric storage heaters and they are certainly much less bulky than they used to be. However, they still seem too big, and lacking in flexibility as far as useful output is concerned.

Solar energy is said to be the fuel of the future. Its adherents claim that it is one of the few methods which make sense from an ecological point of view. One day, when other aspects of this book are still entirely applicable, the solar-heating clan may have finally succeeded in making the sun's rays do all the work which a conventional system does now, so refuting my present reservations. As far as old buildings are concerned they are as follows.

The heat-absorbing panels can be difficult to integrate without looking unsightly; and the heat supplied is not yet sufficient to meet all our everyday needs. For the best results, a building should be insulated to a very high degree of efficiency so that a modest amount of generated heat may have the maximum effect. This could involve, among other things, extensive resort to double or even triple glazing – a costly business which might be extremely damaging to the fenestration of an old house.

Despite my niggardly views about central heating, I am much more enthusiastic when it comes to ventilation. Far too many houses suffer the effects of mildew and decay through its absence. Kitchen smells and fumes linger, when an efficient air extractor/ventilator could remove them without trouble. Amendments to the Building Regulations now demand that kitchens should have both mechanical and background ventilation. If your kitchen is part of a new extension or constitutes a 'material change', you will need to check the best means of satisfying these requirements with the council's building control department. Hotels especially should seriously explore methods of ventilation or air conditioning – since they have such resistance to the idea of opening windows. Technically, it is just a question of procuring suitable cavities to accommodate the necessary ducting. The ventilator or air-conditioning grilles can be quite unobtrusively sited. Ventilation engineers suggest a minimum of eight air changes per hour for restaurants. An air change means that the entire volume of

air in the room has been exchanged for fresh air. I leave you to imagine how many air changes are obtained in the average hotel dining-room, or even in some centrally heated private houses for that matter.

It is recommended that private living-rooms have five changes per hour, private WCs, kitchens and bathrooms between ten and fifteen, and offices six. Recommended air changes can be confusing, since nobody seems to agree about anything except that you probably need more fresh air in a room than you are getting. The 1989 amendments to the Building Regulations, for example, give an option for kitchens of continuous mechanical ventilation at the rate of one air change per hour; but this is *in addition* to the other basic ventilation requirements. No window is better designed to provide generous and easily adjusted ventilation than the traditional double-hung sash. If you have such windows and open them, you can forget about abstruse air-change calculations.

Kitchens

Since new extensions to old houses are frequently required for kitchens, this is probably the moment to consider them. Of all the rooms in an old building, the kitchen is likely to be the one which is most constantly in use, however visually disagreeable it may be.

Although they may take considerable pains to retain the sense of age in other parts of the house, many owners still seem to think that a kitchen should be governed by a quite different set of rules. The result is that many kitchens are about as mellow and reassuring as the supermarket's bacon counter. They abound in smooth synthetic finishes, noisy stainless-steel sinks and draining-boards, and banks of mass-produced cupboards, with all the level surfaces lined up to form continuous counters. This melancholy scene is often illuminated by the even, surrealistic glare of fluorescent lighting.

To put up with such a state of affairs in a modern bungalow is unnecessary, and wilfully to create a kitchen of that kind in an old cottage is unimaginably self-defeating. You would not dream of having a room with a big open hearth, brick floor and old furniture on one side of a sixteenth-century screens passage and a glistening, modern dining-room, like something from a Sixties penthouse, on the other. No more should you duck beneath an ancient timber lintel to emerge in a space-age kitchen: the flavour of the kitchen should accord with that of the building as a whole. Quite how you handle the situation will depend upon the basic structure and proportions of the room involved. One Georgian house could have a flagstoned basement kitchen, little different from that of a seventeenth-century farmhouse, while another might be in a tall, ground-floor room with cornice mouldings.

A kitchen needs to have the atmosphere of a proper room. It should be furnished rather than fitted. Most of the trappings of a thoroughly liveable-in kitchen have almost become clichés since the macrobiotic university graduates of the 1970s, in their long skirts and cheese-cloth blouses, colonised the terraces of darkest Islington and farmhouses in the Welsh Marches. (If they were never right in any other particular, I believe they were so when it came to creating old-fashioned kitchens.) Exposed beams and joists, cheerful dressers, pictures, books, lath-back armchairs, enamel pots and pans, scrubbed pine tables, marble slabs for preparing food, wooden plate-racks and draining-boards, printed cotton curtains, butchers' chopping blocks and rise-and-fall lights with coloured coolie-hat shades are redolent of the 'old-fashioned' kitchen reinterpreted by the affluent middle classes. It will be recalled that many old-fashioned kitchens, presided over by old-fashioned cooks, tended to be wholesome, but distinctly dull.

If you are lucky enough to have a kitchen or scullery with an original timber sink, protect and repair it at all costs. They are very kind to good china, and, while slightly difficult to maintain, have become period pieces. The best kind of sink to fit in a mellow kitchen is the white, glazed earthenware trough known as the Belfast sink. It is thoroughly practical, looks the part, and can be large enough to contain a number of piles of dirty plates, or to use for washing the dog. A particularly adaptable version is broad and fairly shallow, and has a biscuit-coloured exterior, decorated with vaguely Jacobean flutes. Belfast sinks are still manufactured and can also be readily obtained second hand.

With all kitchen sinks – baths and basins too – fit brass taps if you can. These can be purchased new at great expense, or you can find old ones, or strip the chromium plate from second-hand chromium ones.

Not so long ago it was no problem to order your new taps in polished-brass finish, and be charged £2 or £3 extra on the bill. Now, for some reason, brass finish is treated as a luxury for which the frivolous customer must pay through the nose. Brass taps are colourful and restful. You should not lacquer them, just polish when you can muster the energy. They look better than chrome, even if dirty.

I need say nothing about kitchen gadgets, cookers, refrigerators and freezers; most are efficient, ugly and necessary. If you can make their presence less obtrusive all the better; but the main point is to prevent them dominating through sheer volume and number. As far as the size of a kitchen is concerned, you will be restricted by circumstances. But if you are adding a kitchen be as generous as you can; large kitchens are greatly to be preferred. I have lived most of my adult life with inconveniently small ones and speak from experience. It is easy to be beguiled by arguments of economy, or to misread the situation when looking at plans. By the time you have put in all the required fittings and furniture, you may hardly have room to squeeze around the table to reach the cooker. Try to get at least 12×15ft (3.66×4.57m).

All the same, relatively small kitchens can be pleasing and orderly, provided that you do not have to eat your meals in them. Apart from the bathroom, there is no room which can suffer more ill-effects from condensation, so take care to ensure good ventilation and select rather soft lime-plaster wall finishes. If you can afford a fume-extractor hood over the cooker, it (*right*) This is the lean-to dairy of the imaginary cottage shown on page 8, before conversion has started. It is to become the new kitchen; and the walls will be built up to form a slate-hung bathroom extension above. This scheme reduces the cost and simplifies the job by grouping the water and drainage pipework in one part of the building.

(*left*) This is the kind of kitchen which all too many house restorers impose upon gentle old buildings. Such modernisation negates the whole purpose of the enterprise.

(*right*) The final aim is to achieve a kitchen which retains the spirit of an old building. The flagstoned floor has been cleaned up and relaid over a screed and damp-proof membrane. The walls have been replastered but not straightened up, the corners being allowed to wander a little and the arrises blunted. The building inspectors have been persuaded to agree to an open-joisted ceiling and the joists themselves have been made much thicker than the span requires to give half-hour fire resistance. A new sixteen-pane sash window has been made in the west wall and the opening has been splayed to provide a broad window-ledge. A new Belfast sink has been fitted with brass bib taps. There is a hardwood draining-board and the fitted cupboards have very plain boarded doors with small brass handles. The whole room has been furnished, rather than fitted out with a profusion of 'units'.

258

is very well worth thinking about. At least install an extractor fan on an outside wall.

With newly built kitchens, remember to have deep inside sills for the windows, where you can leave things to cool, put down books or place flowers. A stable door, from which you can shake a lettuce or shout for the dog, works well in a kitchen. Have no truck with fully glazed doors: they nearly always look suburban, and present a gloomy spectacle on icy winter afternoons.

POSTSCRIPT

Most of the advice in this book is of a practical nature, but there may well be passages which seem unduly subjective. Others may appear to stray into the realms of hyperbole. However, it would be difficult to convey the spirit of the enterprise without risking such accusations.

Old buildings demand something more than a fitter's manual, and I cannot feel apologetic if these pages do not read like one. The technical and historical aspects are matters of fact, with which I believe a great many specialists will agree. Where choices enter the picture, they are based on the premise that most people would like their house or cottage to retain the restful and rewarding qualities that attracted them in the first place.

It is the variety, beauty and unpredictability of old houses which engage our love and enthusiasm. This is a book of guidance, not a scholastic treatise purporting to say the last word on the subject of repair and conservation. There *is* no last word to say. New techniques are constantly being developed, and historical research obliges us at intervals to revise our most treasured prejudices. Regulations are changed and updated, government policies revised, and fiscal balances adjusted.

Most old buildings are more constant than this and will survive centuries of abuse and neglect, only to succumb beneath the triple assault of indifferent planning officers, hostile committees and greedy developers. Even when they have finally been written off, the allegedly unstable walls often seem consciously to resist the devouring bucket of the bulldozer.

Our job – yours and mine – is to know the practical techniques which can be used to save old buildings with the least disturbance to their structure and architectural features; and since this is not a moral exhortation, we may promise ourselves a reward. That reward will be the pleasure of seeing such buildings about us in our daily life and the joy of living in them.

GLOSSARY

Abutment The junction between one building surface and another, for example, where a roof joins the gable wall of an adjoining house.

Acanthus The leaf of this plant is widely used as a form of decoration in classical architecture, especially for capitals of columns, for arabesques, consoles and cornices.

Adam, Robert (1728–92): Most influential architect and decorator of the second half of the 18th century. His work was chiefly neo-classical.

Adzed Early methods of converting timber involved sawing boards in a saw pit or splitting them with an iron and beetle. The adze was used to clean up the timber after cutting out. It has a broad, stubby, curved blade set flat on the end of a long wooden handle.

Aggregate Stones, chippings, gravel, clinker, etc, used to give bulk and strength to a mix of concrete.

Ancone S-shaped scrolled bracket or 'console', much used in classical architecture, often to support moulded features over doors, windows and fireplaces.

Angle bead Approximately 25mm dia wooden dowel permanently fixed at external angles of walls, against which the plaster finish is struck off.

Angle-tie Timber which joins the wall plates across the corner of a building with a hipped roof. The dragon beam or dragon piece is tusk-tenoned to it.

Anthemion Decorative motif of honeysuckle in stylised form.

Arabesque A flowing and entwining decoration, usually made up of plant tendrils and leaves.

Arch There are numerous types, but in this book mention is particularly made of the **four-centred** kind. This is the most flattened-out form of the pointed arch (normally 16th or early 17th century). The **segmental** arch forms a segment of a circle. **Flat** arches are not arches at all in shape, but conform with traditional arch-type construction.

Architrave A moulding in wood or plaster around the sides and head of a door or window opening. Also serves to decorate and cover the joint between the wall surface and the frame. It is the lowest of three groups of mouldings which comprise a classical entablature.

Archivolt Architrave running around the face of an arch.

Armature A metal reinforcement securing the larger decorative figures and motifs in plaster-work.

Arris The corner formed by the sharp edge where two wooden, stone or plaster surfaces meet: vertical edges of a timber post, for example.

Article 4 Direction Order An order procured by a local authority to enable it to limit specific kinds of development which would otherwise be 'permitted'.

Art Nouveau A style of sinuous, flowing decoration, derived from the works of certain designers of the 1890s and up to World War I.

Arts and Crafts movement An effort to reinstate the guild-orientated standards of craftsmanship and design of medieval Europe. William Morris was one of its leaders.

Asbestos-cement slates Blue-black imitations of machine-cut Welsh slate (usually 12×24in) composed of asbestos fibres and cement.

Ashlar Accurately cut facing stone, constructed with very fine joints.

Ashlaring Framing timbers which continue the vertical face inside a wall to join up with the underside of the common rafters.

Astragal A small rounded moulding like a bead – almost cylindrical. Also used to denote glazing bars of cabinet doors.

Atlantes Male figures used like 'caryatids' to support a classical entablature.

Attic base Base of a column with two torus mouldings separated by 'fillets' and a 'scotia' mould.

Balusters Vertical supports in stone, metal or wood for a staircase handrail, landing balustrade or external balustrade capping.

Balustrade The whole assembly of handrail, 'newels' and 'balusters' which flanks a landing or staircase.

Banister The same as a baluster – common usage.

Barge board Boards fixed to the face of a gable wall, just below the verge slates.

Baroque A style of architecture and decoration which uses the classical language in a grand but lively way. The features have movement; pediments are broken; the eye is led upwards. There are curves and counter curves.

Basket grate A free-standing iron basket, burning logs and, later, coals. Most early ones were fairly plain. Those by Robert Adam, for instance, were most sophisticated. See also Dog grate.

Batter When a wall slopes back from the vertical.

Bay-leaf garland Decorative moulding, usually seen on convex friezes and ceilings. Consists of tightly bundled bay-leaves wound about with ribbon.

Bay window Window protruding from the main wall of a building, having its own side walls and roof. It does not curve like a bow window. The structure must rise from ground level, unlike an oriel.

Bead A small linear convex moulding. Beads are often run in closely spaced groups on such features as architraves.

Bead-and-reel A popular form of enrichment for lengths of rounded moulding. Cylinders alternate with bobbins.

Beam A structural timber of substantial section mainly used to support floor joists, or to tie the feet of principal roof rafters together, thus forming a roof truss.

Bed moulding The mouldings which run below a decorative projection, like the upper part of a classical cornice. The 'cyma recta' and 'corona' are bedded on a group of mouldings underneath.

Belfast sink Trough-shaped white 'stoneware' sink, used in kitchens and sculleries. Typical size: $60 \times 40 \times 25$mm.

Bell flower Popular 18th-century flower decoration strung together to form drops or swags, and also used to enrich convex mouldings.

Bevel To cut the edge off a piece of timber or stone so that it slants at more or less than 45° is to bevel it. A symmetrical cut of 45° is a 'chamfer'.

Binder leads Lead strips used to form the perimeter of a lead-glazed light. The cames are soldered to them.

Blade Principal rafter component of

a roof truss. Also, *cruck blade* which is the curving principal truss timber in cruck construction.

Blinding Sand used to cover, fill and compact site hardcore. Also sharp sand applied to bitumen vertical damp-proofing to key it for plastering.

Boiserie Panelling decorated with carved or moulded designs in the French manner of Louis XIV, XV and XVI.

Bolection moulding Usually used around wall or door panels, but also fireplaces, this moulding stands proud of the framing members and is often shaped in a double curve or 'ogee'.

Bonder Horizontal timbers built into the thickness of old stone or brick walls to provide additional strength to the wall and bearings for joists. They usually rot and cause severe problems.

Bonnet tiles Bonnet-shaped tiles used to cover the hips of roofs.

Boss A carved or moulded ornamental block covering the intersection of beams or ceiling ribs.

Bottle glass Window-panes given this name are really pieces of blown-glass discs, which have been broken off from the blowpipe during manufacture, leaving a 'bullion'. See Crown glass. Also called bull's-eye panes.

Bow window A late 18th- and early 19th-century curved window, especially favoured for shops and taverns.

Braces Timbers fixed at an angle to stiffen structural panels, stud frames, rafters, etc.

Bracket A support, decorative or otherwise, for projecting features like cornices and shelves. Often S-shaped. On the cut string of a staircase a bracket is purely ornamental.

Bressumer Heavy beam supporting a timber-framed jetty, with cantilevered joists mortice-and-tenoned into it; or a similar beam used as a structural lintel across a wide opening.

Bun finial Ornamental bun-shaped projection on top of a newel post.

Cabled fluting A form of fluting, usually of pilasters, in which the lower part of each flute is filled with a convex moulding.

Camber A surface which has a minimal curve is said to be cambered – whether the soffit of an arch or the profile of a road.

Cambered collar Collar joining two principal rafters, with slightly curved form, like a very shallow arch.

Cambered lintel Lintel in the form of a very shallow arch.

Cames Lead strips which hold the glass panes in lead-glazed windows.

Camp ceiling An attic-type ceiling which follows the line of the rafters at the sides and is flat across the middle of the room. Also, and more properly, a similar ceiling with an inward curve suggesting a tent.

Cantilever A projecting beam or slab supported at one end only. Its stability depends upon its own strength and the amount of loading applied to the tethered end

Cap Short form for the word 'capital'. Used to describe the group of mouldings at the top of a feature like a pilaster, pier or newel.

Capital The top part of any column where it spreads out to take the load. It is ornamental with carving or mouldings according to its style and period. Also the top of a pilaster – especially a classical one.

Caracole A curved projection of a wall, often amounting to a turret, built to house a newel staircase.

Cartouche A convex lozenge, often framed with elaborate mouldings in the Baroque manner, used to bear some kind of inscription or heraldry.

Caryatids Female figures used to support a classical entablature.

Casement A window which is side-hung to open outwards in Britain, and inwards in other European countries.

Casting The process of making an object by pouring a material, which will later solidify, into a specially prepared mould.

Cast iron Ironwork made by 'casting' as above.

Cavetto A moulding which curves inwards often in quarter-circle. (See mouldings.)

Cesspit Large, underground, waterproof storage tank, into which drains are run. Tank must be pumped at regular intervals.

Chair rail A moulded wood or plaster decoration which equates with the uppermost moulding of a classical pedestal. It runs round the walls of a room at a level which might prevent the backs of chairs from doing damage to the finishes.

Chamfer See bevel. On beams and lintels etc, usually 17th century or earlier.

Chase A channel cut in a wall to carry pipes or wiring, etc.

Chevron A zig-zag ornament, much used around Norman arches.

Chimney crane A braced iron bar which is hinged to the side of an open hearth and can be swung across the fire to suspend cooking pots.

Chinoiserie Decoration using Chinese motifs, such as pagodas, wooden bridges, exotic birds and dragons. Fashionable in late 17th and 18th centuries. Often includes Rococo influences.

Cladding Weather-proofing and/or decorative surfaces fixed to the structural element of a wall, eg slates, tiles, weatherboards, etc.

Clasped purlin A side purlin which is supported by the collar of each truss at the point where it joins the principal rafter.

Cloam oven Manufactured earthen-ware oven normally built into the wall near or inside the hearth. The oven was pre-heated with a fire of furze, then raked out. The dough was put inside, and the loose door replaced and sealed around with clay.

Closed-string staircase Strings are the main framing members forming the sides of a staircase. In a 'closed-string' type the treads are housed into the strings. In an 'open string' the shape of the stairs is cut out and the treads rested on the string. 'Wall string' is fixed to the wall of the stair-well.

Clout nail A short galvanised round nail with a large head.

Coach bolt Round-headed bolt with a square section just below the head, which digs into the timber, and is thus prevented from turning while it is tightened.

Coach screw A substantial screw with a square bolt head attached to it, often used for fixing steel plates to timber.

Cock's-head hinge Iron or brass 'H'-hinge with ends like cocks' heads. Mainly 17th century.

Cogged joint Timber is notched over the remaining ridge of wood which divides two housings in a piece of timber running at right angles.

Collars Horizontal timbers, spiked to each pair of rafters, checking any tendency for the ends of the rafters to pull apart under load. Attic ceilings are often fixed to them.

Collar beam A tension timber which runs horizontally between the principal rafters of a roof truss. It is normally of the same thickness as the principals and is mortice-and-tenoned into them. It does not always run the full span of the truss.

Collar purlin A roof timber which runs below the centre of every collar in a series of trusses. It was usually supported, in turn, by crown posts.

Composite order Classical Roman order which combines the Ionic capital with the Corinthian one.

Composition concrete tiles Roofing tiles made from a type of close-textured concrete, usually coloured, and sometimes intended to imitate a natural stone. Some are in plain tile form, but many are interlocking.

Concrete Mixture of cement, sand and other aggregates in such proportions as 1:2:4.

Console A type of timber, stone or plaster bracket, usually in the form of an S-scroll; sometimes called 'ancones'.

Continued chimneypiece A large, classically inspired chimneypiece, which extends right up the wall to form an architectural feature with bold cornice or pediment.

Coping A course of fairly substantial stones which usually top a parapet or upstanding gable walls.

Corbel A stone or timber support which projects from a wall to carry the load of a structural member, such as a beam.

Corinthian order The classical order of architecture with column capitals freely decorated with acanthus leaves. The cornice may be identified by its scrolled 'modillion' brackets.

Cornice The uppermost of the three parts of a classical 'entablature'. Composed of several mouldings, usually with a 'fillet' and 'cyma recta' at the top.

Corona Flat vertical band forming middle area of cornice.

Counter battens These are nailed to battens running in the opposite direction, so providing adequate fixing-points for boards or laths.

Courses Bricks, stones, concrete blocks, etc, are laid row upon row in 'courses'.

Cove A substantial concave moulding often used like a cornice to join the walls to the ceiling

Cranked collar A collar which rises to at least half its depth at the centre. The angle so formed is quite pronounced.

Crown glass Glass blown into a disc was cut out to form crown-glass window-panes – the main method until the 1830s.

Crown post In many early roofs, trusses consisted of pairs of rafters joined by tie-beams, from which crown posts supported collar purlins, running under the middle of the collars.

Cruck There are several different types of cruck roof construction. What they have in common are pairs of curved principals, trussed together at bay intervals.

Curtail step The bottom step of a staircase which is projected sideways and curled around, like a 'cur's tail'.

Cusp Gothic tracery often took the form of trefoil or quatrefoil ornament. A quatrefoil is composed of four 'curves. Where they join, they form 'cusps'.

Cut string See 'closed-string'.

Cyma recta This moulding is like an asymmetrical and flattened-out S with the hollow curve uppermost. 'Cyma reversa' has the hollow at the bottom.

Dado That part of a wall's panelling, or other decoration, which equates with the main area between top and bottom mouldings of a classical 'pedestal'. Or the wood panelling running around the lower part of a room's walls.

Dado rail See 'chair rail' above.

Damp-proof course A layer of impervious material built into the thickness of a wall to prevent moisture rising from the ground into the main structure.

Damp-proof membrane Normally a layer of plastic sheeting, laid over the site hardcore of a modern building, to prevent moisture rising from the ground into the floor structure. Also used in restoration work for damp-proofing.

Die (lead glazing) A hard metal block through which lead can be drawn to form it into the section required for making cames.

Dies (joinery) The square blocks of a turned baluster which are jointed to the handrail and to the string or tread, respectively.

Distress To age a material artificially by such means as scoring, denting, chipping or staining.

Dog grate A free-standing iron or steel basket grate incorporating a cast fireback and pronounced front legs. Clearly derived from fire dogs. Mainly 18th century.

Dog-leg staircase The first flight ends at a half-landing and the stairs then double back in a second flight which finishes the ascent. The outer string of the second flight is directly above the first, or very nearly so.

Door case All the joinery work which frames, surrounds and adorns a door opening.

Door hood A projecting feature, usually supported on brackets, which gives visual emphasis to a doorway, and may also serve to keep the weather off those seeking admission.

Doric order The simplest of the three Greek Orders of Architecture.

Dormer window A window which protrudes from the slope of a roof, having its own miniature roofing structure and sides called 'cheeks'.

Dovetail nailing Nails are driven, at opposing angles, through one piece of wood into another, making it difficult for the timbers to be pulled apart.

Down-hearth A fireplace in which the fire is burned upon a bed of ashes actually on the hearthstone, sometimes with the help of fire dogs.

Dragon-tie See Angle-tie.

Dressings Well-cut and finished stones or bricks often used at openings in a wall; sometimes purely ornamental, but sometimes for structural purposes as well.

Dubbing out Filling major cavities and depressions in a wall with mortar before applying first coat of plaster.

Duck's-nest grate Like a hob grate but much lower and less elegant. Sometimes portable and associated with 18th- and early 19th-century kitchens.

Egg-and-dart moulding A row of egg-shaped ornaments, divided up by little arrow or tongue motifs. Used to embellish small convex mouldings.

Empire style The early 19th-century style of design largely derived from neo-classical and Egyptian details, the latter brought back by Napoleon from North Africa.

Encaustic tiles Patterned earthenware tiles, often glazed. Much favoured by Victorian and Edwardian builders.

Entablature The topmost horizontal element in a classical order. Consists of a cornice, frieze and architrave, reading downwards.

Escutcheon In door furniture, the ornamental and protective plate which surrounds the keyhole.

Fanlight Glazed light above a door, often fan-shaped and ornamented; but the term is applied to the rectangular kind of light as well.

Fascia Classical architraves include one or more horizontal flat bands called fascias. Also the board which runs horizontally under the eaves of a building, to which the guttering is normally attached.

Feather-edge boarding Plain weatherboards, tapering in thickness, so that the thick bottom edge of one overlaps the thin top edge of the one below. Old types had a rough, sawn finish.

Fielded panel A wall or door panel with a raised centre area which is sloped off, bevelled or 'fielded' around the sides.

Fillets Plain narrow bands in wood, stone or plaster which divide more shapely mouldings from each other.

Finial Usually a turned or carved ornament in the shape of an urn, ball, bun, spike or figure, often used to decorate staircase newel posts.

Finishes The surfaces which the eye encounters – floor tiles, plaster, paint, gilding, marble.

Fireback A portable cast-iron slab of more or less elaborate design which stands at the back of the hearth to protect the wall and help reflect heat. Long, low shapes tend to be 16th century; the outline becomes increasingly upright, and Baroque in decoration throughout the 17th century.

Fire dogs Have relatively small iron billet-bars, and are three-legged, with upstand ('stauke') at front end. Used for resting logs while they burn. Smaller than andirons. Sometimes called 'brand irons' or 'creepers'.

Fire hood A form of canopy in metal, plaster, brick or stone, to stand right over the fire and funnel smoke into the flue of a big open hearth.

Fire Precautions Act This Act embodies legislation demanding various standards of fire safety in buildings with a public use, like hotels and offices, etc. Enforced and interpreted by local fire officers. (Its requirements are complex and often detrimental to old houses.)

Firrings Tapered pieces of timber nailed to the tops of level joists to give a drainage fall for roof decking. Also used to bring sloping surfaces level, etc.

Fish plate Steel plates bolted on both sides of pieces of timber which have

Flange Horizontal plates at top and bottom of an I-section steel beam. The vertical plate joining the flanges is called the 'web'.

Flashings Strips of waterproof material (usually lead) used to weather the abutments between slates and chimneys; walls and roofs; walls and windows, etc.

Flaunching Domed capping of mortar, trowelled to shape around the base of a chimney-pot, to weather and secure it.

Flitch plate Steel plate slotted into the thickness of a beam to reinforce and/or extend it.

Flush bead moulding A fairly fine bead, usually worked near the edge of a panel or door-framing member. Also at edge of dado lining-boards, etc. The top of the bead is level with the surface of surrounding timber.

Flush door Plain, smooth door constructed with either a solid or hollow core, surfaced with plywood or other laminate.

Fluting Parallel concave channels usually decorating a column, pilaster or architrave. They are sometimes divided by fillets and sometimes by arrises.

Freestone Fine-grained stone which can be worked in any direction, and is relatively easy to carve or cut with a saw. Normally limestone or sandstone.

French windows Pairs of narrow, glazed doors giving access to the outside. They open inwards, as do ordinary windows in France, from which their name derives.

Frieze The middle element of the three which compose a classical entablature. Often decorated with arabesques, etc.

Frog Hollowed-out area of a brick, designed to make it lighter and to provide additional key for bedding mortar.

Gauge Plaster or mortar mixes are said to be 'gauged' when they contain both cement and *lime*.

Gauge The distance between the top of one slating or tiling batten and the top of the one below. It equals the distance between the tail of one slate and the tail of the next down the roof slope. The latter is called the margin.

Geometric staircase Usually a cantilevered staircase which winds up around an open well, the steps being rested upon each other and unsupported at their outer ends. Either circular or elliptical in plan.

Gesso Made from gypsum or chalk with glue to bind it, this is used as a dense white base for the decoration of panelling and for gilded mouldings. Hates damp.

Gilding Gold or other metallic leaf is applied like a transfer to a surface coated with sticky gold-size.

Glazing bars The wooden framing

members in a window sash which divide and contain the glass panes. Early bars (late 17th century and Queen Anne) were wide, flat and rather crudely designed. They became progressively thinner, and often more subtly moulded in the 18th century. Early 19th-century bars were very thin indeed.

Going The total depth of stair tread from front to back, upon which you can place your foot, less the projection of one 'nosing'.

Gothic The architectural style of the Middle Ages, from roughly 1200 to 1500, which is typified by the use of the pointed arch. Much 16th- and even 17th-century work is also Gothic in spirit or detail.

Gothic revival First revival, often called 'Strawberry Hill Gothic', started c1750 (Horace Walpole). Then Regency Gothic, in early 19th century. Finally, Victorian Gothic (Pugin *et al*) from 1840s.

Graining The practice of painting and staining softwoods to make them look like hardwoods with interesting and attractive figuring. Often late 17th century or Victorian and Edwardian.

Greek revival Style of architecture and design which embraced pure classical Greek models rather than those derived from Renaissance Roman forms. Took hold in the late 18th **and** early 19th centuries.

Grip floor Mixture of beaten lime and ash, normally used as a ground floor, but sometimes laid upstairs on laths supported by joists.

Grip handrail Heavily made Elizabethan or Jacobean handrail with a pronounced roll moulding on the top.

Grout Cement and sand slurry with a high cement content and a small amount of sand

Gudgeon pin A round mild-steel pin attached to a backplate or spike, to which a strap hinge is socketed.

Guilloche A decorative design of continuous interlacing circles.

Guttae Drop-like ornaments which occur below the 'triglyphs' in the Doric Order. Sometimes under sills or stair brackets.

Habitable rooms Building Regulations require more exacting standards regarding ceiling heights, etc, for 'habitable rooms', so the exceptions are important. These are: kitchen, scullery, bathroom, WC, porch or conservatory.

Halved joint Many variations exist; but, in principle, half of each piece of timber being jointed together is cut away and the remaining halves are fitted over each other.

Hanging stile A door's vertical framing member on the hinged side.

H-Hinge A hinge used in the 17th and much of the 18th century; it was always nailed, not screwed.

Hip rafter The diagonally placed rafter at the external junction between roof slopes. Internal junctions are valleys.

Hips The external junctions between roof slopes.

H-L hinge Typical 17th- and 18th-century hinge, consisting of two upright steel plates hinged together – one extended to form an L.

Hob grate Cast-iron coal-burning grate. Late 18th century onwards. Three main types.

Holdfast Flat mild-steel plate, tapered to a point, with a screw or nail hole at the wide end. It is driven into masonry or brick bedding joints and is used to secure door frames, etc.

Hood moulding A moulding usually seen over the top of early window openings.

Hopper head Sometimes called a 'bucket head', this is part of a rainwater system which gathers water from various gutters into a downpipe.

Horns Sash horns are continuations of the sash stiles, making it possible to form the joint between stile and rail with a mortice-and-tenon. A Victorian feature.

Hydrated lime Dry, fine, white powdered lime which has been slaked in the works and is sold in bags.

Hydraulic lime Lime made from stones which contain clays. Unlike hydrated lime it sets by chemical action as well as evaporation. Thus it is more like Portland cement; and has various uses in conservation work where additional strength and setting qualities are required. Authentic pointing or bedding of masonry occasionally demands it.

Ingle-nook fireplace A large open fireplace with a 'down-hearth' and a built-in seat, or seats, inside the chimneybreast. Often 16th or 17th century.

Inlaid floor A kind of parquet of wood strips laid with, and against, the grain. Sometimes described as marquetry floor. Seldom is decoration truly inlaid into the actual carcase timber.

In situ For example, a concrete lintel is cast 'in situ' when the operation is carried out in the position where the lintel will remain.

Inspection chamber Commonly called a manhole. It is placed at major junctions of underground drainpipes for cleaning and inspection purposes. Drains run through the chamber in open channels, between walls of smooth mortar 'benching'. Also placed at changes of direction.

Interlocking tiles These, unlike plain tiles, are designed so that their edges fit mechanically one with another, to provide a weather seal.

Intrados The inner curve of an arch.

Intumescent Having the ability to swell up.

Invert The lowest point on the inside of a drainpipe – the point at which the first tiny drop of moisture will gather.

Ionic order The Classical order with bold volutes to the column capitals but no acanthus leaves. Has dentil decoration in the cornice.

Jacobean architecture The English Renaissance style of the early 17th century, embodying a mixture of elaborate decoration, Gothic in spirit, with what are sometimes rather crude efforts at the classical orders.

Jambs The stones, bricks or posts which form the vertical sides of an opening.

Jib door A disguised door fitted within the panelling of a wall, and seeming to be part of it.

Jinny wheel Fairly large pulley-wheel slung from scaffolding, etc, for hauling building materials up to where they are needed.

Joinery The finishing woodwork in a house, eg doors, stairs and panelling. Also the fitting and making of such items.

Ledged door Door of vertical boards nailed to others placed horizontally.

Ledged-and-braced door As above, but with diagonal boards nailed between ledges, to brace the door and prevent sagging.

Linenfold Panel decoration carved to look like stylised folds of fabric. Usually Tudor.

Lintel A structural component, usually of timber, stone or concrete, which carries the load over a window, door or fireplace opening.

Listed building Building listed by the Department of the Environment as of Architectural or Historic Interest. Three grades: I, II star and II, in that order of importance. Buildings are protected by law (inside and out) against unauthorised demolition, or work which adversely affects the 'character'.

Listings Mortar fillets weathering the abutment between a roof covering and a wall or chimney. Also slates, tiles or stones set at an angle for the same purpose.

Lock rail The horizontal framing rail of a panelled door to which the lock is attached.

Lozenge A diamond-shaped decoration, often of Jacobean panelling.

Lump lime See Quick lime.

Manhole See Inspection chamber.

Margin See Gauge. The part of a slate or tile which is exposed to view on the surface of the roof.

Marginal bars Glazing-bars which contain narrow slips of glass around the perimeter of a sash. A glazing pattern of the second quarter of the 19th century.

Mason's mitre Method of returning a moulding around a panel, or similar feature in stone or wood.

Mast newel staircase Spiral stair with timber treads housed into a central post, which rises up through the building or an attached turret or wing.

Matchboarding Thin softwood boards, sometimes with an edge bead, used for lining rooms.

Mechanical bond A structural union between building materials, as opposed to one effected purely by ordinary adhesion or a filling material. Paint has *no* mechanical bond, while plaster, pressed into expanding metal lath, *has*.

Medallion Decorative oval or circle, often containing a head or figure in relief, or sometimes a painting.

Medullary rays Radial markings on the face of quarter-sawn oak boards, formed by the cell structure of the timber.

Mitre At corners, mouldings like architraves are sawn to butt together at 45°.

Modillions Scrolled brackets placed horizontally under the corona of a classical cornice – should be Corinthian or Composite.

Mortar Normally a mixture of cement and sand; but may include lime as well, or be composed of lime and sand only.

Mortice A slot cut out of a framing timber to house a tenon, thus forming an effective joint.

Mouldings Shaped lengths of great variety in plaster, stone or wood, used as ornament, weathering and for the concealment of joints between one building surface and another.

Muff glass Muff or cylinder glass was blown in cylinder form, slit lengthways and unrolled into flat pieces from which window panes could be cut.

Muffle An oversized metal template used to run the first plaster coat of a moulding, before the final coat is run with the template, proper.

Mullions Stone or timber uprights, dividing window openings vertically into separate 'lights'.

Muntins The intermediate upright framing members in doors or panelling.

Muntin-and-plank partition Vertical boards are slotted into the edges of slightly thicker framing pieces, which may be described as 'muntins', 'studs', or 'posts'. In this book 'post-and-plank' describes the thickest type, 'muntin-and-plank', rather less thick; and 'muntin-and-board' the thinner and often later varieties.

Neo-classical The type of design which became popular in the second half of the 18th century. It was derived from ancient Roman and Greek forms observed at first hand by architects, travellers and archaeologists. One of the most important was Johann Winckelmann (1717–68). Also Robert Adam.

Newel A structural post in stone, metal or wood, supporting a balustrade at the top, bottom and changes of direction in a staircase.

Newel stair A spiral staircase in timber or stone.

Niche A wall recess for a statue, bust or shelves. Often has an arched top.

Nogging Short lengths of timber fixed between the vertical studs of a partition to stiffen the construction. In timber-framed houses, bricks used to fill the space between posts are 'brick nogging'.

Norfolk and Suffolk latches Latches consisting of a simple handgrip and thumbplate on one side of the door, and a wooden or metal sneck on the other.

Nosing The part of a stair tread which projects beyond the riser.

Notched joint To join two pieces of timber which are placed at right angles, one is partly cut away so that it fits snugly over the other.

Ogee moulding An S-shaped section – cyma recta or reversa.

Open-well stairs Staircase rising in relatively short flights around a well. Has quarter-landings, and earliest are 16th century.

Orders of Architecture The formal system of classical design in architecture started by the Greeks, adapted by the Romans and re-interpreted in the Italian Renaissance. Without some basic study of the Orders, little sense can be made of architecture and interiors.

Overmantel The upper part of a chimneypiece above the level of the mantelshelf. Usually consists of a panel topped by cornice or pediment, sometimes flanked by volutes or pilasters.

Ovolo Small convex moulding.

Painted floor Type of floor decoration favoured in 18th century; but original examples are very rare.

Palladian Architecture and design based upon Andrea Palladio's 16th-century Renaissance buildings and versions of the Roman Orders. Exceptionally influential in English 17th- and 18th-century work.

Pallets Thin pieces of softwood built into brickwork or masonry as fixing points for carpentry and joinery.

Panelled door Door framed with stiles, rails and muntins and filled in with panels. Earliest tend to be 17th century in Britain.

Panelled spandrel The panelled partition enclosing the space under a staircase.

Pargetting Ornamental exterior plasterwork, either in relief or intaglio. Often found in eastern counties of England.

Parging Soft lime plaster is implied by this term; and it is normally used today for plastering the inside of a flue.

Patera Small disc-like ornament, round or oval, particularly favoured by neo-classical designers, like Robert Adam. Often seen on fireplaces and plaster friezes

Pebbledash Small, rounded stones, like pea gravel, are dashed on to a cement and sand plaster before it has set, giving a densely textured wall finish.

Pedestal The rectangular base structure upon which a classical column sometimes stands. Also for statues.

Pediment A triangular (or sometimes segmental) feature of timber, stone or plaster which forms the topmost element of a classical door, window or fireplace.

Peggies Small slates of random widths, measuring from 200mm to 300mm deep. The width is always at least half the depth. Sold by the ton.

Pellet A small piece of wood, cut to conceal the head of a bolt or screw. Usually circular, it is made to match the grain.

Pendant A form of decorative boss formed as a downward extension of the ribs in Elizabethan and Jacobean ceilings. Also Gothic. Used, also, to decorate bottom of staircase newel.

Pentice A roof with post, pillar or pier supports, covering an outside staircase. Also a gallery with its own roof.

Permitted development Certain types of minor works which may be carried out without planning permission.

Perpends The upright joints between coursed stones or bricks.

Pilaster A flattened-out rectangular version of a column attached to the wall, more for decoration than structure.

Pilaster panel A pilaster without cap or base mouldings. Often first half of 19th century.

Pitch The angle between the 'pitch line' and the floor from which a flight of steps rises. The pitch line is a notional one running across the edge of each nosing.

Planted moulding A wooden moulding applied to a surface rather than one shaped from the framing itself.

Plinth block Block at skirting level which finishes a door or fireplace architrave.

Plugs Wedges or other short sections of timber driven into a wall to provide fixing-points for nails or screws.

Plumb bob A cone-shaped iron weight hung on the end of a length of bricklayer's line, used to obtain a true vertical, either for screeding plaster or to check for inward or outward lean of walls.

Plumb-cut A vertical cut made to provide the ends of rafters with an upright face suitable for the applica-tion of a fascia board – or for any other similar purpose.

Pointing The visible mortar filling between bricks or stones in a wall.

Poling boards Stout boards placed vertically to retain the sides of trenches or as part of a system of temporary support for a wall.

Pontil The iron rod to which a cylinder of molten glass is transferred from the blowing-iron, for the purpose of spinning a disc of 'crown glass'.

Post An upright structural timber of fairly generous section, normally stressed in compression.

Principal rafters The heavy main rafters which carry the whole roof-load, consisting of purlins, common rafters, battens and slates, in a double roof.

Pulvinated frieze Frieze of convex section. Often used for fireplaces and over-doors.

Purlins Normally square-section timbers, running lengthways along the roof, and resting on the principal rafters. Purlins carry the lighter and more numerous common rafters.

Putti Cherubs, much employed for decorative purposes in Baroque and Rococo art and architecture.

Quadrant mould Moulding of quarter-circle section. Also called 'ovolo'.

Quarry tiles Unglazed, burnt clay tiles. Non-porous. Usually red.

Queen post Structural roof timber strutting between a tie-beam and the underside of a through purlin. When this type of strut supports the collar and not the purlin itself it is called a queen strut.

Quick lime Unslaked lime, also called 'lump lime'. The addition of water slakes the lime so it becomes 'hydrated'.

Quoins Squared-up or 'dressed' stones which form the corners of a building.

Rafters (common rafters) These carry the battens or laths from which the roof tiles or slates are hung. Rafters rest on purlins which transfer the load to the roof trusses.

Rag bolts Substantial bolts with enlarged, ragged ends for fixing into mortar, concrete or lead, so they will not pull out.

Rails The horizontal framing members in door or panelling construction.

Raking cut Stones, tiles, slates, bricks, etc, are said to be raking cut when cut at angles other than a right angle.

Randoms Slates between 600mm and 300mm deep but of random widths. No slate is narrower than half the depth. Sold by the ton.

Rebate Step-shaped recess cut from a piece of timber, eg picture frame is 'rebated' to take glass.

Reeding Closely spaced half-round beading often used in 19th-century architraves.

Register grate A cast-iron or sheet-metal plate fills the main fireplace opening. The plate itself has a diminished opening for a relatively small fitted grate. Normally second half of 19th century.

Register plate Metal plate which fits and seals an opening like that of a fireplace or flue.

Reinforced concrete Concrete containing steel bars, which greatly increase its strength.

Relieving arch A relieving arch built into the general brickwork or masonry, above a lintel or the arch of an opening, to direct the vertical forces outwards and down the sides of the jambs.

Rendering First coat of plaster on a solid wall. Sometimes the only coat. When used on laths it should strictly be called a 'pricking-up' coat.

Reticulation Stone or plaster rustica-tion in the form of vaguely circular sinkings, closely placed but not connected with each other.

Return When a wall or moulding changes direction at right angles. Adequate 'returns' are visually vital at the sides of openings.

Reveal The visible exterior part of a window jamb, between frame and main wall surface.

Reveal pin Like an adjustable steel prop, fixed inside a window opening to provide an anchorage for a scaf-folding transom, which is coupled to it.

Ribbon or strap pointing Pointing between stones which protrudes beyond the face of the stone, forming a ribbon-like pattern of mortar.

Ribs A projecting band on a ceiling. Elizabethan and Jacobean plaster-work is usually made up of 'ribbed' designs. Also the structural 'ribs' of vaulting.

Ridge The apex of a pitched roof where one slope meets the other.

Ridge-board and ridge-pole Timber running along the apex of the roof. Ridge-board is sandwiched between the tops of the rafters. Ridge-pole is cradled by ends of rafters in various ways.

Ridge purlin A long piece of timber, of roughly square section, running along the apex of the roof trusses, by which the ends of the common rafters are supported.

Ridge tile Angled or half-round tile used to weather the junction between the roof slope coverings at the ridge.

Rim lock Metal-cased lock screwed or bolted to face of door.

Riven flagstones Flagstones which have been split and show a 'riven' face rather than a sawn one.

Riser The vertical face of a step.

Rococo A style of decoration using freely flowing designs of 'C' and 'S' scrolls, foliage, shells and arab-esques. Popular in England during the middle third of the 18th century. Very elegant and frivolous.

Roman cement A cement derived from clay which is hydraulic (sets chemically as well as by evaporation). It is brown and was thus thought useful for plastic repair of stonework. Ousted by Portland cement, it was much used in the l9th century.

Roof trusses Pairs of rafters, usually tied by a third member, to form a loadbearing triangle of timber or steel.

Rosette Decorative device, normally round, using a stylised flower motif.

Rough cast Wall rendering on to which a mixture of lime and aggregate has been thrown before setting.

RSJ A rolled-steel joist in the smaller version of the I-section universal beam, and has slight differences in the flange. RSJs are from 76 × 76mm to 203 × 102mm

Rubber Soft, fine-grained bricks suitable for rubbing down, carving and cutting for ornamental arches and decorative brickwork.

Sarcophagus grate Early 19th-century cast-iron grate, with marked horizontal emphasis. Has solid back and slab-like ends, which are shaped and decorated with brass mounts in the Empire manner.

Sarking boards Boards nailed to the upper sides of rafters as a base for tiling or slating. They provide an additional thickness to the roof.

Sash The glazed part of a window. It will normally open. Usually suspended with weights rather than hinged.

Sash (box) The cased frame of a double-hung sash window, in which the sash weights are suspended.

Sash frame The frame in which the sashes are suspended, fixed or hinged. Includes stiles, head and sill.

Sash window (double-hung) Window with suspended sashes, opened and closed by vertical movement.

Sash window (horizontal sliding) Window with sash or sashes which open and close by sliding horizontally within the frame.

Scotia mould Concave moulding like a 'cavetto' but less symmetrical.

Screed A layer of mortar used to level up and provide a smooth floor surface upon which to lay finishes. Usually one or two inches thick.

Screens passage Passage formed at the lower end of a medieval hall by screening off the main area from the exterior door. Often minstrels' gallery over.

Septic tank A sealed waterproof tank, usually below ground, in which waste material from a drainage system is biologically decomposed, before finally being part-purified and dispersed through a soakaway. The latter is often constructed in conjunction with the tank.

Shearing This happens when forces act in such a way that components

fail by breaking apart, either horizontally or vertically, like the movement of a pair of scissors being opened.

Shooting (a door) Planing the edges of a door so that they will make an accurate fit within the frame.

Sill The external window member of wood, slate, stone, brick or concrete which aids rainwater in running off the window and clear of the wall face.

Sleeper wall A dwarf wall rising from ground level to support a suspended floor structure of joists and boards.

Slip joint An unjointed or toothed joint in brickwork, concrete or masonry which permits some movement.

Slips Well-finished rectangular slabs which immediately surround the fireplace opening.

Smoke shutter Shutter of steel, glass or some other incombustible material, used to lower the smoke-line of a fireplace opening and prevent smoke leaking into the room.

Smoke hood. See Fire hood.

Snap header A brick of half the normal length, used to make a 'half-brick thick' wall, in stretcher bond for example, look as though it has been built in Flemish bond. Genuinely, such a wall would have to be at least 'one brick thick', ie the *length* of a brick thick.

Sneck Lever for lifting a latch. The sneck-latch is traditional in old houses of many types and periods.

Soakers Sheets of lead inserted between the slates and turned up the side of an abutting wall to weather the junction.

Soffit The underside of window openings, beams, closed staircases, etc.

Soil mechanics The science of establishing the nature and bearing capacity of the subsoil upon which a building stands, or must be placed.

Soldier arch A false arch of bricks placed on end, without either bearings or abutments. It is supported on a mild-steel bar or tied into a structural backing lintel.

Sole plate A short, stout piece of timber used as a base for a principal rafter. It is bedded on the top of the wall.

Spandrel The triangular area formed between an arch and the rectangle of mouldings in which it is placed.

Spatterdash Rough rendering for walls, applied by flicking lumps of wet plaster on to the wall.

Spikes Big strong nails over 4in (100mm) long.

Splat baluster A staircase baluster cut from a thin board to suggest the design of a conventional baluster in one of the classical shapes – barley-sugar twist, vase or column. Often 17th or early 18th century. Recalls the 'splat' of a Queen Anne dining-chair.

Splay When the interior jambs of a window opening are opened out instead of being at right angles to the wall surface, they are called splays.

Steel-trowel finish A smooth plaster finish to a wall or to a mortar floor screed.

Stiles The outermost vertical framing members of a system of panelling, and in the same way of a door.

Stock-lock Like a rim lock it is screwed or bolted to the door, but has a hardwood case.

Stopped chamfer When the lower edge of a beam, for example, ceases to be chamfered off it is said to be 'stopped'. For this purpose various decorative mouldings are used to effect the change from 'chamfer' to plain rectangular section.

Strap hinge Iron band nailed or bolted to a door and curled around at one end over a gudgeon-pin fixed to the jamb or frame.

Strapped A wall which has been provided with lateral or vertical battens as a fixing-ground for laths or sheet material has been strapped.

Strapwork Strictly, banded decoration in the form of leather straps brought to Britain in late 16th and early 17th century pattern-books from the Low Countries and Germany (Vredeman de Vries, *et al*). Loosely used to describe 'Jacobean' ceilings with flat, decorated rib-work.

Stretcher A brick which appears lengthways on the face of a wall. Those that appear end-on are headers.

String See Closed-string Staircase.

String course A band of projecting masonry, brick or plaster, often at or around first-floor sill level, having a mainly decorative purpose.

Strip flooring Narrow floor boards laid in random lengths, often on a solid sub-floor.

Stucco A form of smooth and relatively hard rendering, much favoured for the walls of early 19th-century and Victorian houses.

Studding Upright timbers of sizes such as 4 × 2in which form the main framework of such units as walls and partitions.

Strut In old houses normally a fairly short piece of timber used to hold two other structural timbers apart.

Stuck mouldings Those which are formed on a piece of timber and are therefore integral to it, rather than mouldings which are run in separate lengths and applied afterwards.

Studs Vertical timbers, evenly spaced, to form the frame of a partition or wall.

Sunk panel One with its surface lying below that of the surrounding framework.

Swag A decorative garland of foliage, fruit or flowers, etc., as if suspended between two points.

Swan neck (or Knee and ramp) A

wall coping, pediment or handrail which adopts an ogee curve.

Template A metal plate cut to the exact profile of a plaster moulding, with which the plaster may be shaped. Also a pattern for marking-out materials which are to be cut to a particular outline.

Tenon End of a piece of timber cut away to fit into a mortice. Extensively used framing joint.

Terms Used like a column or pilaster, these supports comprise a human head and torso merging into a downward-tapering pedestal.

Textured paints Paints and similar solutions containing various types of fine, granular material to give texture.

Throating A groove along the underside of a component such as a sill, to form a drip for moisture trying to run back into the wall face.

Thumb latch (or Norfolk latch) Has a grip handle and thumb-piece for pushing up the sneck, which goes through the door and lifts the latch on other side.

Tie-beam Substantial beam, spanning from wall to wall, which ties together the feet of the principal rafters, thus forming a roof truss.

Tile (clay) Baked earthenware roofing and floor units, not to be confused with a slate which is a natural stone, riven and cut to various sizes.

Tilting batten Slating or tiling batten placed near the ridge of a roof. It is slightly thicker than the other battens and chocks up the heads of the top course of slates, preventing the tails from riding up.

Tilting fillet A wedge-shaped batten run down the sides of valley gutters, for example, to tilt the edges of the slates. Its use makes the verge tighter and encourages the water back onto the main roof slope.

Tongue-and-groove boarding Floorboards are typical. The protruding edge of one fits into a slot in the edge of the next.

Toppings Mixture of small stones and fine grit, etc, used by road contractors to consolidate hardcore.

Top rail The uppermost horizontal framing member in a system of panelling.

Torching Traditionally, lime mortar pointing used to seal the heads of slates or tiles inside a roof space. But also, by derivation, describes the all-over plastering of the backs of the slating laths between common rafters.

Torus mould Sometimes called 'bull-nosed', this is a bold convex moulding used on steps, sills and at the bases of columns.

Transom Stone or timber members which divide window openings *horizontally* into separate lights.

Trimmer A floor joist, which carries the ends of other joists which have been cut off to form an opening for a staircase or fire hearth.

Trimming joists Joists of increased thickness which carry the ends of the 'trimmer joist' of a staircase or fire-hearth opening.

Truss (roof) Substantial framework of principal rafters, a tie-beam and/or collar, usually forming either a triangle or an 'A'. It carries the purlins and common rafters. Trusses are positioned at 'bay' intervals along the roof.

Tuck pointing This obnoxious method was used in the 18th and 19th centuries, and involves grooving the soft mortar and filling the grooves with further mortar which stands about 3mm proud of the wall surface. This filling is struck off against a batten to give it straight edges.

Tuscan order The robust-looking Roman simplification of the Doric Order. Has no fluting, no frieze decoration and has a torus moulding between two fillets, instead of an attic base.

Tyrolean finish Wall finish applied by spraying a mixture of plaster grout and fine sharp stone chippings. This is done with a special hand-operated spraying machine.

Universal beam Steel beam of I-section, larger than an RSJ and of marginally different flange detail. Sizes from 203 × 133mm to 305 × 165mm.

Valley The junction at the internal angle between roof slopes. Normally weathered and drained by a lead valley gutter.

Venetian window Window with a round arch head, closely flanked by lower and narrower flat-headed lights. Palladian.

Verges The exposed ends of a sloping roof, where the slates or tiles overhang the gable walls.

Vermiculation Form of rustication in either stone or plaster, consisting of worm-like ridges, outlining hollows in the surface of the material.

Vernacular building Everyday building, usually of indigenous local style.

Vernier Very accurate measuring device, with two scales read in conjunction with each other, so that readings may be taken to $1/100$th part of the appropriate unit. Made in caliper and gauge form.

Vertical crazy paving Unevenly shaped and relatively thin slabs of stone, applied with mortar to a wall surface.

Voussoirs Specially cut stones of wedge shape, which make up the structure or facing of an arch.

Wainscotting Another name for wall panelling.

Waling Timbers placed horizontally along the sides of a trench which are kept firmly in position with struts. Also used for other types of temporary support to walls, etc. A typical section would be 150 × 100mm.

Wall plate Length of timber which runs along the top of a wall to provide a bearing and fixing for the ends of rafters.

Waney-edge weatherboard Weatherboards having a wavy outline along the bottom edge of each board, giving a vaguely rustic appearance.

Weathered Building components which are bevelled or made to a slight fall, so helping rainwater to drain off them. Also commonly employed as a term for other kinds of protection against penetrating damp, and to mean fading or wear of paints and materials due to exposure.

Web An I-section steel beam or RSJ is made up of two components – the flanges at top and bottom (horizontal) and the linking vertical part, the 'web'.

Whimhouse A single-storey building, usually round, which is joined to a barn or granary. In it a horse tramps in a circle hauling a centre-pivoted beam which powers machinery.

Whiting Crushed chalk, used as a white pigment or filler.

Wind bracing Inclined, or diagonally placed, timbers or steel sections employed to stiffen a structure like a roof. Often seen in roofs of 15th- and 16th-century buildings.

Wood float A plastering tool, consisting of a flat piece of wood, with a handle, for texturing renderings.

Wood-float finish Plaster textured to give a slightly roughened effect by being worked over before it sets with a wooden version of the steel trowel.

Wreathed handrail One which curls around to form a decorative termination over the bottom newel post of a staircase. Also, single and double-twist wreaths to carry handrail around a stairwell.

Wrought-iron Ironwork which has been shaped while hot from the fire of a blacksmith's forge. Also a high-quality iron for this purpose.

Zax A long-bladed chopper for cutting slates over a dressing-iron. On the upper edge, it has a spike for holing slates.

FURTHER READING

Barry, R. *The Construction of Buildings* – five volumes (Crosby Lockwood Staples 1969–78)

Beard, Geoffrey *Decorative Plasterwork in Great Britain* (Phaidon 1975)

Billett, Michael *Thatching & Thatched Buildings* (Robert Hale 1979)

Bridgeman, Harriet & Drury, Elizabeth *Needlework – An Illustrated History* (Paddington Press Ltd 1978)

Brown, R. J. *English Farmhouses* (Hale 1982)

Brunskill, R. W. *Vernacular Architecture* (Faber 1978)

Building Research Station Digests – on numerous topics (HMSO 1972)

Clifton-Taylor, Alec *The Pattern of English Building* (Faber 1972)

Clifton-Taylor, Alec & Ireson, A. S. *English Stone Building* (Gollancz 1983)

Coggins, C. R. *Decay of Timber in Buildings* (The Rentokil Library 1980)

Cook, Olive & Smith, Edwin *The English House Through Seven Centuries* (Nelson 1983)

Cordingley, R. A. *British Historical Roof-Types* (from SPAB, see own entry below)

Corkhill, T. *A Concise Building Encyclopaedia* (Pitman 1970)

Cunnington, Pamela *How Old is Your House?* (Collins 1983)

Curl, James Stevens *English Architecture – An Illustrated Glossary* (David & Charles 1977)

Fleming, John, Honour, Hugh & Pevsner, Nikolaus *A Dictionary of Architecture* (Penguin 1975)

Fleming, John & Honour, Hugh *Penguin Dictionary of Decorative Arts* (Penguin 1977)

Fletcher, Sir Banister *A History of Architecture* (Athlone 1975)

Fowler, John & Cornforth, John *English Decoration in the 18th Century* (Barrie & Jenkins 1978)

Fox, Sir Cyril, & Lord Raglan *Monmouthshire Houses* (National Museum of Wales volumes 1–3, 1951–4)

Girouard, Mark *The Victorian Country House* (Yale University Press 1979)

Gloag, John *Early English Decorative Detail* (Academy 1973)

Insall, Donald *The Care of Old Buildings Today* (Architectural Press 1972)

Lander, Hugh *Do's and Don'ts of House and Cottage Conversion* (Acanthus Books 1979)

Lander, Hugh *Do's and Don'ts of House & Cottage Interiors* (Acanthus Books 1982) Use its 286 photographs in conjunction with *House Restorer's Guide*, to identify and select period interior features (available from Lanner, Redruth, Cornwall)

Lloyd, Nathaniel *History of the English House* (The Architectural Press 1975)

Loudon's Encyclopaedia of Cottage, Farm & Villa Architecture (Longman 1842)

McKay, W. B. *Building Construction* – four volumes (Longman 1971–5)

Melville, Ian A. & Gordon, Ian A. *The Repair and Maintenance of Houses* (London, The Estates Gazette Ltd 1973)

Mercer, Eric *English Vernacular Houses* (HMSO 1975)

Mitchell's Building Construction – six volumes (Batsford 1979)

Osborne, Harold *The Oxford Companion to Art* (Oxford 1970)

Owen, Michael *Antique Cast Iron* (Blandford 1977)

Pegg, Brian F. & Stagg, William D. *Plastering – A Craftsman's Encyclopaedia* (Crosby Lockwood Staples 1976)

Powell-Smith, Vincent & Billington, M. J. *The Building Regulations Explained and Illustrated* 7th ed (Granada 1982)

Powys, A. R. *Repair of Ancient Buildings* (from SPAB, see own entry below)

Scott, John S. *A Dictionary of Building* (Penguin 1974)

Seymour Lindsay, J. *An Anatomy of Wrought Iron – 1000 to AD 1800* (Alec Tiranti 1964)

— *Iron & Brass Implements of the English House* (Alec Tiranti 1970)

Sitwell, Sacheverell *British Architects and Craftsmen* (Pan 1973)

SPAB Technical Pamphlets; various titles (from SPAB, 37 Spital Square, London E1 6DY)

Suddards, Roger W. *Listed Buildings – The Law & Practice* (Sweet & Maxwell 1982)

Summerson, John *The Classical Language of Architecture* (UP 1966)

Teynac, Nolot & Vivien *Wallpaper: a History* (Thames & Hudson 1983)

Walker, Philip *Woodworking Tools* (Shire Album 50, 1980)

Whitten, D. G. A. & Brooks, J. R. V. *The Penguin Dictionary of Geology* (Penguin 1972)

Wood, Margaret *The English Mediaeval House* (Dent 1965)

Yarwood, Doreen *The English Home* (Batsford 1956)

INDEX

270

271